Asymptotic Methods in the Buckling Theory of Elastic Shells

SERIES ON STABILITY, VIBRATION AND CONTROL OF SYSTEMS

About the Series

Rapid developments in system dynamics and control, areas related to many other topics in applied mathematics, call for comprehensive presentations of current topics. This series contains textbooks, monographs, treatises, conference proceedings and a collection of thematically organized research or pedagogical articles addressing dynamical systems and control.

The material is ideal for a general scientific and engineering readership, and is also mathematically precise enough to be a useful reference for research specialists in mechanics and control, nonlinear dynamics, and in applied mathematics and physics.

Selected Volumes in Series B

Proceedings of the First International Congress on Dynamics and Control of Systems, Chateau Laurier, Ottawa, Canada, 5–7 August 1999
Editors: A. Guran, S. Biswas, L. Cacetta, C. Robach, K. Teo, and T. Vincent

Selected Topics in Structronics and Mechatronic Systems
Editors: A. Belyayev and A. Guran

Selected Volumes in Series A

Vol. 1 Stability Theory of Elastic Rods
Author: T. Atanackovic

Vol. 2 Stability of Gyroscopic Systems
Authors: A. Guran, A. Bajaj, Y. Ishida, G. D'Eleuterio, N. Perkins, and C. Pierre

Vol. 3 Vibration Analysis of Plates by the Superposition Method
Author: Daniel J. Gorman

Vol. 4 Asymptotic Methods in Buckling Theory of Elastic Shells
Authors: P. E. Tovstik and A. L. Smirinov

Vol. 5 Generalized Point Models in Structural Mechanics
Author: I. V. Andronov

Vol. 6 Mathematical Problems of the Control Theory
Author: G. A. Leonov

Vol. 7 Vibrational Mechanics: Theory and Applications to the Problems of Nonlinear Dynamics
Author: Ilya I. Blekhmam

SERIES ON STABILITY, VIBRATION AND CONTROL OF SYSTEMS

Series A Volume 4

Series Editors: A. Guran, A. Belyayev, H. Bremer, C. Christov & G. Stavroulakis

Asymptotic Methods in the Buckling Theory of Elastic Shells

Petr E. Tovstik
Andrei L. Smirnov

St. Petersburg State University, Russia

Edited by

Peter R. Frise
Windsor University, Canada

Ardéshir Guran
Institute for Structronics, Canada

World Scientific
Singapore • New Jersey • London • Hong Kong

Published by

World Scientific Publishing Co. Pte. Ltd.

P O Box 128, Farrer Road, Singapore 912805

USA office: Suite 1B, 1060 Main Street, River Edge, NJ 07661

UK office: 57 Shelton Street, Covent Garden, London WC2H 9HE

British Library Cataloguing-in-Publication Data
A catalogue record for this book is available from the British Library.

ASYMPTOTIC METHODS IN THE BUCKLING THEORY OF ELASTIC SHELLS

ISBN 981-02-4726-5

Printed in Singapore.

Preface

Many publications both in Russia and abroad are devoted to the analysis of the buckling of thin-walled structures, and the solutions of many important problems have been obtained. In spite of the variety of books that are available on this subject however, there is no one book which contains a general overview of buckling problems and sound base of methodology for the qualitative study of a large number of different types of problems. This gap may be partially filled by the monograph "Asymptotic Methods in Buckling Theory of Elastic Shells", which is a revised and extended edition of the monograph "The Stability of Thin Shells. Asymptotic Methods" by P.E. Tovstik published in 1995 in Russian.

The book contains numerous results developed by the authors and their pupils by means of the application of asymptotic methods to the problem of shell buckling. Previously these methods have been applied to problems of shell statics and vibrations.

There are many new results in this rather compact book. The static stability problem for the case of small membrane deformations and angles of rotation is studied completely. In the general formulation, the principal questions on the development of local buckling modes are studied. The results presented give a clear, qualitative picture which is necessary for further development of modern numerical methods.

One of the unique features of the book is the large bibliography of literature on the application of asymptotic methods to the problems of thin shell buckling. The reference list includes Western as well as the Russian literature on asymptotic methods. The Russian material has generally been translated into English but is still not widely known, despite the high standards prevalent in this body of research.

The book is directed both to researchers working on the analysis and construction of thin-walled structures and continuous media and also to mathematicians who are interested in the application of asymptotic methods to problems of thin shell buckling. The book may also be useful for graduate and post-graduate students of Mathematics, Engineering and Physics.

The present book is a result of the scientific cooperation of the Departments of Theoretical and Applied Mechanics of the Faculty of Mathematics and Mechanics at St. Petersburg State University in Russia and Department of Mechanical and Aerospace Engineering at Carleton University in Ottawa, Canada. The authors would like to express their special thanks to the editors of the book, Prof. Peter R. Frise from the University of Windsor and Prof. Ardeshir Guran from the Institute of Structronics and also to Prof. John Goldak amd Prof. D.R.F. Taylor of Carleton University whose support of the program of scientific cooperation between St. Petersburg and Carleton University made it possible to prepare this manuscript.

This work was supported in part by Russian Foundation for Fundamental Research (grants 9801.01010, 0101.00327) and Soros International Foundation (grant # 54000) and the Natural Sciences and Engineering Research Council of Canada under Dr. Frise's individual research grant.

Our special thanks to Mrs. V. Sergeeva who typeset the main part of the book and drew most of the pictures. We would also like to thank our students N. Kolysheva, Yu. Dyldina, Yu. Balanina, N. Vasilieva, E. Kreis, A. Yambarshev, V. Braulov, and M. Antonov for their help in preparation of the manuscript.

Peter Tovstik and Andrei Smirnov

Contents

Introduction

Buckling analysis is one of the important elements in the modelling of thin-walled structures in many different branches of engineering including shipbuilding, missile-building, and civil and mechanical construction.

The development of numerical methods of solution for the problems of shell buckling (as in the other problems of shell theory) has reached a level such that virtually any problem can be solved numerically. Examples of the methods in use include orthogonal sweep techniques for loaded shells of revolution. In cases that do not permit the separation of variables one can use different variational methods such as the finite element method which has been developed extensively over the last thirty years and is now widely accepted in engineering practice.

Because of the development of numerical methods, it is worthwhile to reexamine the role and importance of analytic methods. In simple cases, analytical methods can give the exact solution to a problem. In many other cases, they provide with adequate precision, an approximate solution and allow us to simplify the numerical solution which can help to speed to numerical process and lower computational costs. Acquaintance with the analytic results also helps to clarify the problem qualitatively and assists in gaining an understanding of the mechanism of buckling.

Asymptotic solution methods using the thin shell approximation have taken the leading place among the available analytical methods.

The rather small class of buckling problems of elastic, smooth, thin shells under conservative surface and edge loads is considered below. The use of the static stability criterion yields the linear boundary value problem which may be solved effectively by means of asymptotic methods.

In order to do this, solution of shell buckling problems which require the application of the dynamic stability criterion, or the solution of non-linear boundary value problems are left aside. In particular, buckling under dynamic loads, buckling of shells in a gas flux, parametric buckling problems and non-linear problems of shallow shell buckling will not be considered in this book. Likewise, the effect of initial imperfections is not taken into account.

Development of approximate asymptotic formulae for expected buckling modes and corresponding critical loads is the basic subject of the book. Shells with various neutral surface geometry and load and support conditions are analyzed.

Significant attention is given to construction of localized buckling modes, in which many small pits appear.

In some cases these pits cover the entire neutral surface while in other cases the localization of buckling modes near some weak lines or points on the neutral surface occurs. This localization is related to non-homogeneity of the initial stress-strain state, or to variability of shell Gaussian curvature and/or of shell thickness. Localization in the neighbourhood of an edge may be related to specific features of the edge support.

In publications on thin shell buckling (see [21, 22, 59, 61, 149, 175]) one can find numerous examples of buckling modes which are localized in neighbourhood of certain lines or points (for example, the buckling problem for a stretched spheroid under external pressure when the pits form along the equator). However, the analytical description of these localized buckling modes that result from the asymptotic solution of the buckling equations is practically absent in shell theory monographs. This fact prompted the authors to write this book.

The methods of solution of the buckling problems used below are based on the methods of the asymptotic solution and qualitative analysis developed by A.L. Goldenveizer and his pupils [51, 52, 57, 70, 88] in connection with the problems of shell static, vibration and buckling.

From the point of view of asymptotic solutions, shell free vibration and buckling have much in common. This is because, in both cases, we come to the boundary value problem with a small parameter in the higher derivatives. The main difference is that in the buckling problems we are interested only in the smallest (or, perhaps only in the smallest few) eigenvalues.

The typical situation for an asymptotic integration of equations with non-constant coefficients is that when in the integration domain, transition lines appear [15, 57, 88, 125, 180], (they are called caustics in the field of acoustics [174]) the region is separated into parts with qualitatively different solution behaviour. In shell buckling problems the transition lines separate the part of the neutral surface which contains the pits which appear under buckling.

The lowest eigenvalue in which we are interested corresponds to the buckling mode which has pits only on a small portion of the neutral surface (i.e. localised in the neighbourhood of the weakest line or point). The peculiarities of asymptotic methods used below are connected with this circumstance.

In the case when the weakest point does not coincide with a shell edge, we have used the method for construction of the buckling mode that was proposed by V.P. Maslov [99].

In cases that permit the separation of variables, the boundary value problem becomes one-dimensional and the turning points play the role of transition lines. Investigation of turning points in shell theory problems was begun by N.A. Alumyae [5]. If the weakest line does not coincide with the shell edge, we approach the case of two close turning points. For construction of the buckling mode the one-dimensional version of V.P. Maslov's method [99] is used. However, if the weakest line coincides with the shell edge then we have the turning point situated near the edge.

For construction of semi-momentless buckling modes for cylindrical and conic shells localized near the weakest generatrix, an algorithm based on an asymptotic separation of variables is suggested. In this case the buckling mode in the circumferential direction is similar to that obtained by V.P. Maslov's method.

To read this book, it is not necessary that the reader have a specific knowledge in a field of asymptotic integration of equations. All of the required information is given in the text. The solutions are presented in the form of formal asymptotic series written in powers of relative shell thickness. The algorithm for the development of the coefficients of the series are indicated and for the first few terms of the series, the explicit expressions are given in many cases. As a rule, the series diverge and by increasing the number of terms, the partial sums of these series do not satisfy the equations and boundary conditions with increasing accuracy.

Proof of the asymptotic character of these series, in other words the statement that the error incurred after substituting the required function by the first few terms of the series is on the order of the first neglected term is beyond the scope of this book.

Let us now briefly describe the structure of the book.

In Chapters 1 and 2 the well-known equations of the two-dimensional theory of thin elastic shells and the buckling equations which come from the static criterion are given.

Chapter 3 contains classical solutions for simple shell buckling problems, such as the buckling of a shallow or cylindrical shell under a homogeneous membrane stress-strain state.

In the remaining chapters (with the exception of the final Chapter 14) the initial stress-strain state is assumed to be membrane and any pre-buckling deformations are not taken into account.

In Chapters 4 and 5 the problems which allow the separation of variables are considered. In Chapter 4, the buckling modes for shells of revolution which are localized in the neighbourhood of the weakest parallel which does not coincide with a shell edge are presented. In Chapter 5, the buckling modes for a cylindrical shell under non-homogeneous axial pressure are developed.

In Chapter 6, the buckling modes for convex and cylindrical shells localized in the neighbourhood of the weakest point which does not coincide with a shell edge are constructed.

The next 4 chapters (from 7 to 10) are devoted to the consideration of semi-momentless buckling modes for cylindrical and conic shells. For such shells the pits formed under buckling stretch along the generatrix. If the stress-strain state varies in the circumferential direction then the localisation of buckling mode near the weakest generatrix will occur. The external normal pressure, torsion, and bending are typical loads which can cause this buckling mode. The effect of boundary conditions on the critical buckling load is also examined.

In Chapter 11 the buckling of shells of revolution with negative Gaussian curvature is investigated. Under buckling, this type of shells generally displays pits which cover the entire neutral surface.

Chapter 12 presents a discussion of the buckling of weakly supported shells. The relationship between the bending of the neutral surface and the buckling of the shell is examined.

The problems for the shells with initially membrane stress-strain state that results

The buckling modes localized near a shell edge are considered in Chapter 13. Again the initial stress-strain state of the shell is assumed to be membrane. In some cases, buckling is connected with a weak edge support while in other cases, the weakest line or point coincides with the edge.

Chapter 14 concludes the book and provides a treatment of buckling problems where the initial stress-strain state is a combination of membrane, bending and edge effects. This type of stress-strain state is referred to as a general stress-strain state in this book. The initial stress-strain state is represented as the sum of a membrane stress-strain state and an edge effect. Problems when the transition to a general formulation slightly changes the buckling load, and problems when the membrane formulation gives incorrect values of the buckling load are considered.

In reading selected chapters it is useful to note that in Chapters 1–3 well-known introductory information is presented, Chapters 4–11 are divided into four parts (weakly connected to each other) according to the method of asymptotic integration used: (i) Sections 4 and 5, (ii) Section 6, (iii) Sections 7, 8, 9 and 10 and (iv) Section 11. Finally Chapters 12, 13 and 14 contain information of general nature and are largely independent of the earlier parts of the book.

Basic notation

Three digit indexing of equations is used in the book. Each equation index number includes the number of the chapter, the number of the section within the chapter, and finally the number of the equation. For brevity the chapter number is omitted if we refer to equations from within the same chapter, and the chapter and section numbers are omitted if we refer to equations from the same section.

Two digit indexing of figures, tables and remarks is used, comprising the number of the chapter and the number of the figure, table or remark.

To designate the order of a function, $\varphi(h_*)$, so as to compare it with another function, $\psi(h_*)$, as $h_* \to 0$ (or $\mu, \varepsilon \to 0$) the symbols O, o, $\sim$ are used.

The notation $\varphi = O(\psi)$ indicates the existence of a constant $A > 0$ which is independent of h_* and $|\varphi| \leq A|\psi|$. Sometimes instead of $\varphi = O(\psi)$ we write $\varphi \lesssim \psi$.

The notation $\varphi = o(\psi)$ means that $\varphi/\psi \to 0$ as $h_* \to 0$. The symbol $\sim$ is used for functions of the same asymptotic order, that is $\varphi \sim \psi$ means that $\varphi = O(\psi)$ and $\psi = O(\varphi)$ simultaneously.

Basic Symbolic Notation

α, β (α_1, α_2) — orthogonal curvilinear coordinates of a point on the neutral surface;

A, B (A_1, A_2) — Lame's coefficients on the neutral surface;

E — Young's modulus;

$\boldsymbol{e}_1$, $\boldsymbol{e}_2$, $\boldsymbol{n}$ — unit vectors of the coordinate system connected with a point on the neutral surface before deformation;

$\boldsymbol{e}_1^*$, $\boldsymbol{e}_2^*$, $\boldsymbol{n}^*$ — unit vectors of the coordinate system connected with a point on the neutral surface after deformation;

ε_{ij} — membrane strains in the neutral surface;

ε_i, ω — membrane strains in the linear approximation;

ε — a small parameter, $\varepsilon^8 = h_*^2/(12(1-\nu_0^2))$ $\varepsilon > 0$;
ε_0 — a small parameter, $\varepsilon^6 = h_*^2/(12(1-\nu_0^2))$, $\varepsilon_0 > 0$;
i — an imaginary unit, $i = \sqrt{-1}$,
h — the shell thickness;
h_0 — the characteristic shell thickness in the case of a shell of non-constant thickness;
h_* — the dimensionless shell thickness, the principal small parameter, $h_* = h/R > 0$;
$K = \dfrac{Eh}{1-\nu^2}$, and $D = \dfrac{Eh^3}{12(1-\nu^2)}$ — the shell stiffnesses;
k_1, k_2 — the principal curvatures of the neutral surface;
L_1, L_2 — sizes of a shallow shell;
λ — load parameter ($\lambda > 0$);
μ — a small parameter, $\mu^4 = h_*^2(12(1-\nu_0^2))$, $\mu > 0$;
M_i, H_i, H — projections of internal stress-couples into the unit vectors $\boldsymbol{e}_i^*$, $\boldsymbol{n}^*$ related to the unit length before the deformation;
P_1, P, M_1, M — external forces and moments on an edge of a shell of revolution (see Figure 1.3);
ν — Poisson's ratio;
p, q — wave numbers;
Π — potential energy of deformation of a shell;
q_1, q_2, q (q_1^*, q_2^*, q^*) — projections of the vector of external load intensity into unit vectors $\boldsymbol{e}_1$, $\boldsymbol{e}_2$, $\boldsymbol{n}$, ($\boldsymbol{e}_1^*$, $\boldsymbol{e}_2^*$, $\boldsymbol{n}^*$) related to the unit area of the neutral surface before the deformation;
R_1, R_2 — the principal radii of curvature of the neutral surface;
R — the characteristic linear size of the neutral surface;
Q_1, Q_2 — shear stress-resultants related to the unit length before the deformation;
T_i, S_i, S — projections of internal stress-resultants into unit vectors $\boldsymbol{e}_1^*$, $\boldsymbol{e}_2^*$ related to the unit length before the deformation;
t_i — membrane dimensionless initial stress-resultants;
t — the index of variation of a stress-strain state (see Section 1.3);
s — the length of an arc generatrix (for shells of revolution and shells of zero curvature);
φ — angle in the circumferential direction;
θ — angle between the axis of rotation and the normal (for a shell of revolution);
u_i, w (u, v, w) — projections of displacement;
ω_i, γ_i — angles of rotation of unit vectors $\boldsymbol{e}_i$, $\boldsymbol{n}$ in the linear approximation;
$\varkappa_i$, τ — bending and torsional strains of the neutral surface in the linear approximation.

Chapter 1

Equations of Thin Elastic Shell Theory

In this chapter we will consider the relations of the theory of surfaces and surface deformations, the equilibrium equations of shell theory, the constitutive relations and some approximations of these equations and relations. The reader can become acquainted with the derivation and detailed discussion of these equations in monographs [12, 24, 28, 33, 51, 52, 87, 111, 119] and others.

1.1 Elements of Surface Theory

Consider the curvilinear coordinate system α, β on the undeformed shell neutral surface. The system axes coincide with the lines of curvature. The position of a point M on the neutral surface is defined by its radius vector $\boldsymbol{r} = \boldsymbol{r}(\alpha, \beta)$.

We introduce local orthogonal coordinate system by means of unit vectors $\boldsymbol{e}_1$, $\boldsymbol{e}_2$, and $\boldsymbol{n}$, where

$$\boldsymbol{e}_1 = \frac{1}{A}\frac{\partial \boldsymbol{r}}{\partial \alpha}, \quad \boldsymbol{e}_2 = \frac{1}{B}\frac{\partial \boldsymbol{r}}{\partial \beta}, \quad \boldsymbol{n} = \boldsymbol{e}_1 \times \boldsymbol{e}_2, A = \left|\frac{\partial \boldsymbol{r}}{\partial \alpha}\right|, \quad B = \left|\frac{\partial \boldsymbol{r}}{\partial \beta}\right|. \tag{1.1.1}$$

The first and second quadratic forms of the surface are:

$$\mathrm{I} = ds^2 = A^2 d\alpha^2 + B^2 d\beta^2, \qquad \mathrm{II} = \frac{A^2}{R_1} d\alpha^2 + \frac{B^2}{R_2} d\beta^2, \tag{1.1.2}$$

where ds is the arc length on the surface, R_1, R_2 are the principal radii of curvature. We also use the notation $k_i = R_i^{-1}$.

Here A, B, R_1, and R_2 are the functions of the coordinates α and β satisfying the Codazzi–Gauss relations.

$$\begin{gathered}
\frac{\partial}{\partial\beta}\left(\frac{A}{R_1}\right)=\frac{1}{R_2}\frac{\partial A}{\partial\beta},\\
\frac{\partial}{\partial\alpha}\left(\frac{B}{R_2}\right)=\frac{1}{R_1}\frac{\partial B}{\partial\alpha},\\
\frac{1}{AB}\left(\frac{\partial}{\partial\alpha}\left(\frac{1}{A}\frac{\partial B}{\partial\alpha}\right)+\frac{\partial}{\partial\beta}\left(\frac{1}{B}\frac{\partial A}{\partial\beta}\right)\right)=-\frac{1}{R_1R_2}.
\end{gathered}\tag{1.1.3}$$

The following formulae for the differentiation of unit vectors are valid:

$$\begin{aligned}
\frac{\partial \boldsymbol{e}_1}{\partial\alpha}&=-\frac{1}{B}\frac{\partial A}{\partial\beta}\boldsymbol{e}_2+\frac{A}{R_1}\boldsymbol{n}, &\frac{\partial \boldsymbol{e}_1}{\partial\beta}&=\frac{1}{A}\frac{\partial B}{\partial\alpha}\boldsymbol{e}_2,\\
\frac{\partial \boldsymbol{e}_2}{\partial\alpha}&=\frac{1}{B}\frac{\partial A}{\partial\beta}\boldsymbol{e}_1, &\frac{\partial \boldsymbol{e}_2}{\partial\beta}&=-\frac{1}{A}\frac{\partial B}{\partial\alpha}\boldsymbol{e}_1+\frac{B}{R_2}\boldsymbol{n},\\
\frac{\partial \boldsymbol{n}}{\partial\alpha}&=-\frac{A}{R_1}\boldsymbol{e}_1 &\frac{\partial \boldsymbol{n}}{\partial\beta}&=-\frac{B}{R_2}\boldsymbol{e}_2.
\end{aligned}\tag{1.1.4}$$

After deformation, a point M occupies position M^*, which is defined by the radius vector $\boldsymbol{r}^*$. We assume that

$$\boldsymbol{r}^*=\boldsymbol{r}+\boldsymbol{U},\qquad \boldsymbol{U}=u_1\boldsymbol{e}_1+u_2\boldsymbol{e}_2+w\,\boldsymbol{n},\tag{1.1.5}$$

where $\boldsymbol{U}$ is the displacement vector and u_1, u_2 and w are its projections on the unit vectors before deformation.

Assuming

$$\begin{aligned}
\frac{1}{A}\frac{\partial\boldsymbol{r}^*}{\partial\alpha}&=(1+\varepsilon_1)\,\boldsymbol{e}_1+\omega_1\boldsymbol{e}_2-\gamma_1\boldsymbol{n},\\
\frac{1}{B}\frac{\partial\boldsymbol{r}^*}{\partial\beta}&=(1+\varepsilon_2)\,\boldsymbol{e}_2+\omega_2\boldsymbol{e}_1-\gamma_2\boldsymbol{n}.
\end{aligned}\tag{1.1.6}$$

we get

$$\begin{aligned}
\varepsilon_1&=\frac{1}{A}\frac{\partial u_1}{\partial\alpha}+\frac{1}{AB}\frac{\partial A}{\partial\beta}u_2-\frac{w}{R_1},\quad \varepsilon_2=\frac{1}{B}\frac{\partial u_2}{\partial\beta}+\frac{1}{AB}\frac{\partial B}{\partial\alpha}u_1-\frac{w}{R_2},\\
\omega_1&=\frac{1}{A}\frac{\partial u_2}{\partial\alpha}-\frac{1}{AB}\frac{\partial A}{\partial\beta}u_1,\quad \omega_2=\frac{1}{B}\frac{\partial u_1}{\partial\beta}-\frac{1}{AB}\frac{\partial B}{\partial\alpha}u_2,\\
\gamma_1&=-\frac{1}{A}\frac{\partial w}{\partial\alpha}-\frac{u_1}{R_1},\quad \gamma_2=-\frac{1}{B}\frac{\partial w}{\partial\beta}-\frac{u_2}{R_2}.
\end{aligned}\tag{1.1.7}$$

After deformation the coordinates α and β become non-orthogonal. We then introduce unit vectors

$$\boldsymbol{e}_1^*=\frac{1}{A_*}\frac{\partial\boldsymbol{r}^*}{\partial\alpha},\qquad \boldsymbol{e}_2^*=\frac{1}{B_*}\frac{\partial\boldsymbol{r}^*}{\partial\beta},\qquad \boldsymbol{n}^*=\frac{\boldsymbol{e}_1^*\times\boldsymbol{e}_2^*}{\sin\theta}.\tag{1.1.8}$$

where θ is the angle between $\boldsymbol{e}_1^*$ and $\boldsymbol{e}_2^*$ $(\boldsymbol{e}_1^*\boldsymbol{e}_2^* = \cos\theta)$.

In the deformed state the first quadratic form of the surface is

$$\mathrm{I}_* = A_*^2\, d\alpha^2 + 2A_*B_* \cos\theta\, d\alpha\, d\beta + B_*^2\, d\beta^2, \tag{1.1.9}$$

where

$$\begin{gathered}\varepsilon_{11} = \varepsilon_1 + \frac{1}{2}\left(\varepsilon_1^2 + \omega_1^2 + \gamma_1^2\right), \qquad \varepsilon_{22} = \varepsilon_2 + \frac{1}{2}\left(\varepsilon_2^2 + \omega_2^2 + \gamma_2^2\right), \\ \cos\theta = \varepsilon_{12} = \omega + \varepsilon_1\omega_2 + \varepsilon_2\omega_1 + \gamma_1\gamma_2, \\ A_*^2 = A^2\left(1 + 2\varepsilon_{11}\right), \quad B_*^2 = B^2\left(1 + 2\varepsilon_{22}\right), \quad \omega = \omega_1 + \omega_2.\end{gathered} \tag{1.1.10}$$

The values ε_{11}, ε_{12} and ε_{22} are the membrane strains of the neutral surface and we assume that they are small compare to unity. The variables ε_1, ω and ε_2 are the membrane strains in the linear approximation. For small deformations ω_1 and ω_2 are the rotation angles of the unit vectors $\boldsymbol{e}_1$, $\boldsymbol{e}_2$ in the tangential plane, and γ_1 and γ_2 are the rotation angles for the normal $\boldsymbol{n}$.

After deformation the second quadratic form of the surface may be represented as

$$\mathrm{II}_* = L_{11} d\alpha^2 + 2L_{12} d\alpha\, d\beta + L_{22} d\beta^2, \tag{1.1.11}$$

where

$$L_{11} = \frac{\partial^2 \boldsymbol{r}^*}{\partial\alpha^2}\boldsymbol{n}^*, \qquad L_{22} = \frac{\partial^2 \boldsymbol{r}^*}{\partial\beta^2}\boldsymbol{n}^* \qquad L_{12} = \frac{\partial^2 \boldsymbol{r}^*}{\partial\alpha\,\partial\beta}\boldsymbol{n}^*. \tag{1.1.12}$$

The expressions for L_{ij} can be found by formulae (4), (6) and (8) which are complicated non-linear functions of the displacements. We can write the approximate formulae in which the terms of the second order with respect to the displacements are neglected

$$\begin{aligned}\frac{1}{R_1^*} &= \frac{L_{11}}{A_*^2} = \frac{1}{R_1} + \varkappa_1 - \frac{\varepsilon_1}{R_1}, \\ \frac{1}{R_2^*} &= \frac{L_{22}}{B_*^2} = \frac{1}{R_2} + \varkappa_2 - \frac{\varepsilon_2}{R_2}, \\ \tau &= \frac{L_{12}}{A_*B_*} = -\frac{1}{B}\frac{\partial\gamma_1}{\partial\beta} + \frac{1}{AB}\frac{\partial B}{\partial\alpha}\gamma_2 + \frac{\omega_1}{R_2} = \\ &\quad -\frac{1}{A}\frac{\partial\gamma_2}{\partial\alpha} + \frac{1}{AB}\frac{\partial A}{\partial\beta}\gamma_1 + \frac{\omega_2}{R_1}, \\ \varkappa_1 &= -\frac{1}{A}\frac{\partial\gamma_1}{\partial\alpha} - \frac{1}{AB}\frac{\partial A}{\partial\beta}\gamma_2, \\ \varkappa_2 &= -\frac{1}{B}\frac{\partial\gamma_2}{\partial\beta} - \frac{1}{AB}\frac{\partial B}{\partial\alpha}\gamma_1.\end{aligned} \tag{1.1.13}$$

With the same precision the derivatives of the unit vectors after deformation have the form [61]

$$\begin{aligned}
\frac{\partial \boldsymbol{e}_1^*}{\partial \alpha} &= \frac{1}{B}\frac{\partial A}{\partial \beta}\,\omega \boldsymbol{e}_1^* + \left(\frac{1}{B}\frac{\partial (B\omega)}{\partial \alpha} - \frac{1}{B_*}\frac{\partial A_*}{\partial \beta}\right)\boldsymbol{e}_2^* + A\left(\frac{1}{R_1} + \varkappa_1\right)\boldsymbol{n}^*,\\
\frac{\partial \boldsymbol{e}_2^*}{\partial \beta} &= \frac{1}{A}\frac{\partial B}{\partial \alpha}\,\omega \boldsymbol{e}_2^* + \left(\frac{1}{A}\frac{\partial (A\omega)}{\partial \beta} - \frac{1}{A_*}\frac{\partial B_*}{\partial \alpha}\right)\boldsymbol{e}_1^* + B\left(\frac{1}{R_2} + \varkappa_2\right)\boldsymbol{n}^*,\\
\frac{\partial \boldsymbol{e}_2^*}{\partial \alpha} &= \left(\frac{1}{B_*}\frac{\partial A_*}{\partial \beta} - \frac{\omega}{B}\frac{\partial B}{\partial \alpha}\right)\boldsymbol{e}_1^* - \frac{\omega}{B}\frac{\partial A}{\partial \beta}\boldsymbol{e}_2^* + A\tau\boldsymbol{n}^*,\\
\frac{\partial \boldsymbol{e}_1^*}{\partial \beta} &= \left(\frac{1}{A_*}\frac{\partial B_*}{\partial \alpha} - \frac{\omega}{A}\frac{\partial A}{\partial \beta}\right)\boldsymbol{e}_2^* - \frac{\omega}{A}\frac{\partial B}{\partial \alpha}\boldsymbol{e}_1^* + B\tau\boldsymbol{n}^*, \qquad (1.1.14)\\
\frac{\partial \boldsymbol{n}^*}{\partial \alpha} &= -A\left(\frac{1}{R_1} + \varkappa_1\right)\boldsymbol{e}_1^* - A\left(\tau - \frac{\omega}{R_1}\right)\boldsymbol{e}_2^*,\\
\frac{\partial \boldsymbol{n}^*}{\partial \beta} &= -B\left(\frac{1}{R_2} + \varkappa_2\right)\boldsymbol{e}_2^* - B\left(\tau - \frac{\omega}{R_2}\right)\boldsymbol{e}_1^*.
\end{aligned}$$

In the linear approximation the relationships for the unit vectors both before and after the deformation are the following

$$\begin{aligned}
\boldsymbol{e}_1^* &= \boldsymbol{e}_1 + \omega_1\boldsymbol{e}_2 - \gamma_1\boldsymbol{n},\\
\boldsymbol{e}_2^* &= \boldsymbol{e}_2 + \omega_2\boldsymbol{e}_1 - \gamma_2\boldsymbol{n}, \qquad (1.1.15)\\
\boldsymbol{n}^* &= \boldsymbol{n} + \gamma_1\boldsymbol{e}_1 + \gamma_2\boldsymbol{e}_2.
\end{aligned}$$

The discussion below is limited to the small displacement case since the application of the approximate formulae (13)–(15) does not allow us to consider problems with finite (but not small) displacements and rotation angles of a shell element. The case of finite displacements and rotation angles is discussed in [87, 26, 131].

1.2 Equilibrium Equations and Boundary Conditions

The equilibrium equations of a deformed element of the shell are

$$\begin{aligned}
\frac{\partial}{\partial \alpha}(B\boldsymbol{F}_1) + \frac{\partial}{\partial \beta}(A\boldsymbol{F}_2) + AB\boldsymbol{q} &= 0,\\
\frac{\partial}{\partial \alpha}(B\boldsymbol{G}_1) + \frac{\partial}{\partial \beta}(A\boldsymbol{G}_2) + A_*B\,\boldsymbol{e}_1^*\times\boldsymbol{F}_1 + A\,B_*\boldsymbol{e}_2^*\times\boldsymbol{F}_2 &= 0.
\end{aligned} \qquad (1.2.1)$$

Here $\boldsymbol{F}_1$ and $\boldsymbol{G}_1$ ($\boldsymbol{F}_2$ and $\boldsymbol{G}_2$) are the stress-resultant and the stress-couple vectors related per the unit length before the deformation acting on the line $\alpha = \text{const}$ ($\beta = \text{const}$) and $\boldsymbol{q}$ is the intensity of the external surface loads per the unit area before the deformation (see Figure 1.1). If the intensity of the external surface loads $\boldsymbol{q}^*$ per the unit area after the deformation is introduced (for example, the hydrostatic pressure) then the last term in the first relation (1) is to be replaced by $A_* B_* \boldsymbol{q}^* \sin\theta$.

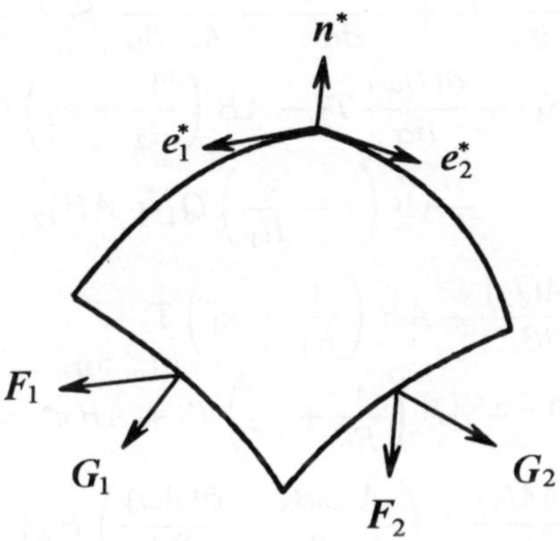

Figure 1.1: The stress-resultants and stress-couples acting on a shell element.

We introduce the projections of these vectors on the basis vectors after deformation

$$\begin{aligned} \boldsymbol{F}_1 &= T_1 \boldsymbol{e}_1^* + S_1 \boldsymbol{e}_2^* + Q_1 \boldsymbol{n}^*, \\ \boldsymbol{F}_2 &= T_2 \boldsymbol{e}_2^* + S_2 \boldsymbol{e}_1^* + Q_2 \boldsymbol{n}^*, \\ \boldsymbol{G}_1 &= H_1 \boldsymbol{e}_1^* - M_1 \boldsymbol{e}_2^*, \qquad \boldsymbol{G}_2 = M_2 \boldsymbol{e}_1^* - H_2 \boldsymbol{e}_2^*, \\ \boldsymbol{q} &= q_1^* \boldsymbol{e}_1^* + q_2^* \boldsymbol{e}_2^* + q^* \boldsymbol{n}^*. \end{aligned} \tag{1.2.2}$$

The scalar equilibrium equations are obtained by projecting equation (1) onto the basis vectors after deformation and using formulae (1.14). Then we can write [61]

$$
\begin{aligned}
&\frac{\partial(BT_1)}{\partial\alpha} - \frac{A}{A_*}\frac{\partial B_*}{\partial\alpha}T_2 + \frac{\partial(AS_2)}{\partial\beta} + \frac{B}{B_*}\frac{\partial A_*}{\partial\beta}S_1 + \frac{\partial A}{\partial\beta}\omega T_1 - \\
&-\frac{\partial B}{\partial\alpha}\omega\,(S_1+S_2) + \frac{\partial(A\omega)}{\partial\beta}T_2 - AB\left(\frac{1}{R_1}+\varkappa_1\right)Q_1 - \\
&\qquad -AB\left(\tau - \frac{\omega}{R_2}\right)Q_2 + ABq_1^* = 0, \\
&\frac{\partial(AT_2)}{\partial\beta} - \frac{B}{B_*}\frac{\partial A_*}{\partial\beta}T_1 + \frac{\partial(BS_1)}{\partial\alpha} + \frac{A}{A_*}\frac{\partial B_*}{\partial\alpha}S_2 + \frac{\partial B}{\partial\alpha}\omega T_2 - \\
&-\frac{\partial A}{\partial\beta}\omega\,(S_2+S_1) + \frac{\partial(B\omega)}{\partial\alpha}T_1 - AB\left(\frac{1}{R_2}+\varkappa_2\right)Q_2 - \\
&\qquad -AB\left(\tau - \frac{\omega}{R_1}\right)Q_1 + ABq_2^* = 0, \\
&\frac{\partial(BQ_1)}{\partial\alpha} + \frac{\partial(AQ_2)}{\partial\beta} + AB\left(\frac{1}{R_1}+\varkappa_1\right)T_1 + \\
&+AB\,(S_1+S_2)\,\tau + AB\left(\frac{1}{R_2}+\varkappa_2\right)T_2 + AB\,q^* = 0, \\
&\frac{\partial(BH_1)}{\partial\alpha} + \frac{\partial(AM_2)}{\partial\beta} + \left(\frac{A}{A_*}\frac{\partial B_*}{\partial\alpha} - \frac{\partial(A\omega)}{\partial\beta}\right)H_2 + \\
&+AB_*Q_2 + AB\omega Q_1 + \left(\frac{\partial B}{\partial\alpha}\omega - \frac{B}{B_*}\frac{\partial A_*}{\partial\beta}\right)M_1 + \\
&\qquad +\frac{\partial A}{\partial\beta}\omega H_1 - \frac{\partial B}{\partial\alpha}\omega M_2 = 0, \\
&\frac{\partial(AH_2)}{\partial\beta} + \frac{\partial(BM_1)}{\partial\alpha} + \left(\frac{B}{B_*}\frac{\partial A_*}{\partial\beta} - \frac{\partial(B\omega)}{\partial\alpha}\right)H_1 + \\
&+BA_*Q_1 + AB\omega Q_2 + \left(\frac{\partial A}{\partial\beta}\omega - \frac{A}{A_*}\frac{\partial B_*}{\partial\alpha}\right)M_2 + \\
&\qquad +\frac{\partial B}{\partial\alpha}\omega H_2 - \frac{\partial A}{\partial\beta}\omega M_1 = 0, \\
&A_*B\,S_1 - AB_*S_2 + AB\left(\frac{1}{R_1}+\varkappa_1\right)H_1 - \\
&\qquad -AB\left(\frac{1}{R_2}+\varkappa_2\right)H_2 - (M_1-M_2)\,AB\tau = 0.
\end{aligned}
\tag{1.2.3}
$$

We consider shells of absolutely elastic isotropic Hookeian material, with Young's modulus E and Poisson's ratio ν. To derive two-dimensional shell theory equations we use the Kirchhoff–Love hypotheses. Let us take the constitutive relations in the form introduced by V. V. Novozhilov and L. I. Balabukh [119]

$$
\begin{aligned}
T_1 &= K(\varepsilon_{11} + \nu\,\varepsilon_{22}), \qquad S_1 = \frac{K(1-\nu)}{2}\left(\varepsilon_{12} + \frac{h^2\tau}{6R_2}\right),\\
T_2 &= K(\varepsilon_{22} + \nu\,\varepsilon_{11}), \qquad S_2 = \frac{K(1-\nu)}{2}\left(\varepsilon_{12} + \frac{h^2\tau}{6R_1}\right),\\
M_1 &= D(\varkappa_1 + \nu\,\varkappa_2), \qquad M_2 = D(\varkappa_2 + \nu\,\varkappa_1),\\
H_1 &= H_2 = H = D(1-\nu)\,\tau,\\
K &= \frac{Eh}{1-\nu^2}, \qquad D = \frac{Eh^3}{12(1-\nu^2)}.
\end{aligned}
\tag{1.2.4}
$$

The error due to these relations is discussed in Section 1.3.

System (3) contains terms of different orders. For elastic shells made of metal, the values of ε_{11}, ε_{12}, ε_{22}, ε_1, ω, ε_2 are small compared to unity. Therefore, in order to simplify system (3) these terms may be neglected compared with 1. It assumed that the functions A and B have equal orders and increase weakly with differentiation. As a result, it may be assumed that in system (3)

$$
A_* = A, \qquad B_* = B \tag{1.2.5}
$$

and the terms containing ω may be omit. After these simplifications system (3) has the form:

$$
\begin{aligned}
&\frac{\partial(BT_1)}{\partial\alpha} - \frac{\partial B}{\partial\beta}T_2 + \frac{\partial(AS_2)}{\partial\beta} + \frac{\partial A}{\partial\beta}S_1 + \\
&\qquad +AB\left[q_1^* - \left(\frac{1}{R_1} + \varkappa_1\right)Q_1 - \left(\tau - \frac{\omega}{R_2}\right)Q_2\right] = 0,\\
&\frac{\partial(AT_2)}{\partial\beta} - \frac{\partial A}{\partial\alpha}T_1 + \frac{\partial(BS_1)}{\partial\alpha} + \frac{\partial B}{\partial\alpha}S_2 + \\
&\qquad +AB\left[q_2^* - \left(\frac{1}{R_2} + \varkappa_2\right)Q_2 - \left(\tau - \frac{\omega}{R_1}\right)Q_1\right] = 0,\\
&\frac{\partial(BQ_1)}{\partial\alpha} + \frac{\partial(AQ_2)}{\partial\beta} + AB\left[\left(\frac{1}{R_1} + \varkappa_1\right)T_1 + \right.\\
&\qquad \left. +\tau(S_1 + S_2) + \left(\frac{1}{R_2} + \varkappa_2\right)T_2 + q^*\right] = 0,
\end{aligned}
\tag{1.2.6}
$$

$$\frac{\partial (BH)}{\partial \alpha} + \frac{\partial (AM_2)}{\partial \beta} - \frac{\partial A}{\partial \beta} M_1 + \frac{\partial B}{\partial \alpha} H + ABQ_2 = 0,$$
$$\frac{\partial (AH)}{\partial \beta} + \frac{\partial (BM_1)}{\partial \alpha} - \frac{\partial B}{\partial \alpha} M_2 + \frac{\partial A}{\partial \beta} H + ABQ_1 = 0.$$

By virtue of relations (4), the last equation (3) is satisfied identically and is not included in system (6).

formulae (2) give the projections of the surface loads on the axes after deformation and they are convenient in the case of the follower loading case. If the load is introduced by its projections on the axes before deformation

$$\boldsymbol{q} = q_1\boldsymbol{e}_1 + q_2\boldsymbol{e}_2 + q\boldsymbol{n}, \tag{1.2.7}$$

then, by virtue of (1.15) in equations (3) and (6)

$$\begin{aligned} q_1^* &= q_1 - \omega_2 q_2 - \gamma_1 q, \\ q_2^* &= q_2 - \omega_1 q_1 - \gamma_2 q, \\ q^* &= q + \gamma_1 q_1 + \gamma_2 q_2. \end{aligned} \tag{1.2.8}$$

Systems (3) or (6) are of the eighth order. Therefore, four boundary conditions must be introduced at each shell edge. First, we consider edge $\alpha = \alpha^0$. If the edge is clamped we introduce the values of

$$u_1, \quad u_2, \quad w, \quad \gamma_1. \tag{1.2.9}$$

If some generalized displacements (9) are not given then we introduce the corresponding generalized forces:

$$T_1, \quad S_1 + \frac{H}{R_2}, \qquad Q_1 - \frac{1}{B}\frac{\partial H}{\partial \beta}, \qquad M_1. \tag{1.2.10}$$

At a free edge all variables (10) must be set.

Instead of (9) and (10) we can also write the expressions for geometric and static variables that must be given at a shell edge which does not coincide with the coordinate line (see [24, 51]).

Let $\boldsymbol{t}$ be the unit vector of a tangent to the edge and $\boldsymbol{m}$ be the unit vector of the normal laid in the tangent plane and directed incide the shell, χ is the angle between the unit vectors $\boldsymbol{e}_1$ and $\boldsymbol{m}$ (see Figure 1.2).

We can write the relations for the generalized displacements and stress-resultants corresponding to them, which are to be introduced at this edge

$$\begin{aligned} u_m &= u_1 \cos\chi + u_2 \sin\chi, \\ v_t &= -u_1 \sin\chi + u_2 \cos\chi, \\ w&, \\ \gamma_m &= \gamma_1 \cos\chi + \gamma_2 \sin\chi; \end{aligned} \tag{1.2.11}$$

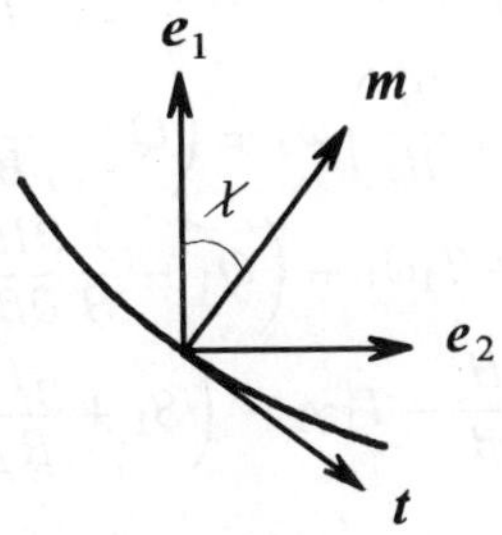

Figure 1.2: The slanted edge.

$$
\begin{aligned}
T_m &= T_1 \cos^2 \chi + (S_1 + S_2) \sin \chi \cos \chi + T_2 \sin^2 \chi - \\
&\qquad - H_{mt} \left(\frac{1}{R_1} - \frac{1}{R_2} \right) \sin \chi \cos \chi, \\
S_t &= (T_2 - T_1) \sin \chi \cos \chi + S_1 \cos^2 \chi - S_2 \sin^2 \chi + \\
&\qquad + H_{mt} \left(\frac{\sin^2 \chi}{R_1} + \frac{\cos^2 \chi}{R_2} \right), \\
Q_m &= \left(Q_1 - \frac{1}{B} \frac{\partial H_{mt}}{\partial \beta} \right) \cos \chi + \left(Q_2 - \frac{1}{A} \frac{\partial H_{mt}}{\partial \alpha} \right) \sin \chi, \\
M_m &= M_1 \cos^2 \chi + 2H \sin \chi \cos \chi + M_2 \sin^2 \chi,
\end{aligned}
\tag{1.2.12}
$$

where

$$
H_{mt} = H \left(\cos^2 \chi - \sin^2 \chi \right) + (M_2 - M_1) \sin \chi \cos \chi \, . \tag{1.2.13}
$$

The stress-resultants in (10) and (12) are introduced by their projections on the unit vectors e_1^*, e_2^*, n^* after deformation and displacements u_i, w are the projections on the unit vectors e_1, e_2, n before deformation. If the external edge load keeps its direction under deformation, it is convenient to use the projections of the stress-resultants on the unit vectors before deformation. Then instead (2), we get

$$
\boldsymbol{F}_1 = T_1' \boldsymbol{e}_1 + S_1' \boldsymbol{e}_2 + Q_1' \boldsymbol{n}. \tag{1.2.14}
$$

Therefore, due to (1.15), instead of generalized forces (10) one must introduce on edge $\alpha = \alpha^0$ the values

$$\begin{aligned} & T_1 + \left(S_1 + \frac{H}{R_2}\right)\omega_2 + \left(Q_1 - \frac{1}{B}\frac{\partial H}{\partial \beta}\right)\gamma_1, \\ & S_1 + \frac{H}{R_2} + T_1\omega_1 + \left(Q_1 - \frac{1}{B}\frac{\partial H}{\partial \beta}\right)\gamma_2, \\ & Q_1 - \frac{1}{B}\frac{\partial H}{\partial \beta} - T_1\gamma_1 - \left(S_1 + \frac{H}{R_2}\right)\gamma_2, \\ & M_1. \end{aligned} \tag{1.2.15}$$

formulae (12) may be transformed in the same way.

In addition to the above assumptions we limit this development to boundary conditions, described as either fixed or free conditions in the generalized directions (9) or (11). We will consider only the simple case when the generalized displacement or generalized force corresponding to it is equal to zero at the edge. In the other words, it is assumed that the boundary stiffness in each direction coinciding with the axes of the trihedron $(\boldsymbol{m}, \boldsymbol{t}, \boldsymbol{n})$, is equal either to infinity or to zero.

Therefore at each edge four boundary conditions should be introduced

$$\begin{aligned} u_m &= 0 \quad \text{or} \quad T_m = 0, \\ v_t &= 0 \quad \text{or} \quad S_t = 0, \\ w &= 0 \quad \text{or} \quad Q_{m*} = 0, \\ \gamma_m &= 0 \quad \text{or} \quad M_m = 0 \end{aligned} \tag{1.2.16}$$

Also we will not consider the cases of elastic support of the type $u_m + \beta\, T_m = 0$ or support by a rib.

We denote variants of the boundary conditions by four-figure numbers of 0's and 1's where 1 indicates the support in the corresponding direction and placing the digits in the same orders as the conditions in (16). For example, 1111 means a clamped edge, 0000 means a free edge and 0110 means a simple support. The boundary conditions are often denoted by C_i and S_i $(i = 1, 2, \ldots, 8)$ [21] (see Table 1.1).

1.3 Errors of 2D Shell Theory of Kirchhoff–Love Type

The entire system of the equations of 2D shell theory of the Kirchhoff–Love type consists of five equilibrium equations (2.6), seven constitutive relations

Table 1.1: The boundary conditions

$u_m = 0$	$v_t = 0$	$w = 0$	$\gamma_m = 0$	1	
$T_m = 0$	$S_t = 0$	$Q_{m*} = 0$	$M_m = 0$	0	
1	1	1	1	C_1	clamped edge
1	1	1	0	S_1	
1	1	0	1	C_5	
1	1	0	0	S_5	
1	0	1	1	C_3	
1	0	1	0	S_3	
1	0	0	1	C_7	
1	0	0	0	S_7	
0	1	1	1	C_2	
0	1	1	0	S_2	simply supported edge (Navier's conditions)
0	1	0	1	C_6	
0	1	0	0	S_6	
0	0	1	1	C_4	
0	0	1	0	S_4	
0	0	0	1	C_8	
0	0	0	0	S_8	free edge

(2.4), nine relations for small strains, angles of rotation and displacements (1.7) and (1.13) and three formulae for the nonlinear membrane deformations (1.10). Therefore the entire system consists of 24 algebro-differential equations in 15 cinematic and 9 force unknowns

$$\begin{gathered} u_1,\, u_2,\, w,\, \varepsilon_1,\, \varepsilon_2,\, \omega_1,\, \omega_2,\, \gamma_1,\, \gamma_2,\, \varkappa_1,\, \varkappa_2,\, \tau,\, \varepsilon_{11},\, \varepsilon_{22},\, \varepsilon_{12}, \\ T_1,\, T_2,\, S_1,\, S_2,\, M_1,\, M_2,\, H,\, Q_1,\, Q_2 \end{gathered} \tag{1.3.1}$$

The solution of this system with four boundary conditions on each edge is called *the general stress-strain state* of a shell. In the cases when the deformations may be ignored the stress-strain state is called shortly *the stress state*.

In this section the error arose under transition from 3D nonlinear equations of the theory of elasticity to 2D equations of the theory of shells is discussed. We also give the classification of the partial stress-strain states of a shell, which not necessarily satisfy the boundary conditions. As a rule for linear problems the general stress-strain state is a sum of the partial stress-strain states.

To make the asymptotic estimates and to compare the orders of the specific terms we introduce, for convenience, a small thickness parameter

$$h_* = \frac{h_0}{R}, \tag{1.3.2}$$

where R is a characteristic linear size of the neutral surface, for example, the radius of the curvature, and h_0 is the characteristic shell thickness. In the case of a shell of a constant thickness we assume $h_0 = h$.

Errors in shell theory equations are caused, in the most cases, by the Kirchhoff–Love hypotheses, which leads to one or another variant of the constitutive relations. At the same time, the expressions for strains and the equilibrium equations generally can be written exactly.

Firstly we discuss the errors of the linear shell theory. The equations of the shell theory contain linear and quadratic terms in unknowns (1). The linear theory is obtained assuming all quadratic terms be equal to zero.

In [121] it was shown that for the linear theory of shells the error Δ of the Kirchhoff–Love hypotheses has the order of the relative shell thickness, i.e.

$$\Delta \sim h_*. \tag{1.3.3}$$

In (see [52, 77]) the similar estimates for the shell stress-strain state unessentially differed from the above one have been obtained.

Let us introduce the index of variation t of stress-strain state as

$$\max\left\{\left|\frac{\partial z}{\partial \alpha}\right|, \left|\frac{\partial z}{\partial \beta}\right|\right\} \sim h_*^{-t} z, \tag{1.3.4}$$

where z is any unknown function which determines this state. We will determine, through $\sim$, the values of the same asymptotic order. A more general error estimation than (3) has the form

$$\Delta \sim \max\left\{h_*, h_*^{2-2t}\right\}, \tag{1.3.5}$$

and it follows from here that estimates (3) and (5) coincide as $t = 1/2$. In [52] it was shown that, by choosing more complicated constitutive relations than (2.4) the error Δ can be reduced to

$$\Delta \sim h_*^{2-2t}, \qquad 0 \leqslant t < 1, \tag{1.3.6}$$

as $t < 1/2$. The advantages and disadvantages of this approach are discussed in [52].

Remark 1.1. The error of the shell theory equations is denoted in formulae (3), (5) and (6) by Δ. The error of the fixed equation is equal to a fraction of the maximum of the omitted terms and the main term of this equation. The error of the system is equal to the maximum of its equation errors. The error of the system depends on the type of solution (mainly on its index t of variation, see (4)). It may occur that the order, Δ_s, of the error of the solution is not equal to the order, Δ, of the error of the system. For example in [56] for the problem of free vibrations of a shell of revolution it was shown that the solution error is larger than the system error. Instead of (6) the estimate $\Delta_s \sim h_*^{2-3t}$ is valid. A similar growth of error may also occur when buckling modes are analyzed. Here we do not consider this problem. For typical buckling modes $t \leqslant 1/2$ and we may expect that $\Delta_s \lesssim h_*^{1/2}$ as $\Delta \sim h_*$. We note also that for the integral solution characteristics, such as the frequency of vibrations or the critical load, the order of their errors Δ_λ does not exceed the order of the system error Δ.

Later in this book, the constitutive relations of type (2.4) will be used. That is why in each equation, terms of order h_* and smaller in comparison with the main term may be neglected without losing accuracy (see [120]).

In linear shell theory the problem of comparing the orders of terms for characteristic partial stress-strain states such as membrane, pure bending, semi-momentless, simple edge effect, or a general stress-strain state which is a combination of some of these stress-strain states and the definition of which are given below has been studied extensively [24, 52, 57].

The order of each of the non-linear terms depends on the level of the external surface and boundary loading. Each buckling problem is characterized by its own level of load and pre-buckling strains.

We denote the maximum value of the membrane pre-buckling strains by ε

$$\varepsilon = \max\{|\varepsilon_{11}|, |\varepsilon_{12}|, |\varepsilon_{22}|\}\,. \tag{1.3.7}$$

Further, in relations similar to (7) we will omit the module sign in estimating the absolute values of the variables. In the book we only consider the pre-buckling stress-strain states for which the following estimate is valid

$$\varepsilon \lesssim h_*, \tag{1.3.8}$$

which means that the order of ε is not larger than the order of h_*. In particular the estimate

$$\varepsilon \sim h_* \tag{1.3.9}$$

is valid for the buckling of a membrane stress-strain state for a convex shell (see Section 3.1) and also for a moderately long cylindrical shell under axial compression (see Section 3.4).

For the buckling of a cylindrical shell under external pressure (see Section 3.5) or torsion (see Section 3.6) one can use the estimate

$$\varepsilon \ll h_*. \tag{1.3.10}$$

Let us now consider one formula of (1.10)

$$\varepsilon_{11} = \varepsilon_1 + \frac{1}{2}\left(\varepsilon_1^2 + \omega_1^2 + \gamma_1^2\right). \tag{1.3.11}$$

Generally, a small value of ε_{11} does not necessarily lead to small values of ε_1, ω_1^2, and γ_1^2. Indeed for pure plate bending, under which the plate turns into a cylindrical shell, $\varepsilon_{11} = 0$, but $\varepsilon_1, \gamma_1 \sim 1$. That is why we assume that both the linear membrane strains and the squares of the angles of rotation are also small variables

$$\max\left\{\varepsilon_i,\ \omega,\ \omega_i^2,\ \gamma_i^2\right\} \lesssim h_*. \tag{1.3.12}$$

Assumption (12) limits the class of buckling problems considered below. In fact, this assumption has already been used in Sections 1.1 and 1.2 to simplify the geometric relations and equilibrium equations.

In some buckling problems characterized by a large bending of the neutral surface estimates (8) does not hold. In particular, in the case of the axisymmetric deformation of a shell of revolution under axial compression the part of the neutral surface has the form close to the mirror reflection from the plane orthogonal to the axis of symmetry OO. In Figure 1.3 the generatrix of the shell of revolution before the deformation (solid line) and after the deformation (dashed line) is shown.

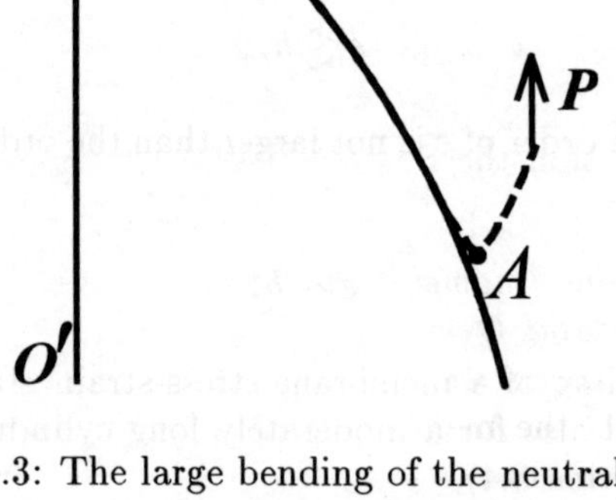

Figure 1.3: The large bending of the neutral surface

In the area of the large bending (near the point A in Figure 1.3) the following estimates are valid

$$\{\varepsilon_1, \varepsilon_2\} \sim h_*^{1/2}, \quad \varkappa_1 \sim R^{-1} h_*^{-1/2}. \tag{1.3.13}$$

The error Δ increases and for its order instead of (3) we have

$$\Delta \sim h_*^{1/2}. \tag{1.3.14}$$

In [167, 168] the constitutive relations for such type of problems have been obtained by means of the asymptotic integration of the 3D nonlinear equations. These relations generalizing equations (2.4) have the relative error of order h_*. In particular, expression for T_1 has the form

$$T_1 = K(\varepsilon_1 + \nu\varepsilon_2 + b_1\varepsilon_1^2 + Rb_2 h_*^2 \varkappa_1^2), \tag{1.3.15}$$

where the additional terms with the multipliers b_1 and b_2 take into account the quadratic dependence of stresses on strains and in their turn the multipliers b_1 and b_2 depend on the nonlinear elastic properties of shell material.

Errors of 2D shell equations and of their simplified forms are discussed in [13].

In the book the problems involving the estimates (13) and (14) are not analyzed. In Section 14.1 one problem in which $\varepsilon \sim h_*^{1/2}$ is pointed out.

The special assumptions on the character of the stress-strain state and the form of the neutral surface lead to further simplification of equations (2.6), (2.4), (1.7), (1.10), (1.13). Below we give the classification of the characteristic stress-strain states of a shell proposed by Goldenveiser [52] and Novozhilov [119]. This classification is based upon the relations between tangential and bending strains (stresses) and the index of variation of the stress-strain state, t. The maximal tensile-compressive stresses, σ_s and bending stresses σ_b are equal to

$$\sigma_s = \frac{1}{h}\max\{T_1, T_2, S_1, S_2\}, \qquad \sigma_b = \frac{12}{h^2}\max\{M_1, M_2, H\}. \tag{1.3.16}$$

Further in this section the relations between the orders of σ_s and σ_b obtained in [52] are given.

1) For *the membrane (momentless)* stress-strain state $\sigma_s \gg \sigma_b$ and $t = 0$. The shear stress-resultants Q_1 and Q_2 may be omit in the first three relations (2.6). In the next chapters we consider membrane pre-buckling stress-strain state. In more details the membrane stress-strain state is discussed in Section 1.4. For the membrane stress-strain state with the index of variation

$t = 0$ the following order relation $\sigma_b \sim h_*^2 \sigma_s$ is valid. If we suppose that $0 < t < 1/2$ then

$$\sigma_b \sim h_*^{2-4t} \sigma_s \qquad \text{as} \qquad 0 < t < 1/2. \tag{1.3.17}$$

2) For *the pure bendings* stress-strain state , the shell neutral surface deforms without tensile and shear, i.e. $\varepsilon_{11} = \varepsilon_{12} = \varepsilon_{22} = 0$. If the nonlinear tangential strains are replaced with their linear approximations, then the stress-strain state with $\varepsilon_1 = \omega = \varepsilon_2 = 0$ is called *infinitesimal pure bendings.* In accordance with assumption (12) we do not make a difference between pure bending and infinitesimal pure bendings. The buckling modes for the weakly supported shell of revolution are close to the pure bendings.

3) If the tensile deformations of the shell neutral surface are small compared to its bending deformations, i.e.

$$\max\{\varepsilon_1, \omega, \varepsilon_2\} \ll R \max\{\varkappa_1, \tau, \varkappa_2\}, \tag{1.3.18}$$

then the corresponding stress-strain state is called *the pseudo-bendings.* Such buckling modes are characteristic for shells of zero and negative Gaussian curvature.

4) The bending of the shell neutral surface generates *the pure moment* stress-strain state, for which the bending deformations and stresses are significantly larger than the tangential deformations. For $0 < t < 1/2$ the relation opposite to (17)

$$\sigma_s \sim h_*^{2-4t} \sigma_b \qquad \text{as} \qquad 0 < t < 1/2. \tag{1.3.19}$$

is valid.

5) Under buckling of shells of zero Gaussian curvature under external pressure and/or torsion the pseudo-bending leads to *the semi-momentless* stress-strain state, for which one of the bending moments may be neglected $M_1 \ll M_2$. For such stress-strain state the indices of variations in axial and circumferential directions (so-called partial indices of variation), $t_1 = 0$ and $t_2 = 1/4$ are different

$$\frac{\partial z}{\partial \alpha} \sim z, \qquad \frac{\partial z}{\partial \beta} \sim h_*^{-1/4} z,$$

where z denotes any of variables (1).

6) For *the combined* stress-strain state the tangential and bending stresses have equal orders $\sigma_s \sim \sigma_b$. As it follows from relations (17) and (19) for this state $t = 1/2$. The detailed description of this state is given in Section 1.5 (see formulae (5.1) and (5.2)). The buckling of convex shells and cylindrical shells under axial compression is accompanied with the combined stress-strain state.

7) The above mentioned stress-strain states do not always satisfy the boundary conditions and the stress-strain state called *the edge effect* may form in the

neighbourhood of the shell edges. Consider the edge $\alpha = \alpha_0$. All variables (1) describing the edge effect decrease exponentially away from the shell edge and have different indices of variations in the edge direction, $t_1 < 1/2$, and in the direction that is orthogonal to the edge, $t_2 = 1/2$:

$$\frac{\partial z}{\partial \alpha} \sim h_*^{-1/2} z, \qquad \frac{\partial z}{\partial \beta} \sim h_*^{-t_2} z, \quad t_2 < 1/2. \tag{1.3.20}$$

If $t_2 = 0$, then the edge effect is called *the simple edge effect.* If the edge effect depends on the pre-buckling stresses it is called *the non-linear edge effect* (see Section 14.1).

8) The partial case of the general stress-strain state is called *the moment* stress-strain state if it is a combination of the membrane (or semi-momentless) stress-strain state and the edge effect. In particular, estimate (3) is valid for moment stress-strain states.

Next we write the simplified equations for the characteristic stress-strain states of a shell.

1.4 Membrane Stress State

This state is characterized by slow variation of unknown functions ($t = 0$). Displacements u_i and w are small variables of equal orders. In the general case, the strains ε_i, ω, γ_i, and $R\varkappa_i$ are also small variables of equal orders. The bending stresses are less than the membrane stresses

$$\{M_i R^{-1}, H R^{-1}, Q_i\} \sim h_*^2 \{T_i, S\}. \tag{1.4.1}$$

The membrane stresses satisfy the non-linear system of equations

$$\begin{aligned}
&\frac{\partial (BT_1)}{\partial \alpha} - \frac{\partial B}{\partial \alpha} T_2 + \frac{\partial (A^2 S)}{A\,\partial \beta} + ABq_1^* = 0, \\
&\frac{\partial (AT_2)}{\partial \beta} - \frac{\partial A}{\partial \beta} T_1 + \frac{\partial (B^2 S)}{B\,\partial \alpha} + ABq_2^* = 0, \\
&(k_1 + \varkappa_1)\, T_1 + (k_2 + \varkappa_2)\, T_2 + 2S\tau + q^* = 0,
\end{aligned} \tag{1.4.2}$$

$$T_1 = K\left(\varepsilon_{11} + \nu\varepsilon_{22}\right), \quad T_2 = K\left(\varepsilon_{22} + \nu\varepsilon_{11}\right), \quad S = \frac{1-\nu}{2} K\varepsilon_{12}.$$

If the curvatures k_i are not equal or close to zero in the entire domain being considered, then in the third equation of system (2) the non-linear terms $\varkappa_i T_i$, τS may be neglected. Further, if the surface deformation in the entire domain is far from the bending $\varepsilon_i = \omega = 0$, then in constitutive relations (2) the non-linear membrane strains may be linearized

$$\begin{aligned} \frac{\partial (BT_1)}{\partial \alpha} - \frac{\partial B}{\partial \alpha} T_2 + \frac{\partial (A^2 S)}{A\,\partial \beta} + ABq_1 &= 0, \\ \frac{\partial (AT_2)}{\partial \beta} - \frac{\partial A}{\partial \beta} T_1 + \frac{\partial (B^2 S)}{B\,\partial \alpha} + ABq_2 &= 0, \\ k_1 T_1 + k_2 T_2 + q &= 0, \end{aligned} \tag{1.4.3}$$

$$T_1 = K(\varepsilon_1 + \nu\varepsilon_2), \quad T_2 = K(\varepsilon_2 + \nu\varepsilon_1), \quad S = \frac{1-\nu}{2} K\omega,$$

i.e. we get linear membrane equations [52, 119].

Let us write the explicit formulae for the membrane stresses for some particular load cases.

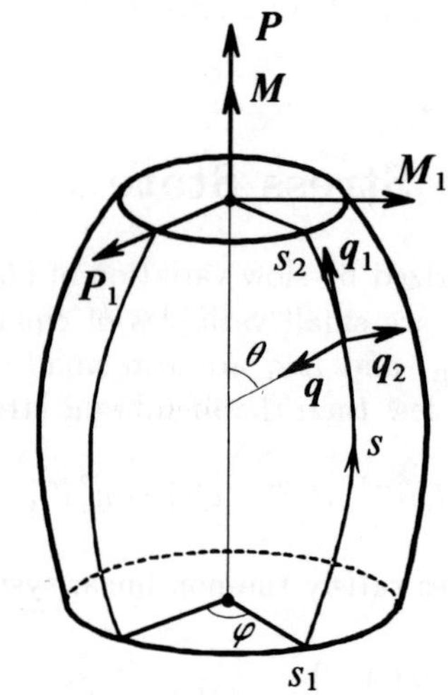

Figure 1.4: Shell of Revolution.

Consider a shell of revolution where as the curvilinear coordinates α, β we introduce the meridian arc length s and the circular angle φ (see Figure 1.4). Then

$$A = 1, \quad B = R_2 \sin\theta, \quad \frac{d\theta}{ds} = -\frac{1}{R_1}, \quad \frac{dB}{ds} = -\cos\theta, \tag{1.4.4}$$

where θ is the angle between the axis of rotation and the normal on the neutral surface. Let the shell be limited by two parallels ($s_1 \leqslant s \leqslant s_2$) and be under the axially symmetric and inversely symmetric (which, in effect, resembles a wind load on a building) surface and edge loads. We assume that

$$\begin{aligned} q_1(s,\varphi) &= q_1^0(s) + q_1^1(s)\cos\varphi, \\ q_2(s,\varphi) &= q_2^0(s) + q_2^1(s)\sin\varphi, \\ q(s,\varphi) &= q^0(s) + q^1(s)\cos\varphi. \end{aligned} \tag{1.4.5}$$

Then we get (see [16]).

$$\begin{aligned} T_1 &= \frac{F_z R_2}{2\pi B^2} - \frac{M_y R_2}{\pi B^3}\cos\varphi, \qquad T_2 = -\frac{T_1 R_2}{R_1} - qR_2, \\ S &= \frac{M_z}{2\pi B^2} + \left(\frac{M_y R_2\cos\theta}{\pi B^3} - \frac{F_x}{\pi B}\right)\sin\varphi, \end{aligned} \tag{1.4.6}$$

where

$$F_x = P_1 - \pi F_q, \qquad F_q = \int_s^{s_2} \left(q_1^1\cos\theta + q_2^1 + q^1\sin\theta\right) B\, ds,$$

$$M_y = M_1 + P_1 z_2 + \pi\int_s^{s_2} \left(B^2 q^1\cos\theta - (F_q + B^2 q_1^1)\sin\theta\right)\, ds,$$

$$F_z = P + 2\pi\int_s^{s_2} \left(q_1^0\sin\theta - q^0\cos\theta\right) B\, ds, \qquad z_2 = \int_s^{s_2} \sin\theta\, ds,$$

$$M_z = M + 2\pi\int_s^{s_2} q_2^0 B^2\, ds.$$

Here forces P and P_1 and moments M and M_1, acting on edge $s = s_2$, are shown in Figure 1.4, and F_x, F_z, M_y, and M_z are the projections of the principal force vector and principal moment vector of the external edge load and surface load acting on the part of a shell between the parallels s and s_2.

We consider also an arbitrary conic or cylindrical shell. As curvilinear coordinates we take the lines of curvature s, φ in such a way that $A = 1$, $R_1 = \infty$. Let the shell be subjected to surface loading q_1, q_2, q, and stresses $T_1 = f_1(\varphi)$, $S = f_2(\varphi)$, be defined at edge $s = s_2(\varphi)$. Then the shell stress-strain state is a membrane stress state with the stress-resultants (see [52], Chapter 13)

$$
\begin{aligned}
T_1 &= \frac{1}{B}\left[B_2 f_1 + \int_s^{s_2}\left(\frac{\partial S}{\partial \varphi} - T_2\frac{dB}{ds} + Bq_1\right) ds\right], \\
T_2 &= -R_2 q, \qquad B_2 = B\left(s_2\right), && (1.4.7) \\
S &= \frac{1}{B^2}\left[B_2 f_2 + \int_s^{s_2}\left(B\frac{dT_2}{d\varphi} + B^2 q_2\right) ds\right]. && (1.4.8)
\end{aligned}
$$

Assuming $s = s_1$ in (7), we find stresses T_i and S at the other shell edge.

1.5 Technical Shell Theory Equations

Here we will examine the mixed stress-strain state for which the membrane and bending stresses are of the same orders. We need to make some assumptions to characterize this state. Let the index of variation t be $t = 1/2$ and let

$$
w \sim h_* R, \qquad k_i \sim R^{-1}, \qquad u_i \ll w. \tag{1.5.1}
$$

Then from the equilibrium equations, the constitutive relations and the geometrical relations we find the orders of all unknown functions

$$
\begin{aligned}
&\max\{u_i\} \sim h_*^{1/2} w; \\
&\max\{\varepsilon_i, \omega_i, \varepsilon_{ij}\} \sim R^{-1} w \sim h_*; \\
&\max\{\gamma_i\} \sim h_*^{-1/2} R^{-1} w \sim h_*^{1/2}; \\
&\max\{\varkappa_i, \tau\} \sim h_*^{-1} R^{-2} w \sim R^{-1}; && (1.5.2) \\
&\max\{T_i, S\} \sim K R^{-1} w \sim K h_*; \\
&\max\{Q_i\} \sim h_*^{1/2} K R^{-1} w \sim K h_*^{3/2}; \\
&\max\{M_i, H\} \sim h_* K w \sim K R h_*^2.
\end{aligned}
$$

Relations (2) give the orders of the maximum values of variables from one group.

We can use the orders of the relations above to simplify equations (2.6),

(2.4), (1.13), (1.10), (1.7). We then obtain

$$\begin{aligned}
\frac{\partial(BT_1)}{\partial\alpha} - \frac{\partial B}{\partial\alpha}T_2 + \frac{\partial(A^2S)}{A\partial\beta} + ABq_1^* &= 0,\\
\frac{\partial(AT_2)}{\partial\beta} - \frac{\partial A}{\partial\beta}T_1 + \frac{\partial(B^2S)}{B\partial\alpha} + ABq_2^* &= 0,\\
\frac{1}{AB}\left(\frac{\partial(BQ_1)}{\partial\alpha} + \frac{\partial(AQ_2)}{\partial\beta}\right) + (k_1+\varkappa_1)T_1 + & \\
+2\tau S + (k_2+\varkappa_2)T_2 + q^* &= 0,\\
\frac{\partial(AM_2)}{\partial\beta} - \frac{\partial A}{\partial\beta}M_1 + \frac{1}{B}\frac{\partial(B^2H)}{\partial\alpha} + ABQ_2 &= 0,\\
\frac{\partial(BM_1)}{\partial\alpha} - \frac{\partial B}{\partial\alpha}M_2 + \frac{1}{A}\frac{\partial(A^2H)}{\partial\beta} + ABQ_1 &= 0,
\end{aligned} \tag{1.5.3}$$

where

$$\begin{aligned}
T_1 &= K(\varepsilon_{11}+\nu\varepsilon_{22}), & T_2 &= K(\varepsilon_{22}+\nu\varepsilon_{11}),\\
M_1 &= D(\varkappa_1+\nu\varkappa_2), & M_2 &= D(\varkappa_2+\nu\varkappa_1),\\
S &= \frac{1-\nu}{2}K\varepsilon_{12}, & H &= (1-\nu)D\tau,\\
\varepsilon_{11} &= \varepsilon_1+\frac{1}{2}\gamma_1^2, & \varepsilon_1 &= \frac{1}{A}\frac{\partial u_1}{\partial\alpha}+\frac{1}{AB}\frac{\partial A}{\partial\beta}u_2-\frac{w}{R_1},\\
\varepsilon_{22} &= \varepsilon_2+\frac{1}{2}\gamma_2^2, & \varepsilon_2 &= \frac{1}{B}\frac{\partial u_2}{\partial\beta}+\frac{1}{AB}\frac{\partial B}{\partial\alpha}u_1-\frac{w}{R_2},\\
\varepsilon_{12} &= \omega+\gamma_1\gamma_2, & \omega &= \frac{B}{A}\frac{\partial}{\partial\alpha}\left(\frac{u_2}{B}\right)+\frac{A}{B}\frac{\partial}{\partial\beta}\left(\frac{u_1}{A}\right),\\
\gamma_1 &= -\frac{1}{A}\frac{\partial w}{\partial\alpha}, & \gamma_2 &= -\frac{1}{B}\frac{\partial w}{\partial\beta},
\end{aligned}$$

$$\begin{aligned}
\varkappa_1 &= \frac{1}{\partial\alpha}\left(\frac{1}{A}\frac{\partial w}{\partial\alpha}\right)+\frac{1}{AB^2}\frac{\partial A}{\partial\beta}\frac{\partial w}{\partial\beta},\\
\varkappa_2 &= \frac{1}{\partial\beta}\left(\frac{1}{B}\frac{\partial w}{\partial\beta}\right)+\frac{1}{BA^2}\frac{\partial B}{\partial\alpha}\frac{\partial w}{\partial\alpha},\\
\tau &= \frac{1}{B}\frac{\partial}{\partial\beta}\left(\frac{1}{A}\frac{\partial w}{\partial\alpha}\right)-\frac{1}{BA^2}\frac{\partial B}{\partial\alpha}\frac{\partial w}{\partial\beta} =
\end{aligned}$$

$$\frac{1}{A}\frac{\partial}{\partial\alpha}\left(\frac{1}{B}\frac{\partial w}{\partial\beta}\right)-\frac{1}{AB^2}\frac{\partial A}{\partial\beta}\frac{\partial w}{\partial\alpha}.$$

System (3) is known as the system of equations of the technical shell theory [32, 61, 109, 171, 175]. All of the simplifications made in defining this system introduce an error of order h_* due to (1) and (2). Thus, this system may be considered to be exact within the limits of the Kirchhoff–Love hypotheses under the assumptions described above (see (1) and $t = 1/2$).

We note some peculiarities of this system. There are no shear stress-resultants in the first two equations. In the third equation all of the terms are equally significant. In the expressions for ε_{ij}, both terms are of the same orders and there are no tangential displacements in the expressions for γ_i, $\varkappa_i$ and τ. System (3) was used as the starting point for many researchers in this field. We will also consider this system further in the book.

Without surface tangential loading ($q_1^* = q_2^* = 0$) system (3) can be written in the compact form ([61])

$$\begin{aligned} -D\Delta\Delta w + (k_1 + \varkappa_1)\, T_1 + 2\tau S + (k_2 + \varkappa_2)\, T_2 + q^* = 0, \\ \frac{1}{Eh}\Delta\Delta\Phi + k_2\varkappa_2 + k_2\varkappa_1 + \varkappa_1\varkappa_2 - \tau^2 = 0, \end{aligned} \tag{1.5.4}$$

where

$$\Delta w = \frac{1}{AB}\left[\frac{\partial}{\partial\alpha}\left(\frac{B}{A}\frac{\partial w}{\partial\alpha}\right) + \frac{\partial}{\partial\beta}\left(\frac{A}{B}\frac{\partial w}{\partial\beta}\right)\right] \tag{1.5.5}$$

is the Laplace operator in the curvilinear coordinates and the membrane stress-resultants may be expressed through the stress function Φ as

$$\begin{aligned} T_1 &= \frac{1}{B}\frac{\partial}{\partial\beta}\left(\frac{1}{B}\frac{\partial\Phi}{\partial\beta}\right) + \frac{1}{A^2B}\frac{\partial B}{\partial\alpha}\frac{\partial\Phi}{\partial\alpha}, \\ T_2 &= \frac{1}{A}\frac{\partial}{\partial\alpha}\left(\frac{1}{A}\frac{\partial\Phi}{\partial\alpha}\right) + \frac{1}{B^2A}\frac{\partial A}{\partial\beta}\frac{\partial\Phi}{\partial\beta}, \\ S &= -\frac{1}{AB}\left(\frac{\partial^2\Phi}{\partial\alpha\,\partial\beta} - \frac{1}{A}\frac{\partial A}{\partial\beta}\frac{\partial\Phi}{\partial\alpha} - \frac{1}{B}\frac{\partial B}{\partial\alpha}\frac{\partial\Phi}{\partial\beta}\right). \end{aligned} \tag{1.5.6}$$

For these T_i, S the first two equilibrium equations are valid with the relative errors of order h_*, but for $k_1k_2 = 0$ systems (3) and (4) are equivalent.

Remark 1.2. Here, equations (4) are obtained assuming that $D = \text{const}$, $Eh = \text{const}$. For shells with variable thickness and elastic material properties with an error of order h_*, the first terms in (4) must be replaced by

$$-\Delta\,(D\Delta w), \qquad \Delta\left(\frac{1}{Eh}\Delta\Phi\right). \tag{1.5.7}$$

It often occurs that the shell stress-strain state is a sum of the main stress-strain state and an edge effect. We consider an edge effect which has the large index of variation ($t = 1/2$) in the neighborhood of line $\alpha = \alpha_0$ in the direction orthogonal to the edge and small index of variation ($t' = 0$) in the edge direction. Such an edge effect spreads in a band, the width of which is of order $Rh_*^{1/2}$ at the line $\alpha = \alpha_0$. To describe the edge effect we may use system (3).

With an error of order $h_*^{1/2}$ we can also use the following equation [61]

$$-D\frac{\partial^4 w}{\partial x^4} + T_1\frac{\partial^2 w}{\partial x^2} - \frac{Eh}{R_2^2}w + T_1\left(\frac{1}{R_1} + \frac{\nu}{R_2}\right) + q^* = 0, \tag{1.5.8}$$

where

$$dx = A\,d\alpha, \quad \frac{\partial T_1}{\partial x} + q_1^* = 0, \quad T_2 = \nu T_1 - \frac{Ehw}{R_2}. \tag{1.5.9}$$

The term $T_1\dfrac{\partial^2 W}{\partial x^2}$ is non-linearly dependent on the loading and that is why equation (8) usually is called the non-linear edge effect equation.

If

$$S = 0, \qquad \frac{\partial B}{\partial \alpha} = 0 \quad \text{at} \quad \alpha = \alpha_0, \tag{1.5.10}$$

the accuracy of equation (8) increases, and the error is of order h_*.

We note that if the load level satisfies estimate (3.9), the first four terms in (8) are of the same orders. For smaller loads, we arrive at the simple edge effect equation

$$-D\frac{\partial^2 w}{\partial x^4} - \frac{Eh}{R_2^2}w = 0. \tag{1.5.11}$$

For shells of revolution equation (8) is discussed in Section 14.1.

1.6 Technical Theory Equations in the Other Cases

We obtained equations (5.3) using three important assumptions

$$t = \frac{1}{2}, \qquad w \sim R^{-1}h_* = h_0, \qquad k_i \sim R^{-1}, \tag{1.6.1}$$

but the equations have wider applicability. Let us discuss the cases when some of assumptions (1) are not fulfilled.

The assumption that $w \sim h_0$ affects only the relative values of the non-linear terms in the third equation (5.3) and in the expressions for ε_{ij}. For $w \ll h_0$ system (5.3) becomes linear after neglecting the non-linear terms.

Let $0 < t < 1/2$, i.e. the stress-strain state has smaller index of variation than the simple edge effect. In this case, systems (5.3) and (5.4) can also be used but their accuracy decreases and the error Δ_1 is of order

$$\Delta_1 \sim h_*^{2t}. \tag{1.6.2}$$

This error is mainly caused by neglecting the tangential displacements in expressions (1.7) for γ_i and by the fact that functions (5.6) do not accurately satisfy the first equilibrium equations.

Cases when the neutral surface deformation is close to bending, i.e. the values of $\varepsilon_i, \omega, \varepsilon_{ij}$ are small due to the reduction of the main terms in the expressions for these values, require special consideration. In this case

$$\varepsilon_i,\ \omega,\ \varepsilon_{ij} \ll R^{-1}w. \tag{1.6.3}$$

and this case will be discussed in Section 12.1.

1.7 Shallow Shells

Let us suppose that $|Rk_i| \ll 1$ (here R is the characteristic size of a shell). In this case the error in equations (5.3) and (5.4) decreases and becomes equal to

$$\Delta_1 \sim \max\{Rk_i\} \cdot h_*^{2t}. \tag{1.7.1}$$

Assuming that A, B, k_1, k_2 are approximately constant, one can simplify the form of these equations. For $A = B = 1$ equations (5.4) turn to [171, 173, 175].

$$\begin{aligned} -D\Delta\Delta w + \frac{\partial^2\Phi}{\partial\beta^2}\left(k_1 + \frac{\partial^2 w}{\partial\alpha^2}\right) + \\ + \frac{\partial^2\Phi}{\partial\alpha^2}\left(k_2 + \frac{\partial^2 w}{\partial\beta^2}\right) - 2\,\frac{\partial^2\Phi}{\partial\alpha\,\partial\beta}\,\frac{\partial^2 w}{\partial\alpha\,\partial\beta} + q^* &= 0, \\ \frac{1}{Eh}\Delta\Delta\Phi + k_1\frac{\partial^2 W}{\partial\beta^2} + k_2\frac{\partial^2 w}{\partial\alpha^2} + \frac{\partial^2 w}{\partial\alpha^2}\,\frac{\partial^2 w}{\partial\beta^2} - \left(\frac{\partial^2 w}{\partial\alpha\,\partial\beta}\right)^2 &= 0. \end{aligned} \tag{1.7.2}$$

For $k_1 = k_2 = 0$ equations (2) become the plate non-linear bending equations [29, 46, 72], the errors of which are of order ε (see (3.7)). The mathematical aspects of existence and uniqueness of solutions of boundary value problems in the nonlinear theory of plates and shallow shells are duscussed in [107, 108, 178].

1.8 Initial Imperfections

Let the unperturbed neutral surface of the shell be defined by the parameters A, B, R_1 and R_2. Function $w^\bullet(\alpha, \beta)$ defines the initial normal deviations of the real surface from the unperturbed one. The projections of displacement which was caused by the deformation of the shell are denoted by u_1, u_2, and w. Then u_1, u_2, $w^\bullet + w$ are projections of the total displacement. We assume that $w^\bullet \lesssim h_0$.

Equations (5.3) and (5.4) must be modified to account for the initial imperfections. The following relations are valid

$$\begin{aligned} \varepsilon_{11} &= \varepsilon_1 + \gamma_1 \gamma_1^\bullet + \frac{1}{2}\gamma_1^2, & \gamma_1^\bullet &= -\frac{1}{A}\frac{\partial w^\bullet}{\partial \alpha}, \\ \varepsilon_{22} &= \varepsilon_2 + \gamma_2 \gamma_2^\bullet + \frac{1}{2}\gamma_2^2, & \gamma_2^\bullet &= -\frac{1}{B}\frac{\partial w^\bullet}{\partial \beta}, \\ \varepsilon_{12} &= \omega + \gamma_1^\bullet \gamma_2 + \gamma_2^\bullet \gamma_1 + \gamma_1 \gamma_2. \end{aligned} \tag{1.8.1}$$

In the third equilibrium equation (5.3) and in equations (5.4) $\varkappa$, τ are replaced by $\varkappa_i + \varkappa_i^\bullet$, $\tau + \tau^\bullet$, where $\varkappa_i^\bullet = \varkappa_i(w^\bullet)$, $\tau^\bullet = \tau(w^\bullet)$. Then equations (5.4) are represented as

$$\begin{aligned} &-D\Delta\Delta w + (k_1 + \varkappa_1 + \varkappa_1^\bullet)\, T_1 + 2\,(\tau + \tau^\bullet)\, S + \\ &\qquad\qquad + (k_2 + \varkappa_2 + \varkappa_2^\bullet)\, T_2 + q^* = 0, \\ &\frac{1}{Eh}\Delta\Delta\Phi + k_1\varkappa_2 + k_2\varkappa_1 + \varkappa_1\varkappa_2 - \tau^2 + \\ &\qquad \varkappa_1\varkappa_2^\bullet + \varkappa_2\varkappa_1^\bullet - 2\tau\tau^\bullet = 0. \end{aligned} \tag{1.8.2}$$

The above relations are valid with relative errors of order h_*, if the index of variation $w^\bullet$ is not larger than 1/2. We will not consider shells with initial imperfections beyond this point.

Buckling of shells with imperfections of the neutral surface and post-buckling behaviour of the imperfect shells are considered in [9, 10, 11, 19, 48, 61, 64, 76, 148, 149]

1.9 Cylindrical Shells

For a cylindrical shell we assume that

$$A = B = R, \quad k_1 = 0, \quad k_2(\beta) = R^{-1} k\,(\beta), \tag{1.9.1}$$

where, for a non-circular cylindrical shell, R is the characteristic radius of curvature. For a circular shell $k(\beta) = 1$.

Many buckling problems involving cylindrical shells can be solved on the basis of systems (5.3) or (5.4) after simplification by means of formulae (1). Some problems however, especially for long cylindrical shells, cannot be described by these equations. An example is that of semi-momentless stress-strain states, for which

$$\frac{\partial^2 w}{\partial \beta^2} \gg \frac{\partial^2 w}{\partial \alpha^2}, \tag{1.9.2}$$

and states for which the contour of the shell-cross section is not deformed (or only slightly deformed). For such stress-strain states, the shear stress-resultants in the first equilibrium equations and the tangential displacements in the expressions for the bending deformations must not be omitted.

We can write a system of equations for a non-circular cylindrical shell

$$\begin{aligned}
&\frac{\partial T_1}{\partial \alpha} + \frac{\partial S}{\partial \beta} + R\, q_1^* = 0, \\
&\frac{\partial T_2}{\partial \beta} + \frac{\partial S}{\partial \alpha} - R\,(k_2 + \varkappa_2)\, Q_2 + R q_2^* = 0, \\
&\frac{\partial Q_1}{\partial \alpha} + \frac{\partial Q_2}{\partial \beta} + R\,[\varkappa_1 T_1 + 2\,\tau S + (k_2 + \varkappa_2)\, T_2 + q^*] = 0, \\
&\frac{\partial H}{\partial \alpha} + \frac{\partial M_2}{\partial \beta} + R Q_2 = 0, \\
&\frac{\partial H}{\partial \beta} + \frac{\partial M_1}{\partial \alpha} + R Q_1 = 0,
\end{aligned} \tag{1.9.3}$$

$$\begin{aligned}
\varepsilon_1 &= \frac{1}{R}\frac{\partial u_1}{\partial \alpha}, & \varepsilon_2 &= \frac{1}{R}\left(\frac{\partial u_2}{\partial \beta} - k w\right), \\
\omega &= \frac{1}{R}\left(\frac{\partial u_1}{\partial \beta} + \frac{\partial u_2}{\partial \alpha}\right), \\
\varkappa_1 &= \frac{1}{R^2}\frac{\partial^2 w}{\partial \alpha^2}, & \varkappa_2 &= \frac{1}{R^2}\frac{\partial}{\partial \beta}\left(\frac{\partial w}{\partial \beta} + k u_2\right), \\
\tau &= \frac{1}{R^2}\frac{\partial}{\partial \alpha}\left(\frac{\partial w}{\partial \beta} + k u_2\right).
\end{aligned}$$

The expressions for ε_{ij} and the constitutive relations we take in the form of (5.3).

There exists a number of different versions in notation for the resolving equations of a circular cylindrical shell [27, 33, 47, 61, 135, 141, 173, 175] and others. Below we will present the equations which are valid descriptions of both the fast varying and the semi-momentless states mentioned above

$$\begin{aligned} -D\Delta\Delta w + \frac{1}{R^4}\frac{\partial^2 w}{\partial \beta^2} + T_1 \varkappa_1 + 2\,S\tau + T_2\left(\frac{1}{R} + \varkappa_2\right) + q^* &= 0, \\ \frac{1}{Eh}\Delta\Delta\Phi + \frac{1}{R}\varkappa_1 + \varkappa_1\varkappa_2 - \tau^2 &= 0, \end{aligned} \tag{1.9.4}$$

where the stress function Φ is introduced by formulae

$$T_1 = \frac{1}{R^2}\frac{\partial^2 \Phi}{\partial \beta^2}, \quad S = -\frac{1}{R^2}\frac{\partial^2 \Phi}{\partial \alpha\,\partial \beta}, \quad T_2 = \frac{1}{R_2}\frac{\partial^2 \Phi}{\partial \alpha^2} + \frac{M_2}{R}. \tag{1.9.5}$$

Here and in (5.4) we assume that $q_1^* = q_2^* = 0$.

With an error of order ε (see (3.7)) we can surmise that in (3) and (4)

$$\varkappa_2 = \frac{1}{R_2}\left(\frac{\partial^2 w}{\partial \beta^2} + w\right). \tag{1.9.6}$$

System (4) is valid for almost all of the stress-strain states considered below. The various stress-strain states are characterized by different main terms of the system. Buckling problems of long cylindrical shells under axial compression are not, however, described by system (4) (see Sections 2.3 and 2.4).

1.10 The Potential Energy of Shell Deformation

We can write an expression for the potential energy of shell deformation as

$$\begin{aligned} \Pi &= \Pi_\varepsilon + \Pi_\varkappa = \\ &\frac{1}{2}\iint\limits_\Omega K\left(\varepsilon_{11}^2 + 2\nu\varepsilon_{11}\varepsilon_{22} + \varepsilon_{22}^2 + \frac{1-\nu}{2}\varepsilon_{12}^2\right) d\Omega + \\ &\frac{1}{2}\iint\limits_\Omega D\left(\varkappa_1^2 + 2\nu\varkappa_1\varkappa_2 + \varkappa_2^2 + 2\,(1-\nu)\,\tau^2\right) d\Omega, \end{aligned}$$

$$d\Omega = AB\,d\alpha\,d\beta, \tag{1.10.1}$$

where the potential energy of shell extension and bending are denoted by Π_ε and $\Pi_\varkappa$ respectively. As in Section 1.2 we neglect the difference in metrics of the neutral surface before and after deformation.

The first variation of the potential energy is equal to

$$\delta\Pi = \iint\limits_{\Omega} \Big(T_1 \delta\,\varepsilon_{11} + T_2 \delta\,\varepsilon_{22} + S\,\delta\,\varepsilon_{12} +$$

$$+ M_1 \delta\,\varkappa_1 + 2\,H \delta\,\tau + M_2 \delta\,\varkappa_2\Big)\,d\Omega, \tag{1.10.2}$$

where

$$S = S_1 - \frac{H}{R_2} = S_2 - \frac{H}{R_1}, \tag{1.10.3}$$

and the other stress-resultants and stress-couples are given by formulae (2.4).

We can also write the elemental work δA of the external surface load $\boldsymbol{q}$ and the edge stress-resultant vectors $\boldsymbol{F}_k$ and the edge stress-couples vectors $\boldsymbol{G}_k$ as

$$\delta A = \iint\limits_{\Omega} \boldsymbol{q}\,\delta\,\boldsymbol{U}\,d\Omega + \int\limits_{\gamma} \Big(\boldsymbol{F}_k\,\delta\,\boldsymbol{U} + \boldsymbol{G}_k\,\delta\,\boldsymbol{\omega}\Big)\,ds, \tag{1.10.4}$$

where $\delta\,\boldsymbol{U}$ and $\delta\,\boldsymbol{\omega}$ are vectors of the elemental displacement and rotation.

The equilibrium equations and static boundary conditions can be obtained from the relation

$$\delta\,\Pi = \delta A \tag{1.10.5}$$

for any virtual displacement $\delta\,\boldsymbol{U}$ of the neutral surface that satisfied the geometric boundary conditions.

1.11 Problems and Exercises

1.1. Find the index of variation for the function $F(x,h) = g(x)\sinh{(z)}$ for $z = h^{-t} f(x)$ as $h \to 0$ assuming $f'(x) \neq 0$ and $t \geq 0$.

Answer The index of variation is equal to t.

1.2. Find the general index of variation, and the direction in which the partial index of variation is minimal as $h \to 0$ for the following functions:

a) $F(x,y,h) = g(x)\sin{(z(x-ay))}$, where $z = h^{-t}$,

b) $F(x,y,h) = \exp{\left(z\frac{x}{y}\right)}$, where $z = h^{-t}$,

assuming $t \geq 0$.

Answer For a) and b) the index of variation is equal to $-t$.

For a) in the direction $x = ay$ the partial index of variation is equal to 0.

For b) in the direction $x = 0$ the partial index of variation is equal to 0.

1.3. Derive relations (1.13) and (1.14) using formulae (1.4–1.10)

1.4. Derive relations (4.6) from the equilibrium equations for the part of a shell of revolution between the parallels s and $s = s_2$. Prove that the obtained stress-resultants satisfy (4.3).

Chapter 2

Basic Equations of Shell Buckling

The various research methods of the buckling of shells under static loading are briefly discussed in this Chapter (see also [12, 15, 21, 33, 45, 48, 61, 62, 95, 123, 166, 149, 170, 175, 178] although only one of these will be employed in the following chapters. The approach presented in this book will permit the examination of the stability of equilibrium states under conservative surface and edge loading by determination of the critical loads using linearized equilibrium equations which will be presented below.

2.1 Types of Elastic Shell Buckling

The equilibrium state or motion of a mechanical system is termed *stable* if the deviation from this state at the moment t is as small as desired for any sufficiently small perturbations at the initial moment of time $t = 0$.

The general method of research in stability is to study perturbed motion in the neighbourhood of unperturbed motion. The method (the dynamic criterion of stability) for conservative mechanical systems was first proposed by Lagrange. A.M. Liapunov later developed an accurate mathematical theory of motion stability [86].

The dynamic criterion can be used for any problem of shell stability but the study of perturbed shell motion is a much more difficult problem than the study of its equilibrium states. That is why, unless it is necessary, the dynamic criterion is rarely used for in the stability analysis of shell equilibrium states. Later on in the book only the problems with the conservative loads are studied, for which the bifurcation (Euler) or the limit point citeria may be used. For these class of problems these criteria are equivalent to the dynamic criterion.

It is important to note, however, that the use of the dynamic criterion is the only possibility in a number of cases. Examples would be problems of shell motion stability under dynamic [85, 146, 147, 176] and non-conservative loads, such as the motion of a shell in a gas flux [124, 176, 177], and parametric shell instability [17, 48, 126]. In this work these problems are not considered and the dynamic criterion of stability is not used.

We will consider a shell loaded with static conservative surface and edge loads. Loads are referred to as conservative, if the work done by them depend only on the end states and do not depend on the way of deformation. In particular, loads which do not change their values and directions, are conservative. However, these loads do not comprise the entire class of conservative loading. Hydrostatic pressure loads, the direction of which depend on the stress-strain state, are also conservative. Note that only the dynamic criterion gives accurate results for cases of static non-conservative loads [17, 126, 182].

In classical problems of linear elasticity where infinitely small deformations are assumed, the equilibrium conditions are assumed to be satisfied by the forces acting on the undeformed elastic system. This assumption which is essential for Kirchhoff's general uniqueness theorem [75] leads to unique solutions for such linear problems. On the other hand, in formulating buckling problems this assumption is dropped and the equilibrium conditions are satisfied by the forces acting on the deformed elastic system. This leads to an essentially nonlinear formulation of such problems in the sense that displacements are not linearly proportional to the externally applied loads, and, in fact, often the deformation of a structure will not be uniquely determined by the applied loading.

According to Kirchhoff's theorem there is only one set of solutions of stresses, strains, and displacements for an elastic body in equilibrium, satisfying all basic equations of linear elasticity for a given body load and boundary conditions. In fact, any two sets of solution for the same body load and boundary conditions, at most may differ only by the rigid body displacement of the system, i.e., the difference in any two sets of solution describes the rigid body motion of the system. Therefore, the solution of the such linearly formulated problem is always stable. The sufficient condition required for satisfying Kirchhoff's theorem is that the potential energy of the elastic system should be a positive definite function. This condition is fulfilled for shells (see Section 1.10).

The physically non-linear formulation of the problem (i.e., the non-linear dependencies of the stresses on the strains) is not considered below. Only buckling, which is caused by the geometric non-linearity of the problem, is studied. The geometric non-linearity is the non-linear dependence of the strains on the displacements and on the non-linear terms, due to the difference of the coordinate systems before and after deformation. On the physically non-linear problem of shell buckling refer to [61, 79, 119, 167, 168].

As with any elastic system it is possible to consider two types of shell buckling from the equilibrium state [1]. The first type is connected with the bifurcation (or branching) of the equilibrium states, and the second is accompanied with the appearance of the limit point.

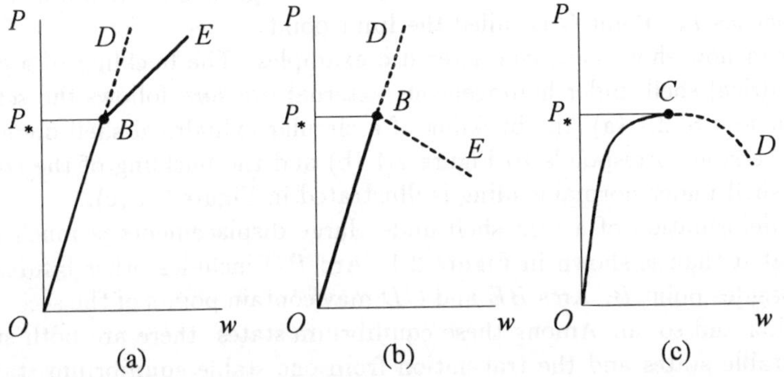

Figure 2.1: Relationship between load and deflection for an ideal shell.

The relationships between the load (P) and the characteristic deflection (w) are plotted schematically in Figure 2.1. Considering Figures 2.1 (a) and (b) let the load increase smoothly from zero, remaining less then P_*. The equilibrium state is then stable, the deflection is uniquely determined (the section OB in Figure 2.1). For $P \geqslant P_*$ one or more adjacent equilibrium states BE, in addition to the basic one BD, appear in the neighbourhood of point B. Point B is called the bifurcation point of the equilibrium states. The equilibrium states in the neighbourhood of point B can be both stable and unstable. In Figure 2.1 the unstable equilibrium states are plotted by the dotted lines. The basic equilibrium state CD is, as a rule, unstable. The stable equilibrium state BE after bifurcation is plotted in Figure 2.1 (a). The case when there are no stable equilibrium states close to point B, is shown in Figure 2.1 (b).

Point B in Figure 2.1 (b) is the buckling point. For $P \geqslant P_*$ the adjacent equilibrium states are unstable.

Rigorously speaking, point B in Figure 2.1 (a) cannot be called the buckling point. But for $P \approx P_*$ a modification of the deflection shape occurs, and the deformation abruptly increases with the small increase in the load. It is clear that the evaluation of P_* is important in engineering analysis. Taking this into account it is reasonable to call P_* the critical load. The same term is used in the cases of Figure 2.1 (b) and 2.1 (c).

Figure 2.1 (c) shows the case when the deflections grow abruptly as the loading increases smoothly to P_*, i.e.

$$\frac{dw}{dP} \to \infty \quad \text{as} \quad P \to P_*. \tag{2.1.1}$$

Buckling occurs at point C. The arc of curve CD consists of the unstable equilibrium states. There are no close equilibrium states for $P > P_*$ and there is no restructuring of the buckling mode, only the growth of its amplitude as P approaches P_*. Point C is called the limit point.

We can now show some characteristic examples. The buckling of a circular cylindrical shell under homogeneous external pressure follows the scheme shown in Figure 2.1 (a); the buckling of a circular cylindrical shell under axial compression corresponds to Figure 2.1 (b) and the buckling of the convex shallow shell under normal loading is illustrated in Figure 2.1 (c).

The deformation of a thin shell under large displacements is much more complicated than is shown in Figure 2.1. Arc BD includes other bifurcation points besides point B. Arcs BE and CD may contain points of the secondary bifurcation and so on. Among these equilibrium states there are both stable and unstable states and the translation from one stable equilibrium state to another can be accomplished continuously (see point B in Figure 2.1 (a)) or by an abrupt jump (points B and C in Figures 2.1 (b) and (c)).

At the present time an understanding of the deformation of thin shells under large deflections is not fully developed even for shells of simple geometry. The problem of shallow shell deformation and the axisymmetric deformation of shells of revolution has been studied in some details [29, 38, 107, 108, 138, 145, 149, 156, 170, 178]. The bifurcation points of the axisymmetric equilibrium state of shells of revolution into non-axisymmetric state have also been obtained mainly by numerical methods [22, 149, 156, 170].

The non-axisymmetric deformation of shells of revolution is considered in [95]. We note also works [132, 133], in which the geometric research methods for studying buckling are used.

It is not normally necessary for the analysis of real structures to obtain a detailed picture of post-buckling shell behaviour because in this case the shell has failed and is no longer of interest. Exceptions to this are shells which work in the large displacement regime.

For engineering analysis it is necessary to determine the critical load P_* accurately. But in many cases the comparison of theoretical and experimental values of P_* reveals their essential divergence. For example, in the problem of a cylindrical shell under compression by an axial force, the theoretical value of P_* exceeds the experimental one by a factor of two or three and this divergence has been systematically observed in many experiments [33, 59, 61].

This flaw in the theory of shell stability was resolved by taking into account small imperfections, mainly of the neutral surface. It becomes clear that the critical load is very sensitive to these imperfections [9, 10, 11, 19, 64, 148].

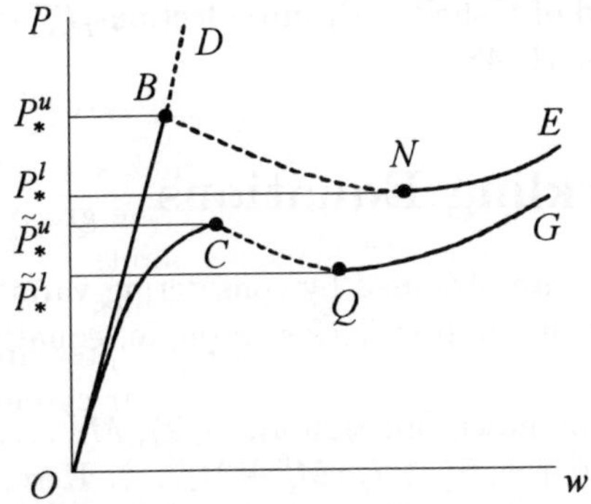

Figure 2.2: The relationship between the load and deflection for an ideal shell and for a shell with imperfections.

The comparison of the "load-deflection" curves for an ideal shell and a shell with imperfections is shown in Figure 2.2. Curve $OBNE$ corresponds to the ideal shell and is similar to the curve plotted in Figure 2.1 (b), the curve $OCQG$ corresponds to a shell with imperfections. Arc OC of this curve corresponds to the pre-buckling stress-strain state. Arcs EN and QG show the stable post-buckling equilibrium states under large deflections.

We will introduce the following notation to assist in the analysis. The ordinates P_*^u and $\tilde{P}_*^u$ of points B and C are called the upper critical loads for the ideal shell and the shell with imperfections. The smallest load under which the post-buckling stable equilibrium state is possible, is called the lower critical load. In Figure 2.2, points N and Q correspond to the lower critical loads P_*^l and $\tilde{P}_*^l$.

Calculation of P_*^u and $\tilde{P}_*^u$ is comparatively simple problem. Explicit analytical expressions for P_*^u have been obtained for many problems of ideal shell buckling. In a number of cases it is possible to evaluate $\tilde{P}_*^u$ by means of one of the numerical methods [62, 95, 115, 170]. The evaluation of P_*^l and $\tilde{P}_*^l$ is, however, a complicated non-linear problem, which is not completely solved at the present time.

A very important question is which load should be taken as the designed load in an actual design. The obvious answer might be to take $\tilde{P}_*^u$, but its evaluation is complicated because, as a rule, the actual imperfections present in the shell are unknown. At the same time it occurs that the approximate values of P_*^l obtained by approximation of the deflection by a series with a few terms (one or two) retained, agrees very well with experimental results. The following studies reveal, however, that P_*^l essentially depends on the accuracy of the deflection approximation and sometimes for a large number of terms it is equal to 1% of P_*^u or it may be negative. That is why at the present time the upper critical load of a shell with imperfections $\tilde{P}_*^u$ is taken as the correct value in actual designs [1, 48, 77].

2.2 The Buckling Equations

Buckling equations are obtained by considering variations of the nonlinear equations, introduced in Chapter 1 (for example, equations (1.2.6) or (1.5.3) or (1.5.4) or (1.9.4)).

To this end, each unknown function u_i, w, T_i, $M_i, \ldots$, in these equations is replaced by $u_i^0+u_i$, w^0+w, $T_i^0+T_i$, $M_i^0+M_i, \ldots$. Here, functions u_i^0; $w^0, \ldots$ describe the initial stress state. The stability of this initial state which satisfies the nonlinear system of equations is to be investigated. Functions u_i, $w, \ldots$ describe the adjacent infinitesimally close equilibrium state. They satisfy the linear homogeneous equations (the buckling equations) and the homogeneous boundary conditions, that are obtained as a result of linearization by u_i, w, ... of the initial non-linear equations.

The existence condition for the non-trivial solution of the buckling equation is used for the evaluation of the critical load. In dealing with buckling problems it is convenient to assume that the load varies proportionally to a loading parameter $\lambda > 0$. Then the pre-buckling state functions (u_i^0, w^0, T_i^0, $M_i, \ldots$) and the coefficients of the buckling equations depend on λ. In this way, the buckling problem is reduced to an eigenvalue problem. The least (positive) eigenvalue is taken as the first critical value $\lambda = \lambda_*$ leading to the corresponding buckling mode. Such an approach is called *equilibrium or Euler analysis of stability* due to L. Euler who in 1744 used this approach to study the stability of axially compressed bars. His paper [36] is considered to be the first work on structural stability.

In the general case the buckling equations are rather complicated (see [60, 61]) and are not written here. Later in this book simplified forms of the buckling equations, applicable under specific cases depending on the character of the initial stress state and buckling mode are presented.

2.3 The Buckling Equations for a Membrane State

The membrane stress state of a shell is described by equations (1.4.3). Let the displacement u_i^0, w^0 and the stress-resultants T_i^0, S^0 characterising this state, be weakly varying functions of α and β (i.e. the index of variation

$t^0 = 0$). Then, due to estimate (1.3.8) and assuming that the deformation of the neutral surface is far from bending, we can conclude, that the neutral surface before and after deformation may be identified. In other words, we may assume, that before buckling occurs the shell is stressed, but not deformed (see [1]).

The form of the buckling equations depends on the expected buckling mode. We will consider the case that the buckling mode can be assumed to satisfy the hypotheses of Section 1.5 that led to equations of the technical shell theory (1.5.3). In other words, the buckling is of a local nature and is accompanied by the formation of a large number of small pits.

Under these assumptions we obtain the buckling equations from (1.5.3)

$$\begin{aligned}
&\frac{1}{AB}\left(\frac{\partial(BT_1)}{\partial\alpha} - \frac{\partial B}{\partial\alpha}T_2 + \frac{1}{A}\frac{\partial(A^2S)}{\partial\beta}\right) + X_1 = 0,\\
&\frac{1}{AB}\left(\frac{\partial(AT_2)}{\partial\beta} - \frac{\partial A}{\partial\beta}T_1 + \frac{1}{B}\frac{\partial(B^2S)}{\partial\alpha}\right) + X_2 = 0,\\
&\frac{1}{AB}\left(\frac{\partial(BQ_1)}{\partial\alpha} + \frac{\partial(AQ_2)}{\partial\beta}\right) + k_1T_1 + k_2T_2 +\\
&\qquad +T_1^0\varkappa_1 + 2S^0\tau + T_2^0\varkappa_2 + Z = 0.
\end{aligned} \tag{2.3.1}$$

The rest of the buckling equations have the form of (1.5.3) and in the elasticity relations one should take $\varepsilon_{ii} = \varepsilon_i$, $\varepsilon_{12} = \omega$. Both equations (1) and (1.5.3) are written in the projections on the unit vectors after deformation. Therefore, in the case of a follower load in (1)

$$X_1 = X_2 = Z = 0. \tag{2.3.2}$$

If the surface load does not change direction (this would be referred to as a "dead" load) and is introduced by its projections (1.2.7), then

$$\begin{aligned}
X_1 &= -\omega_2 q_2 - \gamma_1 q,\\
X_2 &= -\omega_1 q_1 - \gamma_2 q,\\
Z &= \gamma_1 q_1 + \gamma_2 q_2.
\end{aligned} \tag{2.3.3}$$

Therefore, in the case of a follower surface load the load parameter λ is introduced into the stability equations only through stress-resultants T_i^0 and S^0. In the case of a "dead" load λ is also introduced through X_i, and Z. In the case of the follower load $X_1 = X_2 = 0$ and it is possible to introduce stress function Φ by means of formulae (1.5.6). Then the system of buckling equations takes the form

$$\begin{aligned}
-D\Delta\Delta w + \Delta_T^c w + \Delta_k\Phi &= 0,\\
\frac{1}{Eh}\Delta\Delta\Phi + \Delta_k w &= 0,
\end{aligned} \tag{2.3.4}$$

where Δ is the Laplace operator (1.5.5),

$$\Delta_k w = k_1\varkappa_2 + k_2\varkappa_1 = \frac{1}{AB}\left[\frac{\partial}{\partial\alpha}\left(\frac{k_2 B}{A}\frac{\partial w}{\partial\alpha}\right) + \frac{\partial}{\partial\beta}\left(\frac{k_1 A}{B}\frac{\partial w}{\partial\beta}\right)\right], \tag{2.3.5}$$

$$\Delta_T^c = T_1^0\varkappa_1 + 2S^0\tau + T_2^0\varkappa_2.$$

Substituting the expression for $\varkappa_i$, and τ from (1.5.3) into (5) and using the membrane equations for T_i^0, and S^0 we get

$$\Delta_T^c w = \Delta_T w + \Gamma w, \tag{2.3.6}$$

where

$$\begin{aligned} \Delta_T w &= \frac{1}{AB}\left[\frac{\partial}{\partial\alpha}\left(\frac{BT_1^0}{A}\frac{\partial w}{\partial\alpha}\right) + \frac{\partial}{\partial\beta}\left(\frac{AT_2^0}{B}\frac{\partial w}{\partial\beta}\right) + \right. \\ &\qquad \left. + \frac{\partial}{\partial\alpha}\left(S_0\frac{\partial w}{\partial\beta}\right) + \frac{\partial}{\partial\beta}\left(S_0\frac{\partial w}{\partial\alpha}\right)\right], \qquad (2.3.7) \\ \Gamma w &= -\gamma_1 {q_1}^* - \gamma_2 {q_2}^*. \end{aligned}$$

Substituting (6) into the first equation of (4), we obtain the equilibrium equation in the projection on the normal to the neutral surface before deformation.

The operators $D\Delta\Delta w$, $\Delta_k w$, $\Delta_T w$ in (4) and (7) are formally self-adjoint. This means that integrating by parts several times the integral $\iint v\Delta_k u\, d\Omega$ in the domain Ω occupied by a shell, we arrive at the integral $\iint u\Delta_k v\, d\Omega$ and an integral by the boundary contour γ.

The term Γw in (6) makes system (4) not self-adjoint. At the same time the follower surface load with a projection on the plane which is tangent to the neutral surface, is non-conservative. It should be noted that the non-conservative term Γw is relatively small. Indeed, let $t > 0$ be the index of variation of the buckling mode. This term then has order h_*^t compared to the main terms in (4).

Since the method used here is, strictly speaking, not valid for non-conservative loads we suppose further that $\Gamma w = 0$, i.e. considering the follower load we restrict ourselves to a normal follower load (normal pressure). Then system (4) may be rewritten in the form

$$\begin{aligned} -D\Delta\Delta w + \Delta_T w + \Delta_k \Phi &= 0, \\ \frac{1}{Eh}\Delta\Delta\Phi + \Delta_k w &= 0. \end{aligned} \tag{2.3.8}$$

In the case of a "dead" surface load (i.e. one which does not follow the direction of the shell deformation) the terms X_i in equations (1) differ from zero and that prevents the introduction of function Φ by formulae (1.5.6). However by neglecting these terms, we make the error of the same order h_*^{2t}, as that under transition from system (1) to system (8). In the case of "dead" surface loads the non-conservative term Γw is absent and system (8) is suitable for a surface load of any direction.

The errors of systems (1) and (8) have orders h_*^{2t} and for $t = 1/2$ coincide with the error of the original Kirchhoff–Love hypotheses. For $t < 1/2$ the precision of these equations may be insufficient. By identifying the form of the neutral surface before and after deformation we can write the buckling equations

$$\begin{gathered}
\frac{1}{AB}\left(\frac{\partial(BT_1)}{\partial\alpha} - \frac{\partial B}{\partial\alpha}T_2 + \frac{1}{A}\frac{\partial(A^2S)}{\partial\beta}\right) - \frac{Q_1}{R_1} + X_1 = 0, \\
\frac{1}{AB}\left(\frac{\partial(AT_2)}{\partial\beta} - \frac{\partial A}{\partial\beta}T_1 + \frac{1}{B}\frac{\partial(B^2S)}{\partial\alpha}\right) - \frac{Q_2}{R_2} + X_2 = 0, \\
\frac{1}{AB}\left(\frac{\partial(BQ_1)}{\partial\alpha} + \frac{\partial(AQ_2)}{\partial\beta}\right) + k_1T_1 + k_2T_2 + \\
+ T_1^0\varkappa_1 + 2S^0\tau + T_2^0\varkappa_2 + Z = 0,
\end{gathered} \tag{2.3.9}$$

where, unlike (1), the geometric variables γ_i, $\varkappa_i$ and τ should be calculated by the exact formulae (1.1.7) and (1.1.13).

For a initial membrane stress state the buckling equations are equivalent to the variation equation

$$\delta\Pi = 0, \tag{2.3.10}$$

where

$$\Pi = \Pi_\varepsilon + \Pi_\varkappa + \Pi_T,$$

and

$$\begin{aligned}
\Pi_\varepsilon &= \frac{1}{2}\iint_\Omega K\left(\varepsilon_1^2 + 2\gamma\varepsilon_1\varepsilon_2 + \varepsilon_2^2 + \frac{1-\gamma}{2}\omega^2\right)d\Omega, \\
\Pi_T &= \frac{1}{2}\iint_\Omega \left(T_1^0\gamma_1^2 + 2S^0\gamma_1\gamma_2 + T_2^0\gamma_2^2\right)d\Omega,
\end{aligned} \tag{2.3.11}$$

and $\Pi_\varkappa$ is the same as that in (1.10.1). Here the deformations ε_i, ω, $\varkappa_i$, and τ and the rotation angles γ_i are linear functions of the additional displacements originating under transition to the adjacent equilibrium state. We denote the additional potential energy appearing under transition to the adjacent equilibrium state by Π.

For a non-circular cylindrical shell, in notation (1.9.1), the buckling equations may be written as

$$-D\left(\Delta\Delta w + \frac{1}{R^4}\frac{\partial^2 w}{\partial \beta^2}\right) + \frac{T_2}{R} + T_1^0 \varkappa_1 + 2S^0 \tau + T_2^0 \varkappa_2 + Z = 0,$$
$$\frac{1}{Eh}\Delta\Delta\Phi + \frac{\varkappa_1}{R} = 0, \tag{2.3.12}$$

where $\varkappa_i, \tau$ may be evaluated by formulae (1.9.3) and (1.9.6), and the stress-resultants T_i, S are connected with Φ by formulae (1.9.5).

In the problem of axial compression of a long cylindrical shell, system (12) gives an incorrect result. To correct it one must take into account the difference in metrics of the neutral surface before and after deformation, which was neglected above. The value of T_2 in (12) must be calculated by formula [61].

$$T_2 = \frac{1}{R_2}\frac{\partial^2 \Phi}{\partial \alpha^2} + \frac{M_2}{R} + \left(T_1^0 - T_2^0\right)\varepsilon_1. \tag{2.3.13}$$

Now, according to (1.9.3) and (1.9.6) system (12) may be written in the form

$$-D\left(\Delta\Delta w + \frac{2}{R^4}\frac{\partial^2 w}{\partial \beta^2} + \frac{w}{R^4}\right) + \frac{1}{R^3}\frac{\partial^2 \Phi}{\partial \alpha^2} + \frac{T_1^0}{R^2}\left(\frac{\partial^2 w}{\partial \alpha^2} + \frac{\partial u_1}{\partial \alpha}\right) +$$
$$\frac{T_2^0}{R^2}\left(\frac{\partial^2 w}{\partial \beta^2} + \frac{\partial (k u_2)}{\partial \beta} - \frac{\partial u_1}{\partial \alpha}\right) + \frac{2S^0}{R^2}\left(\frac{\partial^2 w}{\partial \alpha \partial \beta} + \frac{\partial (k u_2)}{\partial \alpha}\right) + Z = 0,$$
$$\frac{1}{Eh}\Delta\Delta\Phi + \frac{1}{R^3}\frac{\partial^2 w}{\partial \alpha^2} = 0. \tag{2.3.14}$$

System (14) contains terms, depending on u_1 and u_2, which are essential only if inequality (1.9.2) holds. For such stress states we make an approximation with the assumption that $\varepsilon_2 = \omega = 0$ and get the following relations

$$\frac{\partial u_1}{\partial \beta} + \frac{\partial u_2}{\partial \alpha} = 0, \quad \frac{\partial u_2}{\partial \beta} - kw = 0 \tag{2.3.15}$$

closing system (14).

2.4 Buckling Equations of the General Stress State

Consider two types of the initial stress state: (i) the moment stress state, which is a sum of the membrane stress state and the edge effect and (ii) the general stress state, which is a sum of the membrane stress state and the stress state (may be fast varying) due to local imperfections of the neutral surface. In both cases the the neutral surfaces before and after deformation cannot be identified. Indeed, let the initial deflection $w^0 \sim h$, then for the edge effect the initial changes of curvature $\varkappa_i^0, \tau^0 \sim R^{-1}$, i.e. they are not small.

Let the conditions of applicability of the system of equations of the technical theory (1.5.3) be fulfilled and assume that the initial stress state is the sum of a membrane stress state and an edge effect. The buckling equations we take in the form

$$\begin{gathered}
\frac{\partial BT_1}{\partial \alpha} - \frac{\partial B}{\partial \alpha}T_2 + \frac{1}{A}\frac{\partial (A^2 S)}{\partial \beta} = 0, \\
\frac{\partial AT_2}{\partial \beta} - \frac{\partial A}{\partial \beta}T_1 + \frac{1}{B}\frac{\partial (B^2 S)}{\partial \alpha} = 0, \\
-D\Delta\Delta w + k_1 T_1 + k_2 T_2 - \\
-\frac{1}{AB}\frac{\partial}{\partial \alpha}\left[B(T_1^0\gamma_1 + T_1\gamma_1^0 + S^0\gamma_2 + S\gamma_2^0)\right] - \\
-\frac{1}{AB}\frac{\partial}{\partial \beta}\left[A(T_2^0\gamma_2 + T_2\gamma_2^0 + S^0\gamma_1 + S\gamma_1^0)\right] = 0.
\end{gathered} \tag{2.4.1}$$

This system may be written also through the stress function Φ in the form

$$\begin{aligned}
-D\Delta\Delta w + \Delta_T w + \Delta_k \Phi + \Delta_k^0 \Phi &= 0, \\
\frac{1}{Eh}\Delta\Delta\Phi + \Delta_k w + \Delta_k^0 w &= 0,
\end{aligned} \tag{2.4.2}$$

where

$$\begin{aligned}
\Delta_k^0 \Phi &= \varkappa_1^0 T_1 + \varkappa_2^0 T_2 + 2\tau^0 S, \\
\Delta_k^0 w &= \varkappa_1^0 \varkappa_2 + \varkappa_2^0 \varkappa_1 - 2\tau^0 \tau,
\end{aligned} \tag{2.4.3}$$

and T_i and S are expressed through Φ by formulae (1.5.6). The operators Δ, Δ_k and Δ_T are the same, as in (3.8).

The boundary conditions which are to be introduced to solve systems (3.8), (3.9) and (2) are obtained by linearization of variables (1.2.9)-(1.2.15). We consider edge $\alpha = \alpha^0$ and assume that the external boundary load is conservative.

Then, with an error of order h_* we will equate to zero the generalized additional displacements u_1, u_2, ω, γ_1 or stress-resultants T_1, S, Q_{1*}, M_1, where

$$Q_{1*} = Q_1 - \frac{1}{B}\frac{\partial H}{\partial \beta} - T_1^0\gamma_1 - S^0\gamma_2 - \gamma_1^0 T_1 - \gamma_2^0 S. \tag{2.4.4}$$

If the initial stress state is membrane then the last two terms in (4) may be neglected (with an error of order $h_*^{1/2}$).

Let there exist initial imperfections of the form $w^\bullet$. We can represent the entire deflection in the form $w^\bullet + w^0 + w$, where the term $w^\bullet + w^0$ describes the initial stress state, the stability of which is being considered and w is the additional deflection. Equations (1.8.2) give the buckling equations

$$\begin{aligned} -D\Delta\Delta w + \Delta_T w + \Delta_k \Phi + \Delta_k^0 \Phi + \Delta_k^\bullet \Phi &= 0, \\ \frac{1}{Eh}\Delta\Delta\Phi + \Delta_k w + \Delta_k^0 w + \Delta_k^\bullet w &= 0. \end{aligned} \qquad (2.4.5)$$

The initial imperfections, $w^\bullet$, are included in the operator $\Delta_k^\bullet$, which is obtained from (3) replacing w^0 by $w^\bullet$, and in the term $\Delta_T w$, in which the initial stress-resultants T_i^0, S^0 depend on $w^\bullet$ and may be found from system (1.8.2).

2.5 Problems and Exercises

2.1 Using dynamic criterion investigate the stability of the equilibrium state $w = 0$ for the simply supported bar under axial compression. The bar has a material density of ρ. (Note that, unlike in Euler criterion, using the dynamic criterion one needs to know the mass distribution of the system.)

2.2 The horizontal pipe carries a fluid. The pipe has a length L, modulus of elasticity of E, and the moment of inertia I. The velocity of the flow is V with a mass of m per second flowing through the pipe. Determine the stability of the tube if:

1) both ends of the pipe are simply supported;
2) one end of the pipe is fixed with the other end being free.

2.3 Use the kinetic approach to investigate the stability of the equilibrium state $w = 0$ of a massless bar when subjected to the axial follower load. The bar is clamped at the bottom but it carries a mass m at its top.

2.4 Derive buckling equations (2.3.9) based on variational equation (2.3.10).

Hint. Firstly represent the variations $\delta\Pi_\varepsilon$ and $\delta\Pi_\varkappa$ in the form

$$\delta\Pi_\varepsilon = \iint_\Omega (T_1\delta\varepsilon_1 + T_2\delta\varepsilon_2 + S\delta\omega)\,d\Omega,$$

$$\delta\Pi_\varkappa = \iint_\Omega (M_1\delta\varkappa_1 + M_2\delta\varkappa_2 + H\delta\tau)\,d\Omega.$$

Chapter 3

Simple Buckling Problems

In this chapter we will analyze buckling problems for a membrane homogeneous stress state. Consider a shallow shell and circular cylindrical shell, for which the problem leads to equations with the constant coefficients. The shell edges are assumed to be simply supported which allows us to write the explicit form of the solution. We will also consider the buckling modes for which a shell is covered by a regular pattern of small pits.

In addition, for the case of a non-homogeneous membrane stress state we will determine an estimate for the critical load by means of the energy method and obtain the expected buckling modes (see Section 3.6)

3.1 Buckling of a Shallow Convex Shell

Consider the buckling of a thin, shallow shell under a membrane stress state as determined by the initial stress-resultants T_i^0 and S^0. The metric coefficients A and B, the curvatures k_i and stress-resultants T_i^0 and S^0 are assumed to be constant.

As buckling equations we take system (2.3.8), which, in this case has constant coefficients. We can write the system in the dimensionless form

$$\begin{aligned} \mu^2\Delta\Delta w + \lambda\Delta_t w - \Delta_k\Phi' &= 0, \\ \mu^2\Delta\Delta\Phi' + \Delta_k w &= 0, \end{aligned} \tag{3.1.1}$$

where

$$\begin{aligned} \Delta w &= \frac{\partial^2 w}{\partial x^2} + \frac{\partial^2 w}{\partial y^2}, \\ \Delta_k w &= k_2 \frac{\partial^2 w}{\partial x^2} + k_1 \frac{\partial^2 w}{\partial y^2}, \\ \Delta_t w &= t_1 \frac{\partial^2 w}{\partial x^2} + 2t_3 \frac{\partial^2 w}{\partial x \partial y} + t_2 \frac{\partial^2 w}{\partial y^2}. \end{aligned} \tag{3.1.2}$$

In (1) and (2) the following notation is used:

$$\begin{gathered} \{T_1^0, T_2^0, S^0\} = -\lambda E h \mu^2 \{t_1, t_2, t_3\}, \\ \Phi' = \Phi\left(EhR\mu^2\right)^{-1}, \quad x = \frac{\alpha A}{R}, \quad y = \frac{\beta B}{R}, \\ k_i = \frac{R}{R_i}, \quad \mu^4 = \frac{h^2}{12(1-\nu^2)R^2} = \frac{h_*^2}{12(1-\nu^2)}. \end{gathered} \tag{3.1.3}$$

Here $\mu > 0$ is a small parameter, $\lambda > 0$ is an unknown load parameter. It is assumed, that the loading is one-parametric, i.e. that the initial stress-resultants increase proportional to a parameter λ, under loading. This assumption doesn't violate the generality since, under such loading, any stress state may be obtained. The symbol "minus" in the first group of formulae (3) is introduced for convenience, since buckling is characterised by compressive (negative) stress-resultants T_i^0. Below, the prime symbol by Φ is omitted.

The solution of system (1) we seek in the form

$$\{w, \Phi\} = \{w_0, \Phi_0\} \exp\left\{ i\,\frac{px + qy}{\mu} \right\}, \tag{3.1.4}$$

where w_0, Φ_0, p, and q are unknown constants.

After substitution in (1) we find the equation which wave numbers p and q satisfy. We write this equation in a form, solved with respect to λ

$$\lambda = f(p,q) = \frac{(p^2+q^2)^4 + (k_2 p^2 + k_1 q^2)^2}{(p^2+q^2)^2 (t_1 p^2 + 2t_3 pq + t_2 q^2)} \tag{3.1.5}$$

While not considering the question of satisfying the boundary conditions, we assume, that p and q may attain any real values. The critical value λ_0 of the parameter λ is given by formula:

$$\lambda_0 = \overset{(+)}{\min_{p,q}} f = f(p_0, q_0). \tag{3.1.6}$$

Let

$$p = r\cos\varphi, \qquad q = r\sin\varphi, \tag{3.1.7}$$

then the function f has the form:

$$f = r^2 a(\varphi) + r^{-2} b(\varphi) \tag{3.1.8}$$

and its minimum by r may be found immediately

$$\overset{(+)}{\min_r} f = \frac{2(k_2\cos^2\varphi + k_1\sin^2\varphi)}{t_1\cos^2\varphi + 2t_3\sin\varphi\cos\phi + t_2\sin^2\varphi} \equiv c(\varphi). \tag{3.1.9}$$

Now

$$\lambda_0 = \overset{(+)}{\min_\varphi}\{c(\varphi)\} = c(\varphi_0), \tag{3.1.10}$$

and the minimum is calculated by φ, for which $c(\varphi) > 0$ (the (+) symbol in (10) and also in (6) and (9) remind us of this). Since the shell is convex $k_1, k_2 > 0$, the numerator of $c(\varphi)$ is positive for all φ.

If simultaneously

$$t_1 \le 0, \quad t_2 \le 0, \quad t_3^2 - t_1 t_2 \le 0, \tag{3.1.11}$$

then the denominator of $c(\varphi)$ for all φ is non-positive and the shell will not buckle since there are no compressive initial stresses acting in any direction.

If inequalities (11) are not fulfilled simultaneously, then

$$\lambda_0 = \frac{4k_1k_2}{t_1k_1 + t_2k_2 + \tau_3}, \quad \cot\varphi_0 = \frac{t_1k_1 - t_2k_2 + \tau_3\,\mathrm{sgn}\,t_3}{2k_2t_3}, \qquad \tau_3 = \left((t_1k_1 - t_2k_2)^2 + 4k_1k_2t_3^2\right)^{1/2} \ge 0. \tag{3.1.12}$$

Due to (3) the critical values of the stress-resultants T_i^0 and S^0 satisfy the equality

$$\left((T_1^0R_2 - T_2^0R_1)^2 + 4R_1R_2(S^0)^2\right)^{\frac{1}{2}} - T_1^0R_2 - T_2^0R_1 = \frac{2Eh^2}{\sqrt{3(1-\nu^2)}}. \tag{3.1.13}$$

The above solution was obtained in [139]. Earlier, the problem without shear stress ($S^0 = 0$) was solved in [134]. Similar solutions for some partial cases of loading have been found by geometric methods in [132, 133].

The upper critical loads, given by A.V. Pogorelov in [132, 133] differ from those found by other authors by the factor $\sqrt{1-\nu^2}$. The difference may be explained as follows. The basis is the search for the potential energy of shell deformation under bending. One of the potential energy components is the potential energy of tangential deformation (Π_ε) which is concentrated in the neighbourhood of the fold appearing during the isometric transformation of the surface (see (2.3.12))

$$\Pi_\varepsilon = \frac{k}{2} \iint \left(\varepsilon_1^2 + 2\nu\varepsilon_1\varepsilon_2 + \varepsilon_2^2 + \frac{1-\nu}{2}\omega^2 \right) d\Omega$$

To calculate this energy the author assumed that $\varepsilon_1 \neq 0$ and $\varepsilon_2 = \omega = 0$. Actually the equality $\varepsilon_2 = 0$ introduces a non-existent tie on the surface deformation. If in evaluating Π_ε we assume that $\varepsilon_2 = -\nu\varepsilon_1$ (which corresponds to the minimum of the Π_ε by ε_2) then the results given by A.V. Pogorelov's geometrical method [133] for the upper critical loads coincide exactly with those obtained by the buckling equations.

Let us consider some special cases.

Pure shear.

Let $T_1^0 = T_2^0 = 0$ and $S^0 \neq 0$. Then the critical value of the stress-resultants and the parameters of the buckling mode are the following

$$S^0 = \frac{Eh^2}{\sqrt{3(1-\nu^2)R_1R_2}}, \quad \cot\varphi_0 = \sqrt{\frac{k_1}{k_2}}, \quad r_0^2 = 2\frac{k_1k_2}{k_1+k_2}. \tag{3.1.14}$$

The absence of external normal pressure.

Due to (1.4.3) $t_1k_1 + t_2k_2 = 0$ and formula (13) can be simplified as

$$((T_1^0)^2R_2^2 + (S^0)^2R_1R_2)^{\frac{1}{2}} = \frac{Eh^2}{\sqrt{3(1-\nu^2)}} \tag{3.1.15}$$

The absence of shear stress.

In this case directly from (9) we find

$$\lambda_0 = \frac{2k_2}{t_1}, \qquad \varphi_0 = 0, \qquad T_1^0 = -\frac{Eh^2}{R_2\sqrt{3(1-\nu^2)}} \quad \text{for} \quad t_1k_1 > t_2k_2; \tag{3.1.16}$$

$$\lambda_0 = \frac{2k_1}{t_2}, \qquad \varphi_0 = \frac{\pi}{2}, \qquad T_2^0 = -\frac{Eh^2}{R_1\sqrt{3(1-\nu^2)}} \quad \text{for} \quad t_1k_1 < t_2k_2. \tag{3.1.17}$$

The case when $t_1k_1 = t_2k_2$ is special and is considered in Section 3.3.

3.2 Shallow Shell Buckling Modes

It follows from (1.4) that in all of the cases considered above, the buckling mode is in the form of a system of pits which are highly elongated in one direction. The pits are inclined to axis Y by the angle $-\varphi_0$ and spread from one edge of the shell to the other. The distance between two neighbouring pits is

$$\Delta l = \frac{2\pi\mu R}{r_0} \tag{3.2.1}$$

and is of order $\sqrt{Rh}$. The buckling mode physically resembles a washing board.

In particular, for the case of pure shear

$$\Delta l = \frac{\pi\sqrt{h(R_1+R_2)}}{\sqrt[4]{3(1-\nu^2)}} \tag{3.2.2}$$

and for $S^0 = 0$, $t_1k_1 > t_2k_2$

$$\Delta l = \frac{2\pi\sqrt{hR_2}}{\sqrt[4]{12(1-\nu^2)}} \tag{3.2.3}$$

and the pits are elongated in the Y–direction.

It is clear that the buckling mode constructed satisfies none of the boundary conditions. The simplest boundary condition to satisfy is that of a simple support for a rectangular plane shell without shear. The boundary conditions of a simple support (Navier's conditions) for $x = 0$ and $x = l_1$ have the form

$$u_2 = w = T_1 = M_1 = 0$$

or

$$w = \frac{\partial^2 w}{\partial x^2} = \Phi = \frac{\partial^2 \Phi}{\partial x^2} = 0. \tag{3.2.4}$$

The conditions for $y = 0$ and $y = l_2$ are introduced similarly. Here $l_i = L_i R^{-1}$ where L_i are the shallow shell sizes in a plane.

The functions

$$\{w, \Phi\} = \{w_0, \Phi_0\} \sin\frac{p_m x}{\mu} \sin\frac{q_n y}{\mu}$$

$$p_m = \frac{\mu m \pi}{l_1}, \quad q_n = \frac{\mu n \pi}{l_2}, \quad m, n = 1, 2, \ldots \tag{3.2.5}$$

satisfy these conditions. From (1.5) we find $\lambda_{m,n}$. The critical value of λ is equal to

$$\lambda_0 = \overset{(+)}{\min_{m,n}} \lambda_{m,n}, \tag{3.2.6}$$

where the minimum we seek to those m, n for which $\lambda_{m,n} > 0$.

Let us consider the case when $t_1 k_1 > t_2 k_2$, $t_3 = 0$, for which the pits are elongated along axis Y. It was stated above that for arbitrary p and q the minimum is attained at $p_0 = \sqrt{k_1}$, $q_0 = 0$. That is why minimum (6) is attained at $n = 1$ and p_m which is closest to p_0. We can estimate the difference between λ_0 and $\bar{\lambda}_0$. By taking into account that $| p_m - p_0 | \leq \pi\mu/(2l_1)$ we find that

$$\frac{| \lambda_0 - \bar{\lambda}_0 |}{\lambda_0} \leq \frac{\pi^2}{\sqrt{12(1-\nu^2)}} \left[\frac{R_2 h}{L_1^2} + \frac{2R_2^2 h}{R_1 L_2^2} \left(1 - \frac{T_2^0 R_1}{T_1^0 R_2} \right) \right]. \tag{3.2.7}$$

Therefore if R_i, $L_i \sim R$ error (7) has order h_*.

While it is more difficult to construct the buckling modes for $S^0 \neq 0$ and for the other boundary conditions it is relatively easy to solve the problem for $S^0 = 0$ in the case when two opposite edges of a rectangular plane shell are simply supported and at the other two edges arbitrary boundary conditions are introduced.

The exact construction of a buckling mode was introduced above. Now let us consider the estimate of the critical load under different boundary conditions. Here we can make two statements.

First, since the buckling mode has a local character, increasing the stiffness of the boundary (up to a clamped edge support condition $u_1 = u_2 = w = \gamma_1 = 0$) may lead to an increase of the critical value of the load parameter λ compared to (1.12) only by a small value of order h_*. This statement will be proved at the end of Section 3.6.

Second, there exist 8 variants of weak shell edge support (with arbitrary support of the other edges) for which the parameter λ decreased. The buckling mode does not cover the entire neutral surface but rather it is localized in the neighbourhood of the weakly supported edge. This case will be considered in Section 13.4.

Remark 3.1. Generally speaking, the method of study described above is not valid for shells of negative ($k_1 k_2 < 0$) and zero ($k_1 k_2 = 0$) Gaussian curvature.

Indeed, let $k_2 > 0$ and $k_1 \leq 0$. Then we find from (1.9) and (1.10) that

$$\lambda_0 = 0, \quad r_0 = 0, \quad \cot \varphi_0 = \pm \sqrt{-\frac{k_1}{k_2}},$$

from which it follows that the order of the critical load decreases and the distance between the pits, due to (1), increases. The angle of inclination of the pits φ, depending on the initial stresses, could only be found. An exception to this is the case when $k_1 = T_2^0 = S^0 = 0$ and this will be examined in Section 3.4.

3.3 The Non-Uniqueness of Buckling Modes

For the condition

$$T_1^0 R_2 = T_2^0 R_1 < 0, \quad S^0 = 0 \tag{3.3.1}$$

the critical load could be found simply from

$$\lambda_0 = \frac{2k_2}{t_1} = \frac{2k_1}{t_2},$$
$$T_1^0 R_2 = T_2^0 R_1 = -\frac{Eh^2}{\sqrt{3(1-\nu^2)}}, \tag{3.3.2}$$

but the buckling mode is still indeterminate.

Let p and q in (1.4) be arbitrary values. We can write (1.5) in the form

$$\lambda = \frac{\lambda_0}{2}\left(z + \frac{1}{z}\right), \quad z = \frac{k_2 p^2 + k_1 q^2}{(p^2+q^2)^2}. \tag{3.3.3}$$

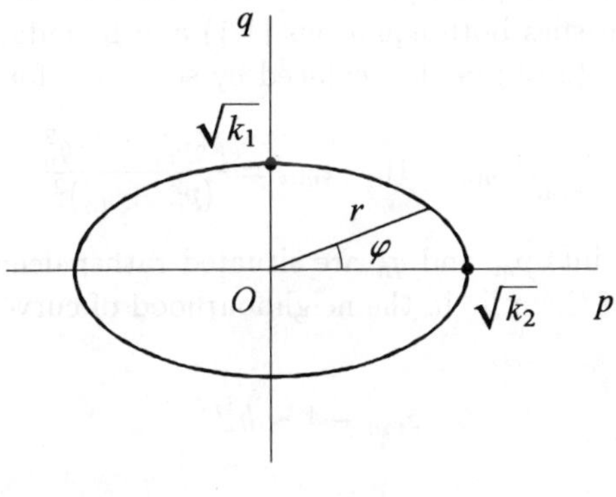

Figure 3.1: Wave numbers for the special case of a convex shell.

It is clear that the critical load is realized for $z = 1$, i.e. for any p and q, lying on the curve

$$(p^2 + q^2)^2 = k_2 p^2 + k_1 q^2 \tag{3.3.4}$$

(see Figure 3.1 or in polar coordinates (1.7))

$$r^2 = k_2 \cos^2\varphi + k_1 \sin^2\varphi. \tag{3.3.5}$$

If p and q are fixed and satisfy equation (4), then the pits (as in Section 3.2) are highly elongated in the same direction. However, now both $\pm p$ and $\pm q$ satisfy equation (4). Composing the linear combination of the functions

$$\exp\left\{\frac{i}{\mu}\left(\pm px \pm qy\right)\right\} \tag{3.3.6}$$

we obtain the possible buckling mode

$$\{w, \Phi\} = \{w_0, \Phi_0\} \sin\frac{p(x-x_0)}{\mu} \sin\frac{q(y-y_0)}{\mu}, \tag{3.3.7}$$

where p and q satisfy equation (4), and x_0 and y_0 are arbitrary.

Remark 3.2. Similar to (1.4) buckling modes (7) are obtained not depending on any boundary condition. The importance of constructing such modes is that they reflect the principal qualitative peculiarities of the buckling modes for real boundary conditions in this, and other more difficult problems (such as those with variable initial stresses, curvatures, etc.). As for the critical load, the value of λ_0 sought here is a good first approximation of the critical load. The following discussion confirms this statement.

If the shell has a rectangular planar form and its edges are simply supported, then mode (2.5) satisfies both equations (1.1) and boundary conditions (2.4). Due to (3), relation (2.6) may be replaced by searching for the minimum

$$\min_{m,n} |z_{mn} - 1|, \quad z_{mn} = \frac{k_2 p_m^2 + k_1 q_n^2}{(p_m^2 + q_n^2)^2}. \tag{3.3.8}$$

For small h_* points p_m and q_n are situated rather densely on (p, q) plane and the majority of them lie in the neighbourhood of curve (4). We note that for

$$z_{mn} - 1 \sim h_*^{1/2} \tag{3.3.9}$$

$$\frac{\lambda_0 - \bar{\lambda}_0}{\lambda_0} \sim h_* \tag{3.3.10}$$

The special case considered here is important in the problem of the buckling of a spherical shell under external pressure. For that $T_1^0 = T_2^0 = -\dfrac{qR}{2}$, $R_1 = R_2 = R$ and by formula (2) we can find the critical external pressure (see [61, 183])

$$q_0 = \frac{2Eh^2}{\sqrt{3(1-\nu^2)}R^2} \tag{3.3.11}$$

Although a complete sphere is not a shallow shell, within each pit, the shell may be considered as shallow and that justifies the use of equations (1.1).

Remark 3.3. It should not be supposed that the buckling of a real shell occurs by a mode for which the difference $|z_{mn} - 1|$ is a minimum. Initial imperfections and other factors not taken into account tend to distort the picture. On the other hand, the fact that eigenvalues λ_{mn} concentrate near λ_0 has a large effect on the sensitivity of the critical load to any initial imperfections. It is known that a spherical shell under external pressure (mentioned above) and a cylindrical shell under axial compression are the most sensitive (see Section 3.4). For both of these problems, buckling with many modes can occur.

The increase in sensitivity is caused by two factors. The more likely is that an occasional imperfection is close to one of the buckling modes and the other is that multiple and adjacent eigenvalues are more sensitive to perturbations (imperfections) than a simple one. This problem is considered in more detail in [14, 78, 145, 150].

3.4 A Circular Cylindrical Shell Under Axial Compression

Let us consider the buckling problem of circular cylindrical shell of radius R and length L_1 under a membrane stress state $T_1^0 < 0$, $T_2^0 = S^0 = 0$. For the case of a shell of moderate length ($L_1 \sim R$) we can use equations (1.1) as the buckling equations, in which

$$k_1 = 0, \quad k_2 = 1, \quad 0 \leq x \leq l_1 = \frac{L_1}{R}, \quad 0 \leq y \leq 2\pi. \tag{3.4.1}$$

If edges $x = 0$, and l_1 are simply supported (2.4), then the specific case considered in Section 3.3 takes place and the buckling mode is defined by formula (3.7), in which p and q satisfy equation (3.4)

$$(p^2 + q^2)^2 = p^2, \tag{3.4.2}$$

and the critical load, by virtue of (3.2) is equal to

$$\lambda_0 = \frac{2}{t_1}, \quad T_1^0 = -\frac{Eh^2}{\sqrt{3(1-\nu^2)}R}. \tag{3.4.3}$$

Formula (3) has been obtained by R. Lorenz [94] and S.P. Timoshenko [154].

The set of values of p and q satisfying (2) which is a couple of circles of radius $1/2$ is shown in Figure 3.2.

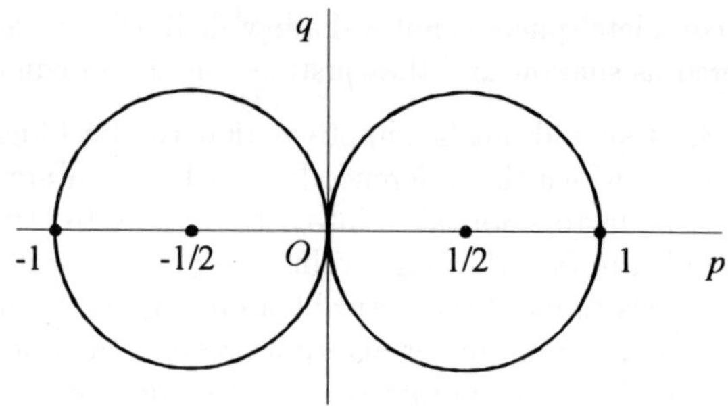

Figure 3.2: Wave numbers for axial compression of a cylindrical shell.

Satisfying the boundary conditions and the periodic conditions by y we find that p and q may take on only the discrete values

$$p_m = \frac{\mu m \pi}{l_1}, \quad q_n = \mu n, \quad x_0 = 0 \tag{3.4.4}$$

($m = 1, 2, \ldots$, $n = 0, 1, 2, \ldots$). The expected buckling mode is possible for a couple of values (m, n) for which the value of

$$z_{mn} = \frac{p_m^2}{(p_m^2 + q_n^2)^2} \tag{3.4.5}$$

is close to unity. Among them are the values $p_m = 1$ and $q_n = 0$, i.e.

$$n = 0, \quad m \simeq \frac{L_1}{\pi \mu R} = \frac{L_1 \sqrt[4]{12(1-\nu^2)}}{\pi \sqrt{Rh}} = l_0, \tag{3.4.6}$$

corresponding to an axially symmetric buckling mode.

Formula (3) is inapplicable for very long and very short shells and so we will now consider them separately.

Consider sufficiently long shells where $l_0 \geq 1$ (see (6)), then we can apply formula (3). If $l_0 < 1$ then we assume $m = 1$, and $n = 0$ and the critical load is then equal to

$$T_1^0 = -\frac{Eh^2}{\sqrt{12(1-\nu^2)}R} \left(l_0^2 + l_0^{-2} \right). \tag{3.4.7}$$

If, besides that, $l_0^4 \ll 1$, then neglecting the first term in (7) we come to the Euler formula

$$T_1^0 = -\frac{Eh^3 \pi^2}{12(1-\nu^2) L_1^2} \tag{3.4.8}$$

for the buckling of a beam-strip cut out from the shell by generators.

Now consider long shells. It follows from Figure 3.2 that among the expected buckling modes there are modes for which p_m and q_n are simultaneously close to zero (i.e. m and n are small). This case corresponds to the smaller variation of the stress state and system of equations (1.1) is not applicable.

For small p_m and q_n $p_m^2 \ll q_n^2$ (see Figure 3.2), i.e. inequality (1.9.2) is fulfilled and we can use systems (2.3.12) and (2.3.13). We seek the solution in the form

$$\{w, \phi\} = \{w_0, \phi_0\} \sin \frac{p_m x}{\mu} \sin ny,$$

$$u_1 = -\frac{p_m w_0}{\mu n^2} \cos \frac{p_m x}{\mu} \sin ny, \qquad u_2 = -\frac{w_0}{n} \sin \frac{p_m x}{\mu} \cos ny. \tag{3.4.9}$$

As a result we get

$$\frac{T_1^0}{Eh} = \min_{m,n} \left\{ \mu^4 \frac{(\lambda_m^2 + n^2)^2 - 2n^2 + 1}{\lambda_m^2 (1 + n^{-2})} + \frac{\lambda_m^2}{(\lambda_m^2 + n^2)^2 (1 + n^{-2})} \right\}, \tag{3.4.10}$$

where $\lambda_m = m\pi / l_1$. By means of this formula we can also get formulae (3) and (8). In the case $\lambda_m^2 \ll n^2$ formula (10) may be simplified

$$\frac{T_1^0}{Eh} = \min_{m,n} \left\{ \mu^4 \frac{(n^2 - 1)^2 n^2}{\lambda_m^2 (n^2 + 1)} + \frac{\lambda_m^2}{n^2 (n^2 + 1)} \right\}. \tag{3.4.11}$$

Minimizing it by λ_m we get the formula developed by P. Southwell and S.P. Timoshenko [61], generalizing (3) for the case of small n

$$T_1^0 = -\frac{Eh^2}{\sqrt{3(1 - \nu^2)} R} \frac{n^2 - 1}{n^2 + 1}, \qquad \lambda_m = \mu n \sqrt{n^2 - 1}, \tag{3.4.12}$$

For very large l_1 the case when $m = n = 1$ takes place. In this case we find from (11)

$$T_1^0 = -\frac{Eh}{2} \left(\frac{\pi}{l_1} \right)^2, \tag{3.4.13}$$

and the buckling of the shell occurs in a manner similar to a rod of circular cross-section.

Comparing the values of T_1^0, given by formula (12) for $n = 2$ and by formula (13) we find that buckling occurs similar to a rod with

$$\frac{L_1}{R} > 3.7 \left(\frac{R}{h} \right)^{1/2}. \tag{3.4.14}$$

More accurately, the effect of the critical loads T_1^0 on the characteristics of the shell are studied in [45, 61].

3.5 A Circular Cylindrical Shell Under External Pressure

We will now consider the buckling problem of a membrane stress state

$$T_1^0 = S^0 = 0, \qquad T_2^0 = -qR \tag{3.5.1}$$

of circular cylindrical shell described in Section 3.4. In the case of simply supported edges the buckling mode is

$$w = w_0 \sin\frac{\pi x}{l_1} \sin ny. \tag{3.5.2}$$

After substitution into equations (2.3.14) we find

$$T_2^0 = -\min_n \frac{Eh}{n^2-1}\left[\mu^4\left((\lambda_1^2+n^2)^2 - 2n^2 + 1\right) + \lambda_1^4(\lambda_1^2+n^2)^{-2}\right], \tag{3.5.3}$$

where $\lambda_1 = \pi/l_1$, $l_1 = L_1/R$.

First, we will consider a moderately long thin shell ($\mu \ll 1$, $\lambda_1 \sim 1$). Minimizing by n and assuming that we can neglect λ_1^2 and 1 compared to n^2 in (3) (i.e. ($n \sim h_*^{-1/4}$)), we approximately get

$$T_2^0 = -\min_n Eh\left(\mu^4 n^2 + \lambda_1^4 n^{-6}\right). \tag{3.5.4}$$

We could also obtain this formula by starting with system (1.1).

Minimizing (4) by n we get the Southwell–Papkovich formula [61, 104, 127, 144]

$$q_0 = -\frac{T_2^0}{R} = \frac{4Eh\mu^4 n_0^2}{3R} = \frac{0.856\,E}{(1-\nu^2)^{3/4}}\left(\frac{h^5}{L_1^2R^3}\right)^{1/2}, \tag{3.5.5}$$

where

$$n_0 = 3^{1/8}\left(\frac{\lambda_1}{\mu}\right)^{1/2} = 2.77\left(1-\nu^2\right)^{1/8}\left(\frac{R^3}{L_1^2 h}\right)^{1/4}. \tag{3.5.6}$$

Buckling occurs for an integer n which is closest to n_0. For these n, the critical value q_0 may be obtained more precisely by means of formula (3). The error of formula (5) is of order $h_*^{1/2}$ and is a result of the inaccuracy of formula (4) and with necessity of transition to an integer value of n.

From formulae (5) and (6) it follows that both q_0 and n_0 decrease with the growth of shell length L_1. The dependence of q_0 on the length of the shell may be also used to explain why buckling mode (2), has one half-wave in the longitudinal direction. The case of a few half-waves is similar to the case of a shorter shell.

The minimum value of n is equal to 2. For sufficiently long shells ($\lambda_1 \lesssim \mu$) we can neglect the second term in (4), and for $n = 2$ we get

$$q_0 = \frac{Eh^3}{4(1-\nu^2)R^3}. \tag{3.5.7}$$

This is the Grashof–Bress formula [18]. The shell length is not introduced in formula (7). This formula can be derived by considering the buckling of a ring made from a shell under an external follower pressure [1].

The dependence of n_0 and q_0 on the shell parameters for small $n_0 = 2$–6 is studied in [61].

Let us now consider the combined loading of a cylindrical shell under an axial force and a normal pressure (T_1^0, $T_2^0 \neq 0$, $S^0 = 0$). We will restrict the discussion to the case of a moderately long shell with simply supported edges and consider system (1.1). It occurs [61], that for $T_1^0 < 0$ (as in the case when $T_1^0 = 0$) that the buckling mode has only one wave in the longitudinal direction and a few waves in the circumferential direction ($m = 1$, $n \sim h_*^{-1/4}$).

With an error of order $h_*^{1/2}$, the critical values of the stresses T_1^0 and T_2^0 can be found from the relation

$$\mu^4(\lambda_1^2+n^2)^4 + \left(\frac{T_1^0}{Eh}\lambda_1^2 + \frac{T_2^0}{Eh}n^2\right)\left(\lambda_1^2+n^2\right)^2 + \lambda_1^4 = 0, \tag{3.5.8}$$

which, with an error of the same order may be written as

$$2\tau_1 + c\xi\tau_2 = \xi^2 + \xi^{-2}, \tag{3.5.9}$$

where

$$\tau_1 = \frac{T_1^0}{T_1^*}, \qquad \tau_2 = \frac{T_2^0}{T_2^*}, \qquad \xi = \mu n^2 \lambda_1^{-1},$$
$$T_1^* = -2Eh\mu^2, \quad T_2^* = -cEh\lambda_1\mu^3, \quad c = 4\cdot 3^{-3/4} = 1.755. \tag{3.5.10}$$

Here τ_1 and τ_2 are the nondimensional stresses related to the critical values obtained in (4.3) and (5). The critical value of τ_2 for fixed τ_1 we get from (9) minimizing it by ξ. This dependence is represented in Figure 3.3.

For $\tau_2 < 0$, (the internal pressure) the value of τ_1 does not depend on τ_2 in the linear approximation and so the buckling mode is axially symmetric. For $\tau_1, \tau_2 > 0$ the curve is close to straight line $\tau_1 + \tau_2 = 1$. The tangents to this curve drawn through points (1,0) and (0,1) are

$$\tau_1 + 0.877\,\tau_2 = 1, \quad 0.866\,\tau_1 + \tau_2 = 1. \tag{3.5.11}$$

This case of combined loading is discussed in more detail in [45, 61, 155].

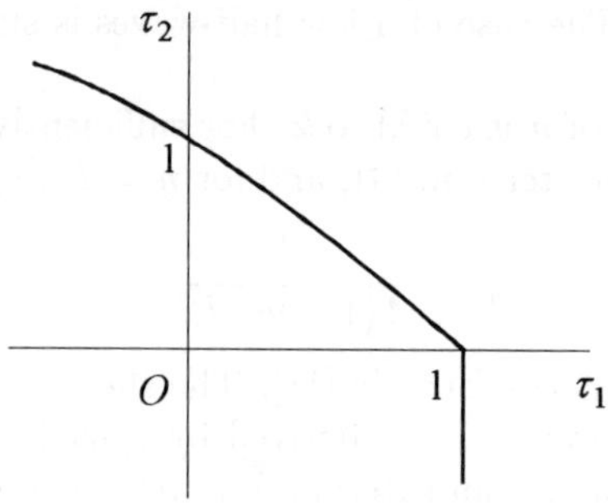

Figure 3.3: Dimensionless stress-resultants for a cylindrical shell under combined loading.

3.6 Estimates of Critical Load

The problems considered in this chapter comprise the class of shell buckling problems which have exact closed form solutions. In subsequent chapters different approximate solutions will be developed. The order of the critical loads for membrane buckling are obtained by energy methods. Generally, the results given below are contained in [54, 55]. A similar approach was used in [57, 158] in which shell free vibrations were studied. Applications of various variational principles to shell theory are discussed in [169].

Membrane stress state buckling may be studied starting from the variation problem which comes from (2.3.10)

$$\lambda = \min{}^{(+)} J, \qquad J = \frac{\Pi'_\varepsilon + \mu^4 \Pi'_\varkappa}{\mu^2 \Pi'_T}, \tag{3.6.1}$$

where

$$\begin{aligned}
\Pi'_\varepsilon &= \frac{\Pi_\varepsilon}{E_0 h_0} = \frac{1}{2}\iint\limits_\Omega \frac{K}{E_0 h_0}\left[\varepsilon_1^2 + 2\nu\varepsilon_1\varepsilon_2 + \varepsilon_2^2 + \frac{1-\nu}{2}\omega^2\right] d\Omega, \\
\Pi'_\varkappa &= \frac{\Pi_\varkappa}{E_0 h_0 \mu^4} = \frac{1}{2}\iint\limits_\Omega \frac{D}{E_0 h_0 \mu^4}\left[\varkappa_1^2 + 2\nu\varkappa_1\varkappa_2 + \varkappa_2^2 + 2(1-\nu)\tau^2\right] d\Omega, \\
\Pi'_T &= \frac{1}{2}\iint\limits_\Omega \left(t_1\gamma_1^2 + 2t_3\gamma_1\gamma_2 + t_2\gamma_2^2\right) d\Omega.
\end{aligned} \tag{3.6.2}$$

Here, $\lambda > 0$ is an unknown load parameter. The minimum in (1) is taken over all possible complementary displacements u_1, u_2, w, for which the value of J in (1) is positive (the (+) sign reminds us of this) and which satisfies the geometric boundary conditions. We assume that the small parameter μ (or h_*) in (1) is the only essential parameter. In other words, we analyze a moderately long shell with smoothly varying metrics, curvature and membrane initial stresses.

The value of J in (1) does not depend on the range of the complementary displacements and therefore, we will assume for convenience that $w \sim R$ and all geometric shell sizes are related to R. Then, all the variables in (1) become non-dimensional.

The research method used below consists of the substitution into (1) of various displacements which satisfy the geometric boundary conditions. For the load parameter, λ, we obtain an estimate of the upper bound. The symbol O is used instead of $\sim$ in the case when the exact order with respect to h_* of the estimated variable, is not found.

It is clear that $\Pi'_\varepsilon \geq 0$, $\Pi'_\varkappa \geq 0$ for any of the displacements u_i, w. The value of Π'_T may have any sign depending on T_i^0 and S^0, but we are interested only in displacements for which $\Pi'_T > 0$. Consideration of the quadratic form in γ_1 and γ_2 in the expression for Π'_T shows that such displacements exist only in cases when inequalities (1.11) are not fulfilled simultaneously. In the other words it is necessary for buckling that at least one of the inequalities be fulfilled

$$T_1^0 < 0 \quad \text{or} \quad T_2^0 < 0 \quad \text{or} \quad (S^0)^2 - T_1^0 T_2^0 > 0. \tag{3.6.3}$$

These inequalities provide the existence of the directions along which the compressive initial stresses act. Condition (3) is not necessary in the entire domain Ω occupied by the shell. It is sufficient that this condition be fulfilled only in a certain portion Ω_1 of this domain.

Let us consider the system of equations

$$\varepsilon_1 = \varepsilon_2 = \omega = 0, \tag{3.6.4}$$

where the tangential deformations ε_i, ω are defined in (1.1.7), (1.1.10). The non-trivial solutions of this system with respect to u_i, w, satisfying the geometric boundary conditions are called infinitesimal surface bending. If, for them, the values of $\varkappa_i$, τ differ from zero then we have non-trivial bending.

We assume that there exists non-trivial bending for which the displacements u_i, w are slowly varying functions (index of variation $t = 0$). Then we obtain from (1) the following estimates

$$\lambda = O(h_*), \quad T^0 = O\left(\frac{Eh_0^3}{R^2}\right), \quad T_0 = \max\left\{T_i^0, S^0\right\}. \tag{3.6.5}$$

The estimate $\lambda \sim h_*$ takes place in the buckling problem for a long or free cylindrical shell under external pressure (see (5.7)), and also in the buckling of a weakly supported shell of negative Gaussian curvature (see Section 12.2).

Later it will be shown (see Chapter 12) that in the case when non-trivial bending exists, functional (1) does not give the asymptotically exact value of the critical load λ as $h_* \to 0$, although estimate (5) is valid.

To explain this, we represent λ in the form $\lambda = a_0 h_*^\alpha + o(h_*^\alpha)$. Then, if bending exists, function (1) gives the correct value of α and in the general case, an incorrect value of a_0. At the same time for the other cases considered in this section, the accuracy of (1) is sufficient for the evaluation of both α and a_0.

Now, we assume that no non-trivial solution satisfying the geometric boundary conditions of system (4) exists. As the test functions for the substitution into (1) take first the displacement field

$$u_1 = u_2 = 0, \quad w = \varphi(\alpha, \beta)\sin z, \quad z = h_*^{-t} f(\alpha, \beta). \tag{3.6.6}$$

The geometric boundary conditions are fulfilled by a special choice of function φ that is assumed to vanish together with the few first derivatives on the boundary of domain Ω.

Function (6) for $\left(\dfrac{\partial f}{\partial \alpha}\right)^2 + \left(\dfrac{\partial f}{\partial \beta}\right)^2 > 0$ has index of variation t. Substitution of (6) into (1) gives $\Pi'_\varepsilon = O(1)$, $\Pi'_\varkappa \sim h_*^{-4t}$. The integrand in Π'_T takes the form

$$h_*^{-2t}\varphi^2\cos^2 z\left[t_1\left(\frac{\partial f}{\partial \alpha}\right)^2 + 2t_3\frac{\partial f}{\partial \beta}\frac{\partial f}{\partial \alpha} + f_2\left(\frac{\partial f}{\partial \beta}\right)^2\right] + O(h_*^{-t}) \tag{3.6.7}$$

and if conditions (3) are fulfilled it is possible to find functions φ and f in (6) for which $0 < \Pi'_T \sim h_*^{-2t}$. The substitution of these estimates into (1) gives $J = O(h_*^{2t-1} + h_*^{1-2t})$ and for $t = 1/2$ we obtain $J = O(1)$.

Hence, independent of the shape of the shell and the method of shell support we obtain under the above assumptions

$$\lambda = O(1), \quad T^0 = O\left(\frac{Eh_0^2}{R}\right). \tag{3.6.8}$$

For "well" supported convex shells ($k_1 k_2 > 0$) estimate (8) is exact i.e. $\lambda \sim 1$ (see Sections 3.1, 12.3, 13.4). Here the shell is called "well supported" if restrictions on the displacements prevent neutral surface bending.

In the case of weakly supported shells, estimate (8) may be improved, however we will not consider this case here (see Chapter 12).

We will, however, try to improve estimate (8) for arbitrarily supported shells of zero and negative Gaussian curvature. Let us consider a shell of zero curvature $k_1 = 0$. By virtue of the Codazzi–Gauss relation (1.1.3) $\dfrac{\partial A}{\partial \beta} = 0$. The displacement field we take in the form [158]

$$
\begin{aligned}
&u_1 = h_*^{2t}\varphi_1 \sin z, && u_2 = h_*^t \varphi_2 \cos z + h_*^{2t}\varphi_3 \sin z, \\
&w = \varphi \sin z, && z = h_*^{-t} f(\beta), \qquad f'(\beta) \neq 0, \\
&\varphi_1 = -\frac{B^2}{Af'}\frac{\partial(\varphi_2/B)}{\partial\alpha}, && \varphi_2 = \frac{k_2 B}{f'}\varphi, \qquad \varphi_3 = -\frac{1}{f'}\frac{\partial \varphi_2}{\partial \alpha}.
\end{aligned}
\tag{3.6.9}
$$

By direct substitution we can verify that $(\varepsilon_1, \varepsilon_2, \omega) \sim h_*^{2t}$. Hence $\Pi'_\varepsilon \sim h_*^{4t}$, $\Pi'_\varkappa \sim h^{-4t}$. Further $\gamma_1 \sim 1$, $\gamma_2 \sim h_*^{-t}$ and by virtue of (2)

$$
\begin{aligned}
&0 < \Pi'_T \sim h_*^{-2t} && \text{for} \quad T_2^0 < 0, \\
&0 < \Pi'_T \sim h_*^{-t} && \text{for} \quad T_2^0 = 0, \quad S^0 \neq 0, \\
&0 < \Pi'_T \sim 1 && \text{for} \quad T_2^0 = S^0 = 0, \quad T_1^0 < 0 \quad \text{and} \\
& && \text{for} \quad T_2^0 > 0, \quad (S^0)^2 - T_1^0 T_2^0 > 0.
\end{aligned}
\tag{3.6.10}
$$

Substituting the above estimates into (1) and minimizing by t in the three cases of (10) we obtain respectively $t = 1/4$ and

$$
\begin{aligned}
&\lambda = O\left(h_*^{1/2}\right), && T^0 = O\left(Eh_0\left(\frac{h_0}{R}\right)^{3/2}\right), \\
&\lambda = O\left(h_*^{1/4}\right), && T^0 = O\left(Eh_0\left(\frac{h_0}{R}\right)^{5/4}\right), \\
&\lambda = O(1), && T^0 = O\left(\frac{Eh_0^2}{R}\right).
\end{aligned}
\tag{3.6.11}
$$

For the particular case of a circular cylindrical shell, the first and third estimates have been given in Sections 3.5 and 3.4. We note that in the case of $\lambda \sim 1$, displacements (6) for $t = 1/2$ lead to the same estimate. Therefore, the improvement of the general estimate (8) takes place only in the first two cases.

Remark 3.4. Estimates (11) remain valid in the case when the inequalities in (10) are fulfilled only in a certain (not too small) portion Ω_1 of domain Ω. To obtain estimate (11) we take the function in (9) which is non-zero only in Ω_1. We deal in a similar way in obtaining estimate (8), if conditions (3) are fulfilled in a portion of domain Ω.

Now we consider a shell of negative Gaussian curvature, i.e. $k_1 k_2 < 0$. We take

$$
\begin{aligned}
u_1 &= h_*^t \varphi_1 \cos z, & \varphi_1 &= A k_1 \left(\frac{\partial f}{\partial \alpha}\right)^{-1} \varphi, \\
u_2 &= h_*^t \varphi_2 \cos z, & \varphi_2 &= B k_2 \left(\frac{\partial f}{\partial \beta}\right)^{-1} \varphi, \\
w &= \varphi \sin z, & z &= h_*^{-t} f(\alpha, \beta),
\end{aligned}
\tag{3.6.12}
$$

where function f satisfies the equation

$$
\frac{k_1}{A^2}\left(\frac{\partial f}{\partial \alpha}\right)^2 + \frac{k_2}{B^2}\left(\frac{\partial f}{\partial \beta}\right)^2 = 0. \tag{3.6.13}
$$

For such displacements we find $\varepsilon_i,\ \omega \sim h_*^t$. Let us turn to the estimation of Π_T' and assume that $k_1 < 0,\ k_2 > 0$.

The method of the construction of displacements (12) imposes complementary restriction (13) for $\dfrac{\partial f}{\partial \alpha}$ and $\dfrac{\partial f}{\partial \beta}$ contained in (7). The requirement that (7) be positive leads to the condition

$$
T_1^0 k_2 - T_2^0 k_1 - \left| S^0 \sqrt{-k_1 k_2} \right| < 0. \tag{3.6.14}
$$

If this condition is fulfilled at least in a part of domain Ω, there are displacements for which $0 < \Pi_T' \sim h_*^{-2t}$.

After substitution into (1) and minimization by t we find

$$
t = 1/3, \qquad \lambda = O\left(h_*^{1/3}\right), \qquad T^0 = O\left(E h_0 \left(\frac{h_0}{R}\right)^{4/3}\right). \tag{3.6.15}
$$

Condition (14) imposes stronger restrictions on the initial stress-resultants T_i^0, S^0 than do conditions (3). If conditions (3) are fulfilled, but (14) are not fulfilled, then buckling occurs. However it is possible to guarantee only estimate (8) for the critical load.

Remark 3.5. The improvement of estimate (8) is connected with the construction of surface pseudo-bending i.e. such surface displacements (9) and (12) for which the tangential deformations are small.

So far we considered neither the cases of weak shell support nor the cases when the Gaussian curvature of the surface changes its sign.

The estimates of the critical load obtained in this section and the expected buckling modes are used below for construction of the asymptotic solutions.

Let us return to the case considered in Sections 3.1–3.2 ($k_1, k_2 > 0$) and prove that the increase of stiffness in the boundary support does not greatly increase the critical value compared to the value given by (1.12). We write the displacements in the form

$$\begin{aligned} &u_1 = \mu\, a\, w_0 \cos z, \qquad u_2 = \mu\, b\, w_0 \cos z, \\ &w = w_0 \sin z, \qquad z = \mu^{-1}(px + qy) + z_0, \\ &a = \frac{(q^2 - \nu p^2)(k_2 p^2 + k_1 q^2)}{p(p^2 + q^2)^2} - \frac{k_1}{p}, \\ &b = \frac{(p^2 - \nu q^2)(k_2 p^2 + k_1 q^2)}{q(p^2 + q^2)^2} - \frac{k_2}{q}, \end{aligned} \tag{3.6.16}$$

where the values of p and q have been found in Section 3.1. Substitution of (16) into (1) for $w_0 \equiv 1$ gives λ equal to (1.12).

Now we take $w_0 = w_0(x, y)$ and require that

$$w_0 = \frac{\partial w_0}{\partial m} = 0, \qquad x, y \in \Gamma, \tag{3.6.17}$$

where Γ is a contour of domain Ω. Then functions (16) satisfy the condition of a clamped support. For each integral Π'_ε, $\Pi'_\varkappa$, Π'_T in (1) we have

$$\Pi'_j = \Pi^0_j \iint_\Omega w_0^2 d\Omega + O(\mu^2), \qquad \mu \to 0, \quad j = \varepsilon, \varkappa, T, \tag{3.6.18}$$

where Π^0_j are such that it is possible to write formula (1.5) in the form $\lambda^0 = (\Pi^0_\varepsilon + \Pi^0_\varkappa)(\Pi^0_T)^{-1}$. It follows from above that $\lambda \leq \lambda^0 + O(\mu^2)$.

3.7 Problems and Examples

3.1. Based on system (2.3.8) derive the system of equations describing the membrane pre-buckling axisymmetric stress-strain state of the shell of revolution. Use the curvilinear coordinates (s, φ), where s is the length of the genetatrix and φ is the angle in the circumferential direction.

Answer The system has the form (3.1.1), where

$$\Delta_t w = \mu^2 \frac{1}{b}(b t_1 w')' - \frac{t_2 \rho^2}{b^2} w$$

$$t_k = -\frac{T_k^0}{\lambda E h \mu^2} \qquad k = 1, 2$$

where $T_k^0(s)$ are the membrane pre-buckling stresses, λ is a loading parameter, $\rho = \mu m$, μ is a small parameter $\mu^4 = \dfrac{h^2}{12(1-\nu^2)R^2}$, $b = \dfrac{B(s)}{R}$, $w(s, \varphi) = w(s)\cos(m\varphi)$, and $\Phi(s, \varphi) = \Phi(s)\cos(m\varphi)$,

3.2. The membrane pre-buckling axisymmetric stress-strain state of the shell of revolution is described by system of equations (3.1.1) (see Problem 3.1) Construct the asymptotic expansions of the solutions of the system

Answer

$$w(s,\mu) \sim \sum_{k=0}^{\infty} \mu^k w_k(s) \exp \frac{1}{\mu} \int_{s_0}^{s} p(s)ds$$

$$\Phi(s,\mu) \sim \sum_{k=0}^{\infty} \mu^k \Phi_k(s) \exp \frac{1}{\mu} \int_{s_0}^{s} p(s)ds$$

where $p(s)$ is a root of equation

$$\left(p^2 - \frac{r^2}{b^2}\right)^4 + \lambda\left(t_1 p^2 - \frac{t_2 r^2}{b^2}\right)^2 \left(p^2 - \frac{r^2}{b^2}\right) + \left(k_2 p^2 - \frac{k_1 r^2}{b^2}\right)^2 = 0$$

and

$$w_0 = \left(p^2 - \frac{r^2}{b^2}\right)\left(b\frac{\partial f}{\partial p}\right)$$

$$\Phi_0 = -\left(k_2\lambda^2 - \frac{k_1 r^2}{b^2}\right)\left(\lambda^2 - \frac{r^2}{b^2}\right)^{-1}\left(b\frac{\partial f}{\partial \lambda}\right)$$

3.3. Derive from (3.6.4) relations for the tangential displacements u_1 and u_2 for the shells of zero and negative Gaussian curvature (3.6.9) and (3.6.12). Derive the estimates (3.6.11) and (3.6.15).

Hint Represent the deflection in the form $w = \varphi \sin z$.

Chapter 4

Buckling Modes Localized near Parallels

The buckling modes of convex shells of revolution under axisymmetric membrane initial stress states are considered in this chapter. Only buckling problems in which the deflections are localized near some parallel (henceforth called the weakest parallel) are studied. It is assumed that the weakest parallel is sufficiently far from the shell edges to avoid edge effects.

4.1 Local Shell Buckling Modes

The buckling modes of thin elastic shells under initial membrane stress states may be divided into two classes. In the first class the buckling concavity occupies all or the greater part of the shell surface. A typical example of this class is the buckling of a shallow convex shell under external pressure. The second class is when the buckling mode consists of many small pits formed on the surface of the shell.

The size and position of the pits and the value of the buckling load essentially depend on some determining functions such as the radii of the curvature of the neutral surface, the shell thickness, the initial membrane stresses, etc. In simple cases when these functions may be assumed to be approximately constant, the buckling pits occupy the entire shell surface (see Section 3.1). This case is important, for example, for the buckling of a circular cylindrical shell under axial compression (see Section 3.4) or under external lateral pressure (Section 3.5) or torsion (Section 9.1). Shells with negative Gaussian curvature also, as a rule, lose their stability by modes for which the pits occupy the entire shell surface (see Chapter 11).

If the determining functions vary from point to point of the neutral surface then localization of the buckling mode will occur.

For the first localization type, the pits are localized near the weakest line on the shell surface. The buckling modes of convex shells of revolution under axisymmetric loading (discussed in this Chapter) and of cylindrical shells under combined action of an axial force and a bending moment (Chapter 5) are problems of this type.

The second localization type is possible for convex shells for which the small buckling pits are concentrated near the weakest point on the neutral surface. The buckling mode of a convex shell of revolution under combined loading may be of this type (see Chapter 6).

Let us note that the for these localization types it is more difficult to find the buckling modes if the weakest point or line is close to an edge of the shell. In this case it is necessary to satisfy the boundary conditions and to take into account the initial bending stresses and pre-buckling deformations (see Chapter 14).

The third localization type is possible for the shells of zero Gaussian curvature (i.e. for cylindrical and conic shells). Buckling is accompanied by the appearance of concavities which are stretched along the shell generatrix from one shell edge to another. The depth of the concavities is maximum near the weakest generatrix and decreases fast away from the generatrix. The buckling modes of non-circular cylindrical and conic shells (and of circular shells with slanted edges) under external pressure or torsion have these forms. This type of buckling occurs for circular cylindrical shells in bending (see Chapters 7 and 9).

We note that for the second and third localization types there exist two buckling modes characterized by very close critical loads.

For the case when the weakest line or point is far from a shell edge the approximate analytical formulae for buckling modes and upper critical loads have been obtained in assumption that the relative shell thickness is small. Unfortunately, if weakest point or line is close to a shell edge, the results are not so simple.

4.2 Construction Algorithm of Buckling Modes

This algorithm is an one-dimensional version of Maslov's algorithm [99]. In shell buckling problems this algorithm was used in [160, 163] (see also [15, 88]).

We describe this algorithm using, as an example, the self-adjoint ordinary differential equation of order $2n$

$$\sum_{k=0}^{n}(-i\mu)^{2k}\frac{d^k}{dx^k}\left(a_k(x)\frac{d^k w}{dx^k}\right)=0, \quad x_1 \leqslant x \leqslant x_2, \tag{4.2.1}$$

where $\mu > 0$ is a small parameter and the coefficients a_k are real and depend linearly upon the parameter $\lambda > 0$

$$a_k = a_k' - \lambda a_k'', \qquad a_k', a_k'' \in C^\infty \tag{4.2.2}$$

and the functions a_k' and a_k'' are infinitely differentiable with respect to x.

At each end of the interval (x_1, x_2) the n self-adjoint boundary conditions are introduced, but we do not satisfy them exactly. We seek eigenvalues, $\lambda^{(m)}$, for the parameter, λ, for which there exist localized solutions of equation (1) which exponentially decay when x approaches the ends of the interval (x_1, x_2). We will study only the case when there exists a point x_0 $(x_1 < x_0 < x_2)$ such that these solutions exponentially decay as $|x - x_0|$ increase. Point x_0 is referred to as the weakest point.

Some shell buckling problems may be reduced to this problem after separation of the variables. As a rule we get a system of equations of type (1) with coefficients regularly depending on the parameter μ. The algorithm described below may be also used in this case without major changes.

If we seek the solution of equation (1) as a formal series

$$w(x,\mu) = \sum_{j=0}^{\infty} \mu^j w_j(x) \exp\left\{\frac{i}{\mu}\int p(x)\,dx\right\}, \tag{4.2.3}$$

then $p(x)$ satisfies the algebraic equation

$$\sum_{k=0}^{n} a_k(x)\, p^{2k} = 0. \tag{4.2.4}$$

Function (3) satisfies the decaying condition mentioned above if there exist points x_1^*, x_2^* such that

$$\begin{aligned} \Im p(x) > 0 \quad &\text{for} \quad x > x_2^*, \\ \Im p(x) = 0 \quad &\text{for} \quad x_1^* \leqslant x \leqslant x_2^*, \\ \Im p(x) < 0 \quad &\text{for} \quad x < x_1^*. \end{aligned} \tag{4.2.5}$$

The simultaneous fulfilment of conditions (5) is possible only in the case when the roots of equation (4) are multiple for $x = x_1^*$ and $x = x_2^*$. In the neighbourhood of points $x = x_1^*$ and $x = x_2^*$ (which are called the turning points) solutions (3) with limited $w_j(x)$ do not exist.

To construct limited asymptotic solutions in this case, the standard equation method [84] has been used in [57, 157]. Further we propose a simpler method, which is based on the assumption that the turning points x_1^* and x_2^* are close to each other and that the coefficients of equation (1) are real. We seek a solution satisfying the decaying condition as a formal series

$$w(x,\mu) = w_* \exp\left\{i\left(\mu^{-1/2} p_0 \xi + (1/2)\, a\xi^2\right)\right\}, \tag{4.2.6}$$

where

$$w_* = \sum_{k=0}^{\infty} \mu^{k/2} w_k(\xi), \quad \xi = \mu^{-1/2}(x - x_0), \tag{4.2.7}$$

and $w_k(\xi)$ are polynomials in ξ. Parameter λ is also expanded into a series in μ

$$\lambda = \lambda_0 + \mu\lambda_1 + \mu^2\lambda_2 + \ldots \tag{4.2.8}$$

Here $w_k(\xi)$, λ_k, x_0, p_0 and a may be found after substitution into equation (1). Satisfying the decay condition for solution (6), we find that p_0 must be real and $\Im(a) > 0$. For that x_0 is the weakest point, parameter p_0 characterizes the oscillation frequency of function w, and $\Im(a)$ gives the rate of decrease of the function amplitude when x deviates from x_0.

To find the parameters x_0 and p_0 we resolve equation (4) in λ

$$\lambda = \frac{\sum\limits_{k=0}^{n} a_k'(x)\, p^{2k}}{\sum\limits_{k=0}^{n} a_k''(x)\, p^{2k}} \equiv f(p,x). \tag{4.2.9}$$

Let the minimum

$$\lambda_0 = \min_{p,x}\{f(p,x)\} = f(p_0, x_0) \tag{4.2.10}$$

exists for real p and $x_1 \leqslant x \leqslant x_2$, and $x_1 < x_0 < x_2$, and the second differential of function f at point (p_0, x_0) is the positive definite quadratic form

$$d^2 f = f_{pp}^0\, dp^2 + 2 f_{px}^0\, dp\, dx + f_{xx}^0\, dx^2 > 0. \tag{4.2.11}$$

Here derivatives f_{pp}^0, f_{px}^0 and f_{xx}^0 are evaluated at point (p_0, x_0).

We rewrite equation (1) in the form

$$H\left(-i\mu\frac{d}{dx}, x, \lambda, \mu\right) w = 0, \tag{4.2.12}$$

where

$$H\,(p,\,x,\,\lambda,\,\mu) = \sum_{k=0}^{n} \sum_{j=0}^{k} (\,-\,i\,\mu)^j\; C_k^j\, \frac{d^j\, a_k}{dx^j}\; p^{2k-j}$$

are polynomials of order $2n$ in p. After the substitution of solution (6) in equation (12) we obtain

$$H\left(p_0 + \mu^{1/2}(a\xi - i\,\frac{d}{d\xi}),\; x_0 + \mu^{1/2}\xi,\; \lambda_0 + \mu\lambda_1 + \ldots,\; \mu\right) \sum_{k=0}^{\infty} \mu^{k/2} w_k\,(\xi) = 0. \tag{4.2.13}$$

The expansion into a power series of $\mu^{1/2}$ leads to equations

$$\sum_{j=0}^{k} H^{(j)}\, w_{k-j} = 0, \quad k = 0, 1, 2, \ldots, \tag{4.2.14}$$

where the first differential operators $H^{(j)}$ are

$$\begin{aligned} H^{(0)} &= H\,(p_0,\, x_0,\, \lambda_0,\, 0)\,, \\ H^{(1)} &= H_p^0 \left(a\xi - i\,\frac{d}{d\xi}\right) + H_x^0\,\xi, \\ H^{(2)} &= \frac{1}{2}\, H_{pp}^0 \left(a^2\xi^2 - 2\,i\,a\,\xi\,\frac{d}{d\xi} - i\,a - \frac{d^2}{d\xi^2}\right) + \\ &\quad + H_{px}^0 \left(a\,\xi^2 - i\,\xi\,\frac{d}{d\xi}\right) + \frac{1}{2}\, H_{xx}^0\, \xi^2 + \lambda_1\, H_\lambda^0 + H_\mu^0; \end{aligned} \tag{4.2.15}$$

$$H_p = \frac{\partial H}{\partial p}, \quad H_{px} = \frac{\partial^2 H}{\partial p\, \partial x}, \;\ldots$$

The superscript ^{0} on the derivatives indicates that they are calculated for

$$p = p_0, \quad x = x_0, \quad \lambda = \lambda_0, \quad \mu = 0.$$

Since, due to (10)

$$H^{(0)} = H_p^0 = H_x^0 = 0, \tag{4.2.16}$$

we examine the equation $H^{(2)} w_0 = 0$. This equation has a solution in the form of a polynomial in ξ, only if the coefficient at ξ^2 is equal to zero, from which we get

$$H_{pp}^0\, a^2 + 2H_{px}^0\, a + H_{xx}^0 = 0. \tag{4.2.17}$$

Taking into account that

$$H_{pp}^0 = -H_\lambda^0 f_{pp}^0, \qquad H_{px}^0 = -H_\lambda^0 f_{px}^0,$$
$$H_{xx}^0 = -H_\lambda^0 f_{xx}^0, \qquad H_\mu^0 = -\frac{i}{2} H_{px}^0,$$

we obtain

$$a = \frac{-f_{px}^0 + i\,r}{f_{pp}^0}, \quad r = \sqrt{D} > 0, \quad D = f_{pp}^0 f_{xx}^0 - \left(f_{px}^0\right)^2. \tag{4.2.18}$$

According to (11) $D > 0$.

Equation $H^{(2)} w_0 = 0$ may be written in the form

$$\frac{d^2 w_0}{d\xi^2} - c\left(2\xi \frac{dw_0}{d\xi} + w_0\right) + \lambda_1 d\, w_0 = 0, \tag{4.2.19}$$

where $c = r/f_{pp}^0$, $d = 2/f_{pp}^0$.

For

$$\lambda_1 = \lambda_1^{(m)} = r\left(\frac{1}{2} + m\right), \quad m = 0, 1, 2, \ldots \tag{4.2.20}$$

equation (19) has a solution in the form of Hermite's polynomial [69]

$$w_0(\xi) = H_m(\theta), \quad \theta = \sqrt{c}\,\xi. \tag{4.2.21}$$

The Hermite polynomials satisfy the equation

$$M_m\left[H_m(\theta)\right] \equiv \frac{d^2 H_m}{d\theta^2} - 2\theta \frac{dH_m}{d\theta} + 2m\, H_m = 0, \tag{4.2.22}$$

and three first polynomials are the following

$$H_0(\theta) = 1, \quad H_1(\theta) = 2\theta, \quad H_2(\theta) = 4\theta^2 - 2.$$

The polynomials $H_m(\theta)$ are either even or odd.

Let us construct the following terms in series (6) and (8). Function $w_1(\xi)$ due to (14) may be found from the equation $H^{(2)} w_1 + H^{(3)} w_0 = 0$, which after substitution $\theta = \sqrt{c}\,\xi$, has the form

$$M_m\left[\tilde{w}_1(\theta)\right] = P_{m1}(\theta), \tag{4.2.23}$$

where $P_{m1}(\theta)$ is a known polynomial in θ. Equation (23) has no polynomial solution for an arbitrary right hand side but only in the case when the orthogonality condition

$$(P_{m1}, H_m) \equiv \int_{-\infty}^{\infty} P_{m1}(\theta)\, H_m(\theta)\, e^{-\theta^2/2}\, d\theta = 0 \tag{4.2.24}$$

is satisfied.

We note the following property of operators (15), $H^{(n)}$. If the polynomial $P(\xi)$ is even then the polynomial $H^{(n)}P(\xi)$ is even for even n and odd for odd n.

If the polynomial $P(\xi)$ is odd then the polynomial $H^{(n)}P(\xi)$ is odd for even n and even for odd n. It follows that condition (24) is always fulfilled. We chose the solution of equation (23), satisfying the condition $(\tilde{w}_1, H_m) = 0$.

The following approximation leads to the equation

$$H^{(2)}w_2 + H^{(3)}w_1 + H^{(4)}w_0 = 0,$$

which may be rewritten in the form

$$M_m\left[\tilde{w}_2(\theta)\right] = P_{m2}(\theta)\,. \tag{4.2.25}$$

Polynomial $P_{m2}(\theta)$ contains the term $2r^{-1}\lambda_2 H_m(\theta)$, where the value of λ_2 may be found from the condition $(P_{m2}, H_m) = 0$. This process may be continued infinitely. At the even step $2k$ the value of correction λ_k for the eigenvalue of λ may be found.

The critical value λ_* may be obtained for $m = 0$ in formula (20) and

$$\lambda_* = \lambda_0 + \mu\frac{r}{2} + O\left(\mu^2\right), \quad w_0(\xi) \equiv 1. \tag{4.2.26}$$

Above we also studied the case when $m \neq 0$, since it does not require additional computations. The information about the distance between the critical value and the other eigenvalues may be used to estimate the sensitivity of the critical value to perturbations (imperfections) and also for the development of numerical algorithms for the calculation of λ_*.

When we seek the minimum of (10), two essentially different cases are possible

$$\begin{aligned} p_0 &= 0 \qquad \text{(case } A\text{)},\\ p_0 &\neq 0 \qquad \text{(case } B\text{)}. \end{aligned} \tag{4.2.27}$$

Remark 4.1. Case A may only take place if $f_{px}^0 = 0$. In this case eigenfunctions (6) are real and eigenvalues (8) are simple. They may be expressed by formulae

$$\begin{aligned} w^{(m)}(x) &= \exp\left(-\frac{c\,(x-x_0)^2}{2\mu}\right)\left[H_m\left(\frac{\sqrt{c}\,(x-x_0)}{\sqrt{\mu}}\right) + O\left(\mu^{1/2}\right)\right],\\ \lambda &= \lambda_0 + \mu\left(m + \frac{1}{2}\right)r + O\left(\mu^2\right), \end{aligned} \tag{4.2.28}$$

where

$$\lambda_0 = \frac{a_0'}{a_0''}, \quad f_{pp}^0 = \frac{2\,(a_1' a_0'' - a_0' a_1'')}{(a_0'')^2}, \quad f_{xx}^0 = \frac{d^2}{dx^2}\left(\frac{a_0'}{a_0''}\right),$$
$$r = \left(f_{xx}^0\, f_{pp}^0\right)^{1/2}, \quad c = \left(f_{xx}^0\right)^{1/2}\left(f_{pp}^0\right)^{-1/2}, \tag{4.2.29}$$

All of the functions in formulae (29) are calculated for $x = x_0$. The eigenfunction for $m = 0$ is plotted in Figure 4.1.

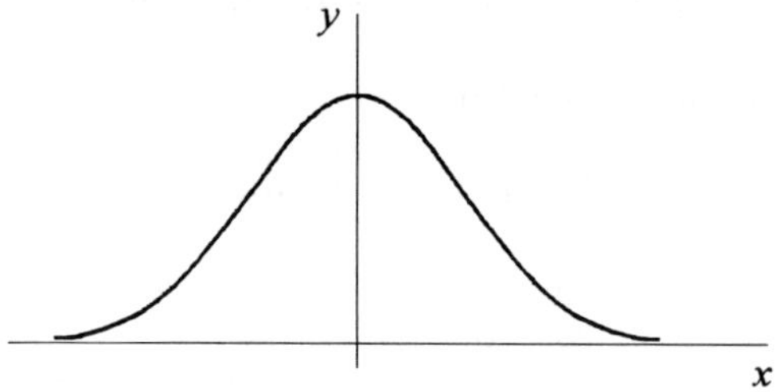

Figure 4.1: The eigenfunction when $p_0 = 0$.

Remark 4.2. In case B eigenfunctions (6) are complex. Since equation (1) is real, both the real part of function (6) and the imaginary part are eigenfunctions. It may seem that the corresponding eigenvalue (8) is double, but this is incorrect, since series (6) and (8) are asymptotic (and not convergent) series and they do not exactly represent the eigenvalues and eigenfunctions. Two real eigenfunctions correspond to each eigenvalue (8), the asymptotic representations of which are the following

$$w^{(j)} = (\Re\, w_* \cos z_j - \Im\, w_* \sin z_j) \exp\left\{-\frac{\Im\, a\,(x - x_0)^2}{2\mu}\right\}, \tag{4.2.30}$$

where

$$z_j = \frac{p_0(x - x_0)}{\mu} + \frac{\Re\, a\,(x - x_0)^2}{2\mu} + \theta_j, \quad j = 1, 2.$$

The initial phase, θ_j, has quite definite values θ_1 and θ_2 ($0 \leqslant \theta_1, \theta_2 < 2\pi$).

Eigenfunctions (30) for $m = 0$ are shown in Figure 4.2. Moreover, the exact eigenvalues corresponding to eigenfunctions (30) are different. We denote them as $\lambda^{(m,1)}$ and $\lambda^{(m,2)}$. The following estimate is valid for any N

$$\Delta\lambda = \lambda^{(m,1)} - \lambda^{(m,2)} = O\left(\mu^N\right)$$

Henceforth, we will call such eigenvalues asymptotically double which refers to the fact that they are nearly (but not exactly) the same and approach each other.

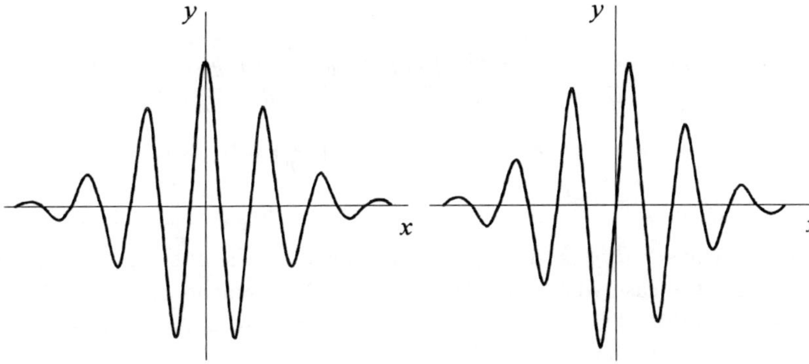

Figure 4.2: The even and odd eigenfunctions when $p_0 \neq 0$.

The method of the asymptotic integration used here does not permit the determination of the difference $\Delta\lambda$ and the values of the initial phases θ_1 and θ_2. For one particular case these values will be found by numerical integration in Section 5.3.

4.3 Buckling Modes of Convex Shells of Revolution

We will now introduce curvilinear coordinates on the neutral surface of a shell of revolution. As the first coordinate we chose the generatrix length s' ($s'_1 \leqslant s' \leqslant s'_2$) and as the second, the circumferential angle φ. We start analysis with system (2.3.8) rewritten in a dimensionless form as in (3.1.1)

$$\begin{aligned} \mu^2\Delta\,(d\,\Delta\,w) + \lambda\,\Delta_t w - \Delta_k \Phi &= 0, \\ \mu^2\Delta\left(g^{-1}\Delta\,\Phi\right) + \Delta_k w &= 0, \end{aligned} \tag{4.3.1}$$

where, for a shell of revolution

$$\begin{aligned} \Delta\,w &= \frac{1}{b}\,\frac{\partial}{\partial s}\left(b\,\frac{\partial w}{\partial s}\right) + \frac{1}{b^2}\,\frac{\partial^2 w}{\partial \varphi^2}, \\ \Delta_k w &= \frac{1}{b}\,\frac{\partial}{\partial s}\left(b\,k_2\frac{\partial w}{\partial s}\right) + \frac{k_1}{b^2}\,\frac{\partial^2 w}{\partial \varphi^2}, \\ \Delta_t w &= \frac{1}{b}\,\frac{\partial}{\partial s}\left(b\,t_1\frac{\partial w}{\partial s}\right) + \frac{t_2}{b^2}\,\frac{\partial^2 w}{\partial \varphi^2} + \frac{1}{b}\,\frac{\partial}{\partial s}\left(t_3\frac{\partial w}{\partial \varphi}\right) + \frac{t_3}{b}\,\frac{\partial^2 w}{\partial s\,\partial \varphi}. \end{aligned} \tag{4.3.2}$$

Here

$$
\begin{aligned}
&\mu^4 = \frac{h_0^2}{12\,(1-\nu_0^2)\,R^2}, \quad d = \frac{E\,h^3\,(1-\nu_0^2)}{E_0 h_0^3\,(1-\nu^2)}, \quad g^{-1} = \frac{E_0 h_0}{E\,h}, \quad b = \frac{B}{R}, \\
&k_i = \frac{R}{R_i}, \quad \{s_i', s'\} = R\,\{s_i, s\}, \quad \{T_i^0, S^0\} = -\lambda T\,\{t_i, t_3\}, \\
&T = E_0 h_0 \mu^2, \quad i = 1, 2.
\end{aligned}
\tag{4.3.3}
$$

The functions b, d, g, k_i and t_i do not depend on the angle φ. Systems (1) is written for the case when parameters E, ν and h depend on s (see Remark 1.1). Typical value of these parameters are denoted by E_0, ν_0 and h_0. If E, ν and h are constant then $d = g = 1$.

We seek a solution of system (1) with n waves in the circumferential direction in the form of (2.6)

$$
w\,(s, \varphi, \mu) = w_*\exp\left\{i\left(\mu^{-1/2}p_0\xi + 1/2\,a\,\xi^2\right) + i\,n\,(\varphi - \varphi_0)\right\}, \tag{4.3.4}
$$

where $\xi = \mu^{-1/2}(s - s_0)$, and w_* has the form of (2.7).

We seek the function Φ in the same form (4).

The function f in formula (2.9) is

$$
f\,(p, \rho, s) = \frac{d\,(p^2+q^2)^4 + g\,(k_2p^2 + k_1q^2)^2}{(p^2+q^2)^2\,(t_1p^2 + 2\,t_3p\,q + t_2q^2)}, \tag{4.3.5}
$$

where $q = \rho\,b^{-1}$, $\rho = \mu\,n$. Function (5) differs from function (3.1.5) only by multipliers d and g. As in Section 3.1, we assume that $k_1k_2 > 0$ and also that conditions (3.6.3) are satisfied. In domain $-\infty < p < \infty$, $0 \leqslant \rho < \infty$, $s_1 \leqslant s \leqslant s_2$ we seek the minimum of the function f

$$
\lambda_0 = \min_{p,\rho,s}{}^{(+)}\{f\} = f\,(p_0, \rho_0, s_0), \tag{4.3.6}
$$

assuming that parameter ρ changes approximately continuously. We then obtain

$$
\begin{aligned}
\lambda_0 &= \min\nolimits_s{}^{(+)}\{\gamma\,(s)\} = \gamma\,(s_0), \\
\gamma(s) &= \frac{4\,k_1k_2(dg)^{1/2}}{t_1k_1 + t_2k_2 + \tau_3}, \qquad \tau_3 = \left((t_1k_1 - t_2k_2)^2 + 4\,k_1k_2t_3^2\right)^{1/2}, \\
\rho_0 &= \frac{b\,(k_2r_0 + k_1)^2}{r_0^2 + 1}\left(\frac{g}{d}\right)^{1/4}, \qquad p_0 = \frac{\rho_0 r_0}{b}, \\
r_0 &= (2\,k_2t_3)^{-1}\,(t_1k_1 - t_2k_2 + \tau_3)\,.
\end{aligned}
\tag{4.3.7}
$$

Here, the values p_0, ρ_0 and r_0 are evaluated for $s = s_0$. The parallel $s = s_0$ in which the function $\gamma(s)$ attains its minimum is called the weakest parallel. The buckling pits are localized near parallel $s = s_0$. We assume that the shell edge supports do not cause buckling. In Chapter 13 we will list the cases of weak edge support for which buckling occurs near the shell edge.

We can write the expressions for initial membrane stress-resultants T_i^0 and S^0 for a shell under axial force P and torsional moment M, applied to edge $s = s_2$, and under external surface loadings $q_i(s)$ and $q(s)$ applied to the shell surface (see (1.4.6))

$$
\begin{aligned}
T_1^0 &= \frac{P\,R_2}{2\pi\,B^2} - \frac{R_2}{B^2}\int_{s_2'}^{s'}\left(q_1\,\frac{B^2}{R_2} + q\,B\,\frac{d\,B}{ds'}\right)ds', \\
T_2^0 &= -\frac{P\,R_2^2}{2\pi\,B^2 R_1} - R_2 q + \frac{R_2^2}{R_1 B^2}\int_{s_2'}^{s'}\left(q_1\,\frac{B^2}{R_2} + q\,B\,\frac{d\,B}{ds'}\right)ds', \qquad (4.3.8) \\
S^0 &= \frac{M}{2\pi\,B^2} - \frac{1}{B^2}\int_{s_2'}^{s'} q_2 B^2 ds'.
\end{aligned}
$$

The stress resultants may be transformed to the dimensionless stress-resultants t_i due to (3). The extraction of the multiplier λ (the load parameter) from the initial stress resultants T_i^0 and S^0 is arbitrary and the choice of parameter λ depends on the problem at hand. The finite results do not depend on that choice.

Let us study some particular cases of shell loading. We are going to use the algorithm described in Section 4.2 which is the reason that the discussion is limited to cases when the weakest parallel is far from a shell edge.

Let the initial shear be absent ($t_3 = 0$). In this case formulae (7) are not convenient and may be replaced by

$$\gamma\,(s) = \frac{2\,k_1}{t_2}\,(gd)^{1/2}, \quad p_0 = 0, \quad \rho\,(s) = b\left(\frac{g\,k_1^2}{d}\right)^{1/4}, \quad \rho_0 = \rho(s_0) \qquad (4.3.9)$$

for $t_2 k_2 > t_1 k_1$ and

$$\gamma\,(s) = \frac{2\,k_2}{t_1}\,(gd)^{1/2}, \quad p_0 = \left(\frac{g\,k_2^2}{d}\right)^{1/4}, \quad \rho_0 = 0 \qquad (4.3.10)$$

for $t_1 k_1 > t_2 k_2$. Here, as in (7), evaluating ρ_0 and p_0 we assume that $s = s_0$.

If $t_1 k_1 = t_2 k_2$ for $s = s_0$ then as in Section 3.3 numerous buckling modes are possible.

Remark 4.3. It follows from the results of this Section that the often used method of coefficient "freezing" allows us to correctly find the value of λ_0 in the expansion of (2.8) and parameters p_0 and n in buckling mode (4). To find the correction λ_1 and the parameter a in (4) determining the buckling mode localization it is necessary to use the more detailed study given below.

4.4 Buckling of Shells of Revolution Without Torsion

Let us first study the case $t_2 k_2 > t_1 k_1$ in which buckling is mainly determined by the hoop stress-resultants, T_2^0. This case is described by formulae (3.9) and was called case A in Section 4.2. The critical load and buckling mode are given by formulae (2.28) and (2.29).

For $m = 0$ we get

$$\begin{aligned} \lambda &= \lambda_0 + \mu\,\lambda_1 + O\,(\mu^2), \\ w\,(s,\varphi) &= \exp\left\{-\frac{c\,(s-s_0)^2}{2\mu}\right\}\cos n\,(\varphi-\varphi_0)\,\left[1+O\,(\mu^{1/2})\right], \end{aligned} \tag{4.4.1}$$

where

$$\begin{aligned} &\lambda_0 = \min_s \gamma\,(s), \qquad \lambda_1 = 1/2\left(f_{pp}^0\, f_{ss}^0\right)^{1/2}, \qquad c = \left(f_{ss}^0/f_{pp}^0\right)^{1/2}, \\ &n \simeq \rho_0\mu^{-1}, \qquad \gamma\,(s) = \frac{2\,k_1}{t_2}\,(gd)^{1/2}, \qquad \rho\,(s) = b\left(\frac{gk_1^2}{d}\right)^{1/4}, \\ &f_{ss}^0 = \frac{d^2\gamma}{ds^2} + 4\,\gamma\left(\frac{1}{\rho}\,\frac{d\rho}{ds}\right)^2, \qquad f_{pp}^0 = 4\,d\,\frac{k_2t_2 - k_1t_1}{k_1t_2^2}. \end{aligned} \tag{4.4.2}$$

The values of f_{ss}^0 and f_{pp}^0 in formulae (2) are evaluated for $s = s_0$. The initial phase φ_0 in (1) is arbitrary and therefore the critical load is double. Due to inequalities $f_{ss}^0 > 0$ and $f_{pp}^0 > 0$ inequality (2.11) is satisfied. The transition from a non-integer value of $n = \rho_0\mu^{-1}$ to the closest integer number of waves in the circumferential direction n does not change the terms given by (1) for λ.

The buckling mode shown in Figures 4.1 and 4.3 illustrates the fact that the buckling pits are elongated in the meridional direction. The tangential displacements u_1 and u_2 are of order μ compared with the normal deflection w. Now let us study some particular loading cases.

Let a shell which has the form of a cupola or spheroid be under the normal

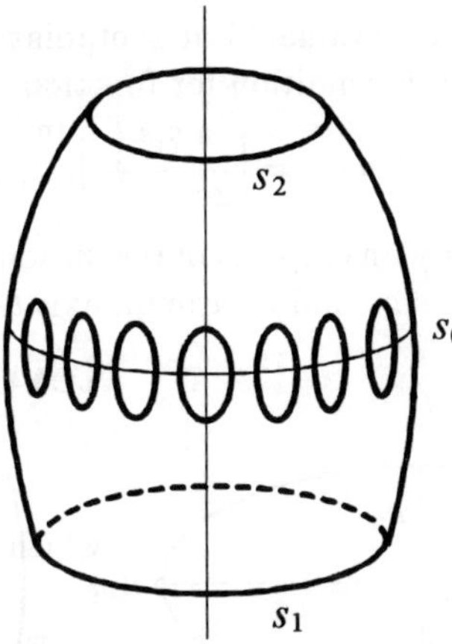

Figure 4.3: The buckling mode of a convex shell of revolution at $t_2 k_2 > t_1 k_1$.

homogeneous pressure q. Then, due to formulae (3.8) we get

$$T_1^0 = -\frac{1}{2} R_2 q, \quad T_2^0 = \left(\frac{R_2^2}{2R_1} - R_2 \right) q, \quad S^0 = 0. \tag{4.4.3}$$

The case $T_2^0 R_1 < T_1^0 R_2$ which we are examining corresponds to external pressure ($q > 0$) for $R_2 < R_1$ and to internal pressure for $R_2 > 2R_1$. If we assume that

$$|q| = \lambda T R^{-1}, \qquad T = E_0 h_0 \mu^2 \tag{4.4.4}$$

then in formulae (2)

$$t_1 = \frac{q_0}{2k_2}, \quad t_2 = \frac{2k_2 - k_1}{2k_2^2} q_0, \quad q_0 = \operatorname{sgn} q. \tag{4.4.5}$$

Assuming that the weakest parallel $s = s_0$ is far from a shell edge, one may use formulae (1) to evaluate the critical load and buckling mode. If parameters E, ν and h are constant then $d = g = 1$ and the expression for the critical load may be rewritten in the form

$$q = \frac{2 E h^2}{\sqrt{3(1-\nu^2)}\,(2R_1 R_2 - R_2^2)} \left\{ 1 + \mu \left(\frac{k_2(k_1 - k_2)}{k_1^2(k_1 - 2k_2)} \right)^{1/2} \right.$$
$$\left. \left[\frac{k_1 - 2k_2}{k_1 k_2^2} \frac{d^2}{ds^2} \left(\frac{k_1 k_2^2}{k_1 - 2k_2} \right) + \frac{1}{k_1^2 b^4} \left(\frac{d}{ds}(b^2 k_1) \right)^2 \right]^{1/2} + O(\mu^2) \right\}. \tag{4.4.6}$$

Here all of the functions are evaluated at the point where $s = s_0$, which may be found from the minimum condition for function

$$\gamma(s) = \left|\frac{4\,k_1 k_2^2}{2k_2 - k_1}\right|. \tag{4.4.7}$$

As an example we will study an ellipsoid of revolution generated by the rotation of an ellipse with semi-axes a_0 and b_0 around axis b_0 (see Figure 4.4).

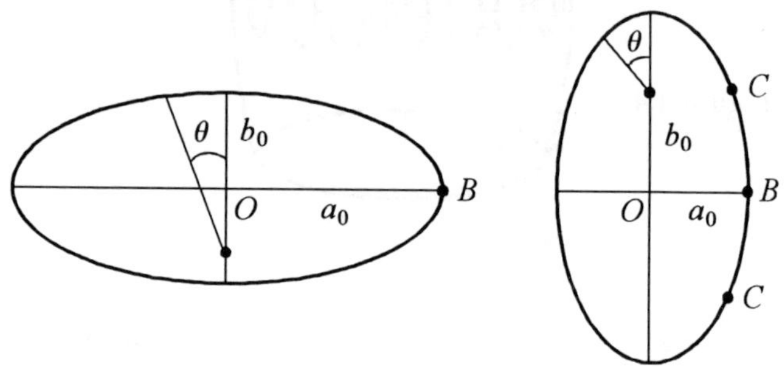

Figure 4.4: An oblate ($\delta < 1$) and a prolate ($\delta > 1$) ellipsoid of revolution.

As a characteristic size we take $R = a_0$. Then

$$\begin{aligned} &k_2 = \left(\sin^2\theta + \delta^2\cos^2\theta\right)^{1/2}, \qquad k_1 = \delta^{-2}k_2^3, \\ &b = k_2^{-1}\sin\theta, \qquad \delta = b_0 a_0^{-1}, \qquad \frac{d}{ds} = k_1\frac{d}{d\theta}, \end{aligned} \tag{4.4.8}$$

where θ is an angle between the axis of revolution and a normal to the shell surface.

First we will consider an prolate ellipsoid ($\delta > 1$) under external pressure. The weakest parallel is the equator (point B in Figure 4.4). By using formulae (6) and (2), we obtain

$$\begin{aligned} q = &\frac{2\,Eh^2}{\sqrt{3(1-\nu^2)}\,a_0^2\,(2\delta^2-1)}\Bigg[1+ \\ &\left(\frac{h}{a_0}\right)^{1/2}\frac{(\delta^2-1)\,(4\delta^2-1)^{1/2}}{\sqrt[4]{3(1-\nu^2)}\,(2\delta^2-1)} + O(h_*)\Bigg], \\ n \simeq &\frac{\sqrt[4]{12(1-\nu^2)}}{\delta}\left(\frac{a_0}{h}\right)^{1/2}, \qquad c = \delta^{-4}\left(2\delta^2 - 1/2\right)^{1/2}. \end{aligned} \tag{4.4.9}$$

Under internal pressure, ellipsoid bucking may occur for $2\delta^2 < 1$. In this case the critical values of parameters q and n may be found by different formulae according to the sign of $2\delta - 1$.

Indeed, function (7) attains its minimum at the weakest parallel. For $1 < 2\delta < \sqrt{2}$ this minimum is attained at the equator and the critical pressure q and wave number n may be found by the same formulae (9) as for external pressure. For $2\delta < 1$ function (7) attains its local maximum at the equator and minimum at

$$\theta_1 = \arcsin\sqrt{\frac{3\delta^2}{1-\delta^2}}, \quad \theta_2 = \pi - \theta_1 \tag{4.4.10}$$

(point C in Figure 4.4), i.e. the pits move away from the equator.

The critical values of q and n are equal to

$$\begin{aligned} q &= -\frac{16\,E\,h^2\delta^2}{a_0^2\,\sqrt{3(1-\nu^2)}}\Bigg[1+ \\ &\quad \left(\frac{h}{a_0}\right)^{1/2}\left(\frac{193\,(1-4\delta^2)}{16\delta\,\sqrt{12(1-\nu^2)}}\right)^{1/2} + O\,(h_*)\Bigg], \\ n &= \left(\frac{6\delta}{1-\delta^2}\right)^{1/2}\sqrt[4]{12(1-\nu^2)}\left(\frac{a_0}{h}\right)^{1/2}. \end{aligned} \tag{4.4.11}$$

For $2\delta = 1$ we have $f_{ss}^0 = 0$. In this case, the condition $f_{ss}^0 > 0$ is not satisfied and formula (1) for the buckling mode must be made more precise, since it does not provide for the decrease of the deflection away from parallel $s = s_0$. However, for estimating q and n, one may use formulae (9) and (11).

The problem of ellipsoid buckling under homogeneous pressure is also considered in [30, 110, 175] .

We will now study the buckling of a shell of revolution under an axial tensile force ($P > 0$). In this case, by virtue of (3.8)

$$T_1^0 = \frac{P\,R_2}{2\,\pi\,B^2}, \quad T_2^0 = -\frac{P\,R_2^2}{2\,\pi\,B^2 R_1}, \quad S^0 = 0. \tag{4.4.12}$$

Assuming that $\lambda T = P/(2\pi\,R)$ we get

$$t_1 = -\frac{1}{b^2 k_2}, \quad t_2 = \frac{k_1}{b^2 k_2^2}, \quad \gamma\,(s) = 2\,(gd)^{1/2}\sin^2\theta, \tag{4.4.13}$$

where the angle θ is the same as in formula (8) (see Figure 4.4).

If parameters E, ν and h are constant ($d = g = 1$) then function $\gamma(s)$ attains its minimum at one of the edges (we recall that the shell is convex).

Let the shell thickness now depend on s. Then function $\gamma(s) = \dfrac{2h(s)}{h_0}\sin^2\theta$ and cases when this function attains its minimum inside the shell are possible.

We consider an ellipsoid of revolution (8) which has its minimum wall thickness at its diameter

$$h(s) = h_0\left[1 + \alpha(\theta - \pi/2)^2\right], \qquad \alpha > 0, \tag{4.4.14}$$

For $\alpha > 1$ the diameter is the weakest parallel and the critical values of P and n are equal to

$$\begin{aligned} P &= \frac{2\pi E h_0^2}{\sqrt{3(1-\nu^2)}}\left[1 + \left(\frac{h_0}{a_0}\right)^{1/2}\frac{a_0^2}{b_0^2}\frac{(\alpha-1)^{1/2}}{(3(1-\nu^2))^{1/4}} + O(h_*)\right], \\ n &\simeq \frac{\sqrt[4]{12(1-\nu^2)}}{\delta}\left(\frac{a_0}{h_0}\right)^{1/2}. \end{aligned} \tag{4.4.15}$$

The buckling mode described by formula (1) is shown in Figure 4.3.

Now we consider the combined loading by an axial force P and a homogeneous pressure q and find the cases in which the buckling modes are the same as in Figure 4.3. For an ellipsoid of revolution of constant thickness described above we assume

$$\begin{aligned} &P = 2\pi a_0 T P_0 \lambda, \qquad q = T a_0^{-1} q_0 \lambda, \quad q_0 = \operatorname{sgn} q, \\ &t_1 = \frac{q_0}{2k_2} - \frac{P_0}{b^2 k_2}, \qquad t_2 = \frac{q_0(2k_2 - k_1)}{2k_2^2} + \frac{P_0 k_1}{b^2 k_2^2}. \end{aligned} \tag{4.4.16}$$

Then buckling occurs near the equator if

$$\begin{aligned} &\text{I:} \quad 1 - \delta^2 < \frac{2P_0}{q_0} < (1-\delta^2)(1-4\delta^2), \qquad q_0 = 1 \\ &\text{II:} \quad (1-\delta^2)(1-4\delta^2) < \frac{2P_0}{q_0} < 1 - 2\delta^2, \qquad q_0 = -1. \end{aligned} \tag{4.4.17}$$

The domains determined by these inequalities are shown in Figure 4.5.

Due to formulae (1), (2) and (3.3) the critical values of P and q may be determined from the relation

$$\begin{aligned} &\frac{P}{\pi a_0^2} + q(2\delta^2 - 1) = \frac{2Eh^2}{\sqrt{3(1-\nu^2)}\, a_0^2}\Bigg[1 + \\ &\left(\frac{h((\delta^2-1)q_0 + 2P_0)((4\delta^2-1)(\delta^2-1)q_0 - 2P_0)}{\sqrt{3(1-\nu^2)}\, a_0 \left((2\delta^2-1)q_0 + 2P_0\right)^2}\right)^{1/2} + \\ &+ O(h_*)\Bigg], \end{aligned} \tag{4.4.18}$$

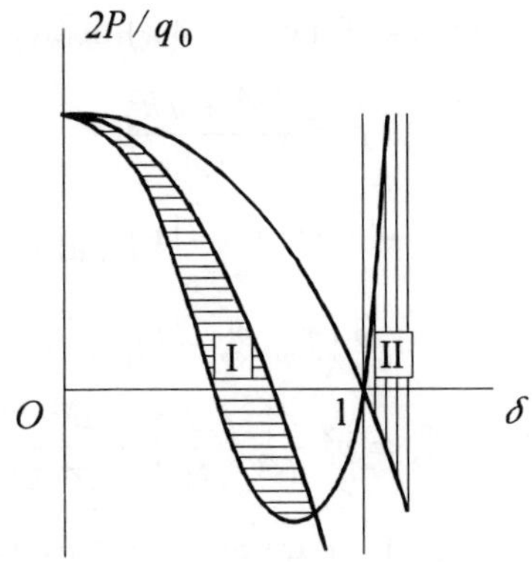

Figure 4.5: Critical value domains for the buckling of an ellipsoid near an equator.

which generalizes formula (9) for the case when $P \neq 0$.

In all of the cases considered above the hoop stress-resultant T_2^0 was the main stress determining buckling. Now we come to the case when

$$t_1 k_1 > t_2 k_2, \tag{4.4.19}$$

and the principal compressive stress-resultant T_1^0 is that on the meridian. In this case, according to formula (3.10), the load parameter is equal to

$$\lambda_0 = \min_s{}^{(+)} \left\{ \frac{2k_2}{t_1} (gd)^{1/2} \right\}. \tag{4.4.20}$$

Let the parameters E, ν and h be constant ($d = g = 1$). Then, in the case considered of combined loading or in cases when the shell is under either an axial force P or a normal pressure q, function (20) attains its minimum at one of the edges.

Let us now consider a shell of non-constant thickness, $h = h(s)$. Let minimum (20) be attained at $s = s_0$ ($s_1 < s_0 < s_2$). Then, due to (3.10) the buckling mode is axisymmetric and is described by function (3.4) for $n = 0$. The buckling mode is the set of axisymmetric concavities the depth of which decreases away from parallel $s = s_0$

$$\begin{aligned} w &= \exp\left\{ i \left(\mu^{-1/2} p_0 \xi + 1/2\, a\, \xi^2 \right) \right\} \left(1 + O\left(\mu^{1/2} \right) \right), \\ \lambda &= \lambda_0 + \mu\, \lambda_1 + O\left(\mu^2 \right). \end{aligned} \tag{4.4.21}$$

Instead of (3.5) we can take function (3.5) in the form

$$f = \frac{d\,p^4 + g\,k_2^2}{t_1 p^2}. \tag{4.4.22}$$

For the evaluation of the parameters in (21) we then obtain

$$\begin{gathered} p_0 = \left(\frac{g\,k_2^2}{d}\right)^{1/4}, \quad \lambda_1 = 1/2\left(f_{pp}^0\,f_{ss}^0 - \left(f_{ps}^0\right)^2\right)^{1/2}, \quad f_{pp}^0 = \frac{8d}{t_1}, \\ f_{ps}^0 = 4\,p_0\,\frac{d}{ds}\left(\frac{d}{t_1}\right), \quad f_{ss}^0 = p_0^2\,\frac{d^2}{ds^2}\left(\frac{d}{t_1}\right) + \frac{1}{p_0^2}\,\frac{d^2}{ds^2}\left(\frac{g\,k_2^2}{t_1}\right) \end{gathered} \tag{4.4.23}$$

(if $f_{ps}^0 = 0$ then $f_{ss}^0 = \gamma_{ss}^0$). Parameter a may be calculated by formula (2.18). Here all of the variables are evaluated at $s = s_0$. The critical load is asymptotically double (see Remark 4.2) and the corresponding buckling modes which are nearly identical for close values of load are shown in Figure 4.2.

As an example consider the same ellipsoid of revolution (14) with variable thickness under an axial compressive force ($P < 0$). Assuming that $P = -2\pi a_0 T \lambda$, we obtain

$$\gamma\,(s) = \frac{2\,k_2^2 b^2 h\,(s)}{h_0} = \frac{2\,h\,(s)}{h_0}\,\sin^2\theta. \tag{4.4.24}$$

Buckling near the equator occurs for $\alpha > 1$ and the asymptotically double critical load is equal to

$$P = -\frac{2\,\pi\,E\,h_0^2}{\sqrt{3\,(1-\nu^2)}}\left[1 + \left(\frac{h_0}{a_0}\right)^{1/2}\frac{a_0^2}{b_0^2}\,\frac{(\alpha-1)^{1/2}}{\left(3\,(1-\nu^2)\right)^{1/4}} + O\,(h_*)\right]. \tag{4.4.25}$$

We note that the absolute value of the critical force found above coincides with the absolute value of the critical load in the case of tensile loading of the shell (see (15)), but that the buckling modes are essentially different.

Remark 4.4. The formulae for the critical load (see (1), (6), (9), (11), (18) and (25)) all have the same structure. They consist of a main term which may be found by the method of freezing coefficients and of small corrections of the relative order $h_*^{1/2}$. Due to (2.28) the next eigenvalue of the load parameter may be obtained by increasing this correction by a factor of three.

4.5 Buckling of Shells of Revolution Under Torsion

Let us now study the buckling of shells of revolution under homogeneous pressure q, axial force P and torsional moment $M > 0$. We introduce the load parameter as

$$\lambda = \frac{M}{2\,\pi\,R^2 T}. \tag{4.5.1}$$

Then due to (3.3) and (3.8) we get

$$\begin{aligned} t_1 &= \frac{q_0}{2k_2} - \frac{P_0}{b^2 k_2}, \quad t_2 = \frac{q_0\,(2k_2 - k_1)}{2k_2^2} + \frac{P_0 k_1}{b^2 k_2^2}, \\ t_3 &= -\frac{1}{b^2}, \qquad P_0 = \frac{P\,R}{M}, \qquad q_0 = \frac{2\,\pi\,R^3 q}{M}. \end{aligned} \tag{4.5.2}$$

Assuming that conditions (3.6.3) are satisfied, we find by formulae (3.7)

$$\begin{aligned} \lambda_0 &= \min_s{}^{(+)}\gamma\,(s) = \gamma\,(s_0), \\ \gamma\,(s) &= 2\,b^2 k_1 (gd)^{1/2} \left[\frac{q_0 b^2}{2k_2} + \left(A^2 + \frac{k_1}{k_2}\right)^{1/2}\right]^{-1}, \qquad \tan\Psi_0 = -r_0, \\ r_0 &= A - \sqrt{A^2 + \frac{k_1}{k_2}}, \quad A = \frac{P_0 k_1}{k_2^2} + \frac{q_0 b^2}{2k_2}\left(1 - \frac{k_1}{k_2}\right). \end{aligned} \tag{4.5.3}$$

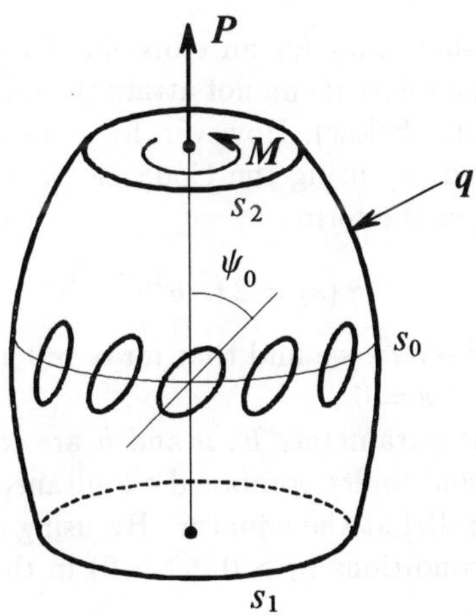

Figure 4.6: The buckling mode of a convex shell of revolution under torsion.

In the case when the weakest parallel s_0 is far from a shell edges (see Figure 4.6) the buckling mode is given by (3.4) which we can write in the form

$$w = e^{-c\xi^2/2}\left[\cos\left(\frac{p_0\xi}{\mu^{1/2}} + \frac{a_1\xi^2}{2} + n\,(\varphi - \varphi_0)\right) + O\,(\mu^{1/2})\right], \tag{4.5.4}$$

where $\xi = \mu^{-1/2}(s - s_0)$ and $a = a_1 + ic$. The values of p_0 and n were found in Section 4.3 and a and λ_1 are evaluated for $m = 0$ by means of formulae (2.18) and (2.20), in which the function f must be taken in the form of (3.5).

As in (4.1), concavities (4) are elongated (their sizes are of order $Rh_*^{1/2} \times Rh_*^{1/4}$) and they are inclined to the meridian by the angle Ψ_0 ($\tan\Psi_0 = -r_0$).

Let us now study some particular cases. First, we assume that the torsional moment $M > 0$ and $P = q = 0$ and that the parameters E, ν and h are constant. We also assume that function $b(s)$ is an even function in $s - s_0$. Then, for $s_1 < s_0 < s_2$ we find from (3), (3.7), (2.18), and (2.20) that the load parameter λ and the coefficients in (4) are the following

$$\begin{aligned}
\lambda &= \lambda_0 + \mu\,\lambda_1 + O\left(\mu^2\right), \quad \lambda_0 = \gamma\left(s_0\right), \\
\gamma\left(s\right) &= 2\,b^2\sqrt{k_1 k_2}, \quad \lambda_1 = 1/2\left(f_{ss}^0\, f_{pp}^0\right)^{1/2}, \quad f_{ss}^0 = \frac{d^2\gamma}{ds^2}, \\
f_{pp}^0 &= \frac{2\,b^2}{\sqrt{k_1^3 k_2}}\left(5\,k_1^2 - 2\,k_1 k_2 + k_2^2\right), \quad \tan\Psi_0 = \sqrt{\frac{k_1}{k_2}}, \\
c &= \left(\frac{f_{ss}^0}{f_{pp}^0}\right)^{1/2}, \quad n = \frac{\sqrt{2\,k_1}\,b\,k_2}{\mu\left(k_1 + k_2\right)}, \quad p_0 = -\frac{\sqrt{2\,k_1}}{k_1 + k_2}, \quad a_1 = 0.
\end{aligned} \tag{4.5.5}$$

The calculations show that for an ellipsoid of revolution, function $\gamma(s)$ is convex upwards and therefore it can not attain the minimum within the interval (s_1, s_2) (see formulae (7) below). However, for some convex shells of revolution this is possible. Indeed, by using the Codazzi–Gauss relations (1.1.3) we can rewrite function $\gamma(s)$ in the form

$$\gamma\left(s\right) = 2\left(-b^3 b''\right)^{1/2}. \tag{4.5.6}$$

Taking $b(s) = 1 - \alpha s^2 - \beta s^4$, we find that for $\alpha > 0$ and $2\beta > \alpha^2$ function $\gamma(s)$ attains its minimum at $s = 0$.

Assuming that the parameters E, ν and h are constant, we try to find in which cases the ellipsoid under combined simultaneous loading of M, P and q has the weakest parallel at the equator. By using designations (1), (2), and (4.8) we rewrite the conditions ($\gamma > 0$, $\gamma'' > 0$) in the form

$$\begin{aligned}
&q_0\delta^2 + A_1 > 0, \\
&A_1 = \left[\left(2P_0 + \left(\delta^2 - 1\right)\, q_0\right)^2 + 4\delta^2\right]^{1/2} > 0, \\
&-2q_0 A_1 \delta^2\left(\delta^2 - 1\right) + \\
&+\left(2P_0 + \left(\delta^2 - 1\right)\, q_0\right)\left(2P_0 + \left(\delta^2 - 1\right)\left(1 - 2\delta^2\right)\, q_0\right) + 4\delta^2 < 0.
\end{aligned} \tag{4.5.7}$$

For buckling at the equator these three inequalities must be satisfied simultaneously. For that, the critical values of dimensional parameters M, P and q satisfy relations

$$\pi q a_0 b_0^2 + \sqrt{\left(a_0 P + \pi q a_0 (b_0^2 - a_0^2)\right)^2 + b_0^2 a_0^{-2} M^2} = 4\pi a_0^2 E_0 h_0 \mu^2 \left(1 + O(\mu)\right). \tag{4.5.8}$$

Let us consider inequalities (7) in more detail. It is clear that for $q_0 = 0$ the last of inequalities (7) is not satisfied. Let now $q_0 \neq 0$ and $P_0 = 0$. The domain of parameters q_0 and δ in which buckling at the equator is possible is shown by the cross-hatching in Figure 4.7.

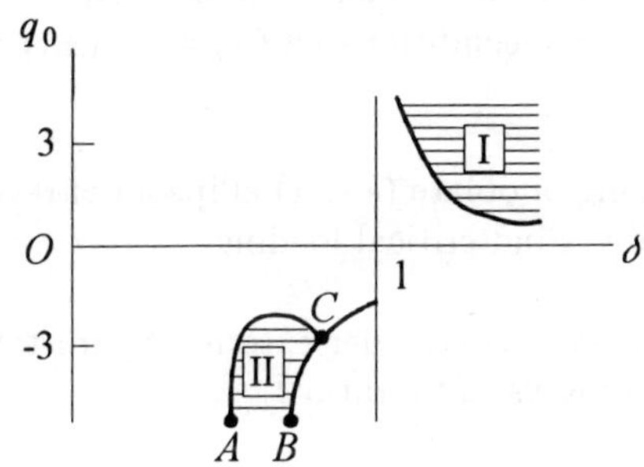

Figure 4.7: The domains of the critical values in the buckling of an ellipsoid near an equator.

The transition of the point (q_0, δ) through the bound of the stability domain I ($q_0 > 0$) and through the part AC of the bound of the domain II ($q_0 < 0$) means that the weakest parallel deviates away from the equator. In approaching the domain II boundary BC the critical values of q and M increase (as in formula (4.9) as $2\delta^2 \to 1$), and after the transition through this boundary shell buckling does not occur (at the equator) since inequalities (3.6.3) are not satisfied.

4.6 Problems and Exercises

4.1 Consider the boundary value problem

$$\mu^4 w'''' + 2\mu^2 \lambda w'' + a_0(x) w = 0, \qquad a_0(x) = 2 - \cos x, \quad \mu \ll 1$$

with the boundary conditions $w(\pm 1) = w'(\pm 1) = 0$.

i) Derive an approximate relation for the asymptotically double eigenvalues

Answer
$\lambda = 1 + \mu\sqrt{2}(\frac{1}{2} + m) + O(\mu^2)$.

ii) For $\mu = 0.1$ and $\mu = 0.01$ find numerically several lower eigenvalues and compare them to each other and to their asymptotic values obtained before.

Hint Under numerical integration seek λ for odd and even eigenfunctions separately solving the boundary value problems with the boundary conditions $w'(0) = w'''(0) = w(1) = w'(1) = 0$ and $w(0) = w''(0) = w(1) = w'(1) = 0$ correspondingly.

iii) Verify that the eigenvalues weakly depend on the boundary conditions solving numerically the initial boundary value problem and the boundary value problem with the boundary conditions $w(\pm 1) = w''(\pm 1) = 0$ for $\mu = 0.1$ and $\mu = 0.01$.

4.2 Consider buckling of oblate ($\delta < 1$) ellipsoid of revolution under external hydrostatic pressure. Find critical loading.

4.3 For ellipsoid of revolution under external hydrostatic pressure plot the function of critical loading vs. axes ratio δ.

4.4 Find the critical load for a shell of revolution if the torsional moment $M > 0$, $P = q = 0$ and the parameters E, ν and h are constant, assuming function $b(s)$ be an even function in $s - s_0$.

Chapter 5

Non-homogeneous Axial Compression of Cylindrical Shells

This Chapter will present a treatment of buckling problems for a cylindrical shell of moderate length under a membrane stress state and a non-homogeneous axial pressure which can be solved by separating the variables. The edges of the shell are assumed to be simply supported. It is also assumed that the determining functions (the axial compression, shell thickness and elastic characteristics of the material) do not depend on the longitudinal coordinate but may depend on the circular coordinate. Buckling modes localized in the neighbourhood of the weakest generatrix are considered, and are developed by means of the algorithm described in Section 4.2.

The buckling of a cylindrical shell under non-homogeneous axial pressure, particularly under a bending moment, has been considered in many previous studies (see reviews [59, 61, 149]). In [68], the Bubnov–Galerkin method was applied and the deflection was approximated by a double trigonometrical series. In [151, 153] the method of asymptotic integration described below was used.

Long cylindrical shells subjected to a bending moment can experience flattening of the cross-section – this is called the Dubjaga–Karman–Brazier effect and it must be taken into account (see [12, 61]) in most analyses. The flattening effect on the critical load for moderately long simply supported shells examined below is, however, not significant and will not be considered here.

5.1 Buckling Modes Localized near Generatrix

Under the assumptions made above we can write the system of buckling equations in the form (see (2.3.8), (4.3.1))

$$\begin{aligned}
&\mu^2\Delta\,(d\,\Delta\,w) + \lambda\,t_1\,\frac{\partial^2 w}{\partial x^2} - k_2\,\frac{\partial^2 \Phi}{\partial x^2} = 0,\\
&\mu^2\Delta\,(g^{-1}\,\Delta\,\Phi) + k_2\,\frac{\partial^2 w}{\partial x^2} = 0, \quad \Delta = \frac{\partial^2}{\partial x^2} + \frac{\partial^2}{\partial y^2},
\end{aligned} \tag{5.1.1}$$

where dimensionless coordinates x and y, similar to those of Section 3.4 are introduced,

$$T_1^0 = -\lambda\,T\,t_1, \qquad T = E_0 h_0 \mu^2, \tag{5.1.2}$$

and the other notation (μ, d, g and k_2) is the same as in (4.3.3). Here, t_1, k_2, d and g depend on y and are assumed to be infinitely differentiable.

The shell occupies the region $0 \leqslant x \leqslant l = L/R$, $y_1 \leqslant y \leqslant y_2$, where L is the shell length, and R is the characteristic radius, taken as the length unit. The simple support conditions (3.2.4) are introduced at the curvilinear edges $x = 0, l$. The boundary conditions at edges $y = y_k$ may be arbitrary. For a shell which is closed in the circumferential direction, the periodicity conditions by y must be fulfilled.

Separating the variables as

$$\begin{aligned}
&w\,(x,y) = w_m(y)\,\sin\frac{p_m x}{\mu}, \qquad \Phi\,(x,y) = \Phi_m(y)\,\sin\frac{p_m x}{\mu},\\
&p_m = \frac{m\,\pi\,\mu}{l}, \qquad m = 1, 2, \ldots,
\end{aligned} \tag{5.1.3}$$

we get the system of ordinary differential equations for unknown functions $w_m(y)$ and $\Phi_m(y)$

$$\begin{aligned}
&\Delta_m\,(d\,\Delta_m w_m) - \lambda\,t_1 p_m^2 w_m + k_2 p_m^2 \Phi_m = 0,\\
&\Delta_m\,(g^{-1}\,\Delta_m \Phi_m) - k_2 p_m^2 w_m = 0, \quad \Delta_m = p_m^2 + (-i\mu)^2\frac{d^2}{dy^2}.
\end{aligned} \tag{5.1.4}$$

As in Section 4.2, we seek a solution of system (4) in the form (note that the subscript m in w_m and Φ_m is omitted)

$$\begin{aligned}
w\,(y,\mu) &= w_*\,\exp\left\{i\left(\mu^{-1/2}q_0\zeta + 1/2\,a\,\zeta^2\right)\right\},\\
w_* &= \sum_{k=0}^{\infty}\mu^{k/2}w_k(\zeta), \quad \zeta = \mu^{-1/2}(y - y_0),\\
\lambda &= \lambda_0 + \mu\,\lambda_1 + \cdots
\end{aligned} \tag{5.1.5}$$

We seek function Φ in the same form as (5). For $\Im a > 0$ solution (5) approaches zero away from the weakest generatrix $y = y_0$. To find function Φ instead of the boundary conditions on the edges $y = y_k$ or the periodicity conditions the assumptions $\Im a > 0$ and $y_1 < y_0 < y_2$ should be used.

For the evaluation of λ_0 we have

$$\begin{aligned} \lambda_0 &= \min_{q,y,m} \{f\} = f(q_0, y_0), \\ f &= \frac{d\,(p_m^2 + q^2)^4 + g\,k_2^2 p_m^4}{t_1 p_m^2 (p_m^2 + q^2)^2}. \end{aligned} \tag{5.1.6}$$

Here the minimum is evaluated for $0 \leqslant q^2 < \infty$, $y_1 \leqslant y \leqslant y_2$, $m = 1, 2, \ldots$ Let the minimum of (6) be attained for $q = q_0$, $y = y_0$ and $m = m_0$. First, we assume that m (and hence p_m) is fixed and also that $k_2 > 0$.

Now we rewrite (6) to the form

$$f(q, y) = \frac{\gamma(y)}{2}\left(z + \frac{1}{z}\right), \quad z = \frac{k_2(p_m^2 + q^2)^2}{p_m^2}\left(\frac{g}{d}\right)^{1/2}, \quad \gamma = \frac{2\sqrt{gd}k_2}{t_1}. \tag{5.1.7}$$

As in Section 4.2, there are possible two cases here: $q_0 = 0$ (case A) and $q_0 \neq 0$ (case B).

The minimum of f is attained for $z = 1$ in which case it is equal to

$$\lambda_0 = \min_y \{\gamma(y)\} = \gamma(y_0) \tag{5.1.8}$$

or for $q_0 = 0$ and it then is equal to

$$\lambda_0' = \min_y \{\gamma'\} = \gamma'(y_0'), \quad \gamma' = \frac{d\,p_m^4 + g\,k_2^2}{t_1 p_m^2}. \tag{5.1.9}$$

Case A occurs if $\lambda_0' < \lambda_0$, otherwise if $\lambda_0 < \lambda_0'$, then the case B occurs.

If $\lambda_0 = \lambda_0'$, then we have a special case which does not fit the scheme of Section 4.2 and which will be considered in Section 5.2. We will consider all of these cases, since each of them may occur under specific conditions depending on the relationship of the various parameters to each other.

After the weakest generatrix y_0 (or y_0') is found, it is useful to choose the values of E_0, ν_0, h_0, R and t_1 such, that

$$d = g = k_2 = 1, \quad t_1 = 2 \quad \text{for} \quad y = y_0 \quad (y = y_0'). \tag{5.1.10}$$

For that, due to (8) and (9)

$$\lambda_0 = 1, \qquad \lambda_0' = \frac{1}{2}\left(p_m^2 + p_m^{-2}\right). \tag{5.1.11}$$

Hence, case A takes place if $p_m > 1$, and case B takes place if $p_m < 1$. If the shell is short and

$$L < \pi\, \mu\, R = \pi \left(\frac{R^2 h_0^2}{12\,(1-\nu^2)} \right)^{1/4}, \tag{5.1.12}$$

then due to (3) $p_1 > 1$ for $m = 1$ and case A takes place.

Let inequality (12) be fulfilled, then due to (4.2.28), for the load parameter eigenvalues and buckling modes we have

$$\begin{aligned} w^{(m,p)} &= e^{-c(y-y_0')^2/(2\mu)} \left[H_p \left(\sqrt{\frac{c}{\mu}}(y-y_0') \right) + O(\mu^{1/2}) \right] \sin \frac{p_m x}{\mu}, \\ \lambda^{(m,p)} &= \frac{1}{2}\left(p_m^2 + p_m^{-2}\right) + \mu \left(p + \frac{1}{2}\right) \left(2\left(1 - p_m^{-4}\right) \frac{d^2 \gamma'}{d{y_0'}^2} \right)^{1/2} + O(\mu^2), \end{aligned} \tag{5.1.13}$$

where $m = 1, 2, \ldots$, $p = 0, 1, \ldots$ and γ' is given by (9),

$$c = \left(\frac{1}{2\left(1 - p_m^{-4}\right)} \frac{d^2 \gamma'}{d{y_0'}^2} \right)^{1/2} \tag{5.1.14}$$

Note that the eigenvalues $\lambda^{(m,p)}$ are simple.

Parameter $\lambda^{(m,p)}$ attains its minimum for $m = 1$, $p = 0$. For that, $H_0 \equiv 1$, the buckling mode is a single pit, and $\lambda^{(1,0)}$ gives a critical load close to (3.4.7) for a short circular cylindrical shell.

Now we pass to case B. Let $p_m < 1$ and assume that relations (10) are valid. Then $\lambda_0 = 1$ and condition $z = 1$ gives

$$q_0^2 = p_m - p_m^2. \tag{5.1.15}$$

If we differentiate (7), we find

$$\begin{aligned} f_{qq}^0 &= \frac{16\,(1-p_m)}{p_m}, \\ f_{qy}^0 &= 0, \\ f_{yy}^0 &= \frac{d^2\gamma}{dy_0} + \left(\frac{d}{dy_0} \left(k_2 \sqrt{\frac{g}{d}} \right) \right)^2. \end{aligned} \tag{5.1.16}$$

Now we can determine the buckling mode and asymptotically double critical value of parameter λ by using the formulae of Section 4.2

$$w^{(j)}(x,y) = e^{-c(y-y_0)^2/(2\mu)} \left[\cos\left(\frac{q_0(y-y_0)}{\mu} + \theta_j\right) + O\left(\mu^{1/2}\right)\right] \sin\frac{p_m x}{\mu},$$

$$\lambda = 1 + \mu\left(\frac{4(1-p_m)}{p_m} f^0_{yy}\right)^{1/2} + O(\mu^2), \tag{5.1.17}$$

$$c = \left(\frac{p_m f^0_{yy}}{16(1-p_m)}\right)^{1/2}, \qquad j = 1, 2.$$

Here, as in (4.2.30), the phases θ_1 and θ_2 are not found. The buckling mode is a system of pits, whose depth decreases away from the weakest generatrix.

5.2 Reconstruction of the Asymptotic Expansions

Consider the case when m varies. Then due to (1.3), p_m takes on a sequence of discrete values. We will examine the dependence of λ on p_m. Let $p_m < 1$, then due to (1.11) and (1.17) $\lambda_0 = 1$ and does not depend on p_m, and the term of order μ decreases as p_m increases and vanishes for $p_m = 1$.

Nevertheless, for $p_m \simeq 1$, (1.17) are not applicable, since $f^0_{qq} = 0$ despite assumption (4.2.11). If $p_m = 1$ then $c = \infty$ and the term $\mu^2\lambda_2$, which is not written out in (1.17) but for one particular case is given in (5.3.3), also approaches infinity.

For $p_m > 1$, parameter λ shows a stronger dependence on p_m. Due to (1.11) the value of λ_0 increases with p_m. For $p_m \simeq 1$ (1.13) are not applicable for the same reason as (1.17) (see also (3.6)).

From the above development it follows that parameter λ reaches its minimum for $p_m \simeq 1$, hence the case when $p_m \simeq 1$ deserves special consideration. Let us assume that

$$p_m = 1 + \varepsilon p'_m, \qquad \lambda = 1 + \varepsilon^2\lambda', \qquad y - y_0 = \varepsilon\eta,$$
$$\varepsilon = \mu^{2/3} = \left(\frac{h_*^2}{12(1-\nu_0^2)}\right)^{1/6} \tag{5.2.1}$$

and seek a solution of system (1.4) in the form

$$w = \sum_{k=0}^{\infty} \varepsilon^k w^{(k)}(\eta), \qquad \Phi = \sum_{k=0}^{\infty} \varepsilon^k \Phi^{(k)}(\eta). \tag{5.2.2}$$

In the zeroth approximation after substituting (2) into (1.4) we obtain the fourth order equation

$$4\,\frac{d^4w^0}{d\eta^4} + (\alpha_1\eta - 8\,p'_m)\,\frac{d^2w^0}{d\eta^2} + \alpha_2\,\frac{d\,w^0}{d\eta} + \\ + \left(a\,\eta^2 + \alpha_3\eta + 4\,p'^2_m - 2\,\lambda'\right)\,w^0 = 0, \qquad (5.2.3)$$

where coefficients α_i and a depend on the first and second derivatives of the functions t_1, k_2, d and g for $y = y_0$ and do not depend on ε. In the case when the first derivatives of these functions are equal to zero for $y = y_0$, we have $\alpha_i = 0$ and equation (3) is simplified

$$\begin{aligned} 4\,\frac{d^4w^0}{d\eta^4} - 8\,p'_m\,\frac{d^2w^0}{d\eta^2} + \left(a\,\eta^2 + 4\,p'^2_m - 2\,\lambda'\right)\,w^0 &= 0, \\ a = \frac{d^2\gamma}{dy_0^2} &> 0. \end{aligned} \qquad (5.2.4)$$

Here γ is given by (1.7).

We seek the values of λ', for which the non-trivial solutions of equations (3) or (4), converging to zero as $\eta \to \pm\infty$, exist. These equations are not integrable in the known functions, however, the fact that the coefficients of these equations do not depend on μ, provides the opportunity to make the following conclusions.

For $p_m \simeq 1$ the critical load differs from $\lambda_0 = 1$ by a quantity of order $\mu^{4/3}$ (for $p_m < 1$ this difference has order μ due to (1.17)). For $p_m \simeq 1$, the index of variation of the solution in the y direction is equal to $t = 1/3$ (see (1)), for $p_m < 1$ it is equal to $t = 1/2$, and for $p_m > 1$ it is equal to $t = 1/4$. In other words, in the neighbourhood of $p_m = 1$ buckling mode reconstruction takes place.

Let us consider equation (4). Applying Fourier transform

$$w^0(\eta) = \frac{1}{\sqrt{2\,\pi}} \int_{-\infty}^{\infty} w^F(\omega)\,e^{i\,\omega\,\eta}\,d\omega \qquad (5.2.5)$$

we come to a second order equation for function $w^F(\omega)$.

$$a\,\frac{d^2w^F}{d\omega^2} + \left[2\,\lambda' - 4\,(p'_m + \omega^2)^2\right]\,w^F = 0. \qquad (5.2.6)$$

We transform this equation to a form with only two parameters Λ and k,

$$\frac{d^2w^F}{dx^2} + \left(\Lambda - (x^2 + k)^2\right)\,w^F = 0, \qquad (5.2.7)$$

where

$$w = \left(\frac{a}{4}\right)^{1/6} x, \quad k = p'_m \left(\frac{4}{a}\right)^{1/3}, \quad \Lambda = \frac{\lambda'}{2}\left(\frac{4}{a}\right)^{2/3}. \tag{5.2.8}$$

For each k $(-\infty < k < \infty)$ there exists the countable set $\Lambda_i(k)$ of values of Λ, for which there exist non-trivial solutions of equation (7) converging to zero as $x \to \pm\infty$. Functions $\Lambda_0(k)$ and $\Lambda_1(k)$ are given in Figure 5.1.

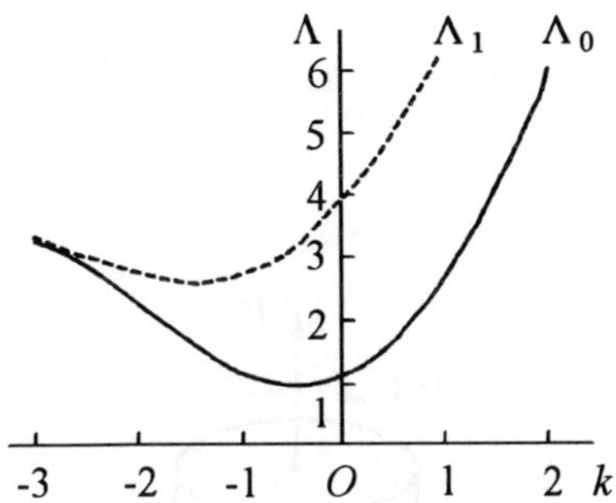

Figure 5.1: Functions $\Lambda_0(k)$ and $\Lambda_1(k)$.

By virtue of (1) and (8) we get

$$\begin{aligned} \lambda^{(i)} &= 1 + 2\,\mu^{4/3}\left(\frac{a}{4}\right)^{2/3}\Lambda_i(k) + O\left(\mu^2\right), \\ k &= \mu^{-2/3}(p_m - 1)\left(\frac{4}{a}\right)^{1/3}. \end{aligned} \tag{5.2.9}$$

Function $\Lambda_0(k)$ attains its minimum $\Lambda_0 = 0,905$ for $k = -0.44$. Hence, the critical load is minimal for

$$p_m = 1 - 0.44\,\mu^{2/3}\left(\frac{a}{4}\right)^{1/3} \tag{5.2.10}$$

and is equal to

$$\lambda_{\min} = 1 + 1.81\left(\frac{a}{4}\right)^{2/3}\mu^{4/3} + O\left(\mu^2\right). \tag{5.2.11}$$

For other $p_m < 1$ the critical load differs from this value by an amount of order μ (see (1.17)).

It should be noted that for $k < 0$, the curves of Λ_0 and Λ_1 approach each other as $|k|$ increases (see Figure 5.1). This means that the eigenvalues are asymptotically double for $p_m < 1$.

5.3 Axial Compression and Bending of Cylindrical Shell

Let us consider a circular cylindrical shell with constant parameters h, E and ν under axial force P and bending moment M_1 (see Figure 5.2). The curvilinear shell edges are simply supported and the initial stress state is assumed to be membrane and defined by the stress

$$T_1^0 = \frac{P}{2\,\pi\,R} + \frac{M_1 \cos y}{\pi\,R^2}. \tag{5.3.1}$$

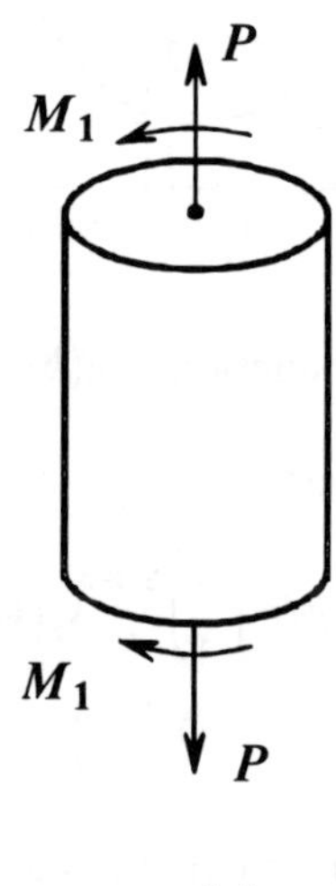

Figure 5.2: The non-homogeneous axial compression of a cylindrical shell.

For $M_1 < 0$, $PR + 2M_1 < 0$, the generatrix $y_0 = 0$ is the weakest one. Let us assume that in (1.2)

$$t_1 = \frac{2\,(\alpha + \cos y)}{\alpha + 1}, \qquad \alpha = \frac{P\,R}{2\,M_1} > -1. \tag{5.3.2}$$

Assuming that $d = g = k_2 = 1$ in the formulae of Section 5.1 we now find that for $p_m < 1$ the buckling mode is given by expression (1.17). In this case due to symmetry it is possible to define angles θ_1 and θ_2, namely $\theta_1 = 0$, $\theta_2 = \pi/2$, that correspond to the even and odd buckling modes by y. The method used here gives the same critical load for these modes [151]

$$
\begin{aligned}
\lambda &= 1 + 2\mu \left(- \frac{t_1''(0)\,(1-p_m)}{2\,p_m} \right)^{1/2} + \\
&+ \mu^2 \left[\frac{(1-p_m)\, t_1^{IV}(0)}{4\,p_m t_1''(0)} + \frac{t_1''(0)\,(-6\,p_m^2 + 21\,p_m - 14)}{64\,p_m(1-p_m)} \right] + \\
&+ O\,(\mu^3),
\end{aligned} \tag{5.3.3}
$$

$$
M_1 = -\frac{2\,\pi\,R^2 \lambda\, T}{1+\alpha}, \quad P = \frac{2\,M\,\alpha}{R}, \quad T = E\,h\,\mu^2,
$$

$$
t_1''(0) = -\frac{2}{1+\alpha}, \qquad t_1^{IV}(0) = \frac{2}{1+\alpha}.
$$

In [151] to analyze the differences in critical load corresponding to even and odd buckling modes, system of equations (1.4) has been integrated numerically. The following boundary conditions correspond to the even mode

$$
w' = w''' = \Phi' = \Phi''' = 0 \quad \text{at} \quad y = 0, \pi, \tag{5.3.4}
$$

and to the odd mode

$$
w = w'' = \Phi = \Phi'' = 0 \quad \text{at} \quad y = 0, \pi. \tag{5.3.5}
$$

For problems with parameters $\dfrac{h}{R} = 0.01$, $l = \dfrac{L}{R} = 2$, $\alpha = 1$ and $\nu = 0.3$, the values of λ are given in Table 5.1 for various values of p_m.

Table 5.1: Critical buckling loads and wave numbers for a cylindrical shell under combined loading

I	*II*	*III*	*IV*
p_m	λ^{even}	λ^{odd}	λ^a
0.173	1.180667	1.180671	1.1706
0.518	1.076492	1.076494	1.0734
0.691	1.051833	1.051847	1.0486
0.864	1.024585	1.031065	1.0257
0.951	1.010245	1.031077	1.0177

In columns *II* and *III* the values of λ^{even} and λ^{odd} corresponding to the even and odd buckling modes found by integrating of system (1.4) are given. The values of λ^a found by formula (3) are presented in column *IV*. It follows from Table 5.1 that for $p_m < 1$ the values of λ^{even} and λ^{odd} are rather close, hence the treatment of λ^a, as if it is asymptotically double is justified. The values λ^{even} and λ^{odd} diverge as p_m approaches unity. In each case the smaller critical load corresponds to the even buckling mode.

Now let $p_m > 1$. The buckling mode is given by expression (1.13) for $p = 0$, and parameter λ is equal to

$$\lambda = \frac{1}{2}\,(p_m^2 + p_m^{-2}) + \mu\, p_m^{-3}\left(-\frac{1}{2}\,t_1''(0)\,(p_m^8 - 1)\right)^{1/2} + \mu^2\left[\frac{3\,t_1^{\text{IV}}(0)\,(p_m^4 - 1)}{4\,p_m^4 t_1''(0)} - \frac{t_1''(0)}{64\,p_m^4\,(p_m^4 - 1)}\,(95\,p_m^8 - 64\,p_m^6 + 172\,p_m^4 + 64\,p_m^2 + 45)\right] + O\,(\mu^3). \tag{5.3.6}$$

Here P and M_1 are determined by the same formulae (3). If $m = 1$ for a short shell (see (1.12)) the buckling mode is in the form of a single pit. By virtue of (1.13)

$$w^{(1,0)} = \exp\left\{-\frac{c\,y^2}{2\,\mu}\right\}\,\sin\frac{p_1 x}{\mu}\left(1 + O\,(\mu^{1/2})\right), \tag{5.3.7}$$

where

$$c = \left[2\,(1 - p_1^{-4})\,(1 + \alpha)\right]^{1/2}, \qquad p_1 = \frac{\pi\,R}{L}. \tag{5.3.8}$$

For the next (odd according to y) buckling mode $w^{(1,1)} \simeq y w^{(1,0)}$ corresponds to

$$\lambda = \frac{1}{2}\,(p_m^2 + p_m^{-2}) + 3\,\mu\, p_m^{-3}\left(-\frac{1}{2}\,t_1''(0)\,(p_m^8 - 1)\right)^{1/2} + O\,(\mu^2). \tag{5.3.9}$$

For $p_m \simeq 1$ we can use the results of Section 5.2. We represent the critical load in the form

$$\begin{aligned} &\lambda = 1 + 2\,\mu^{4/3}\Lambda_i(k)\big(4\,(1 + \alpha)\big)^{-2/3} + O\,(\mu^2), \\ &k = \mu^{-2/3}4^{1/3}(p_m - 1), \end{aligned} \tag{5.3.10}$$

where functions $\Lambda_0(k)$ and $\Lambda_1(k)$ are shown in Figure 5.1.

Consider an example from [152]. Let $\dfrac{h}{R} = 0.001$, $l = 2$, $\alpha = 0$ and $\nu = 0.3$. In Figure 5.3 parameters λ^{even} and λ^{odd} obtained numerically for even and odd buckling modes respectively are shown.

For $p_m < 1$, the values λ^{even} and λ^{odd} approach each other. A comparison of numerical and asymptotic results is presented in Table 5.2. For even and

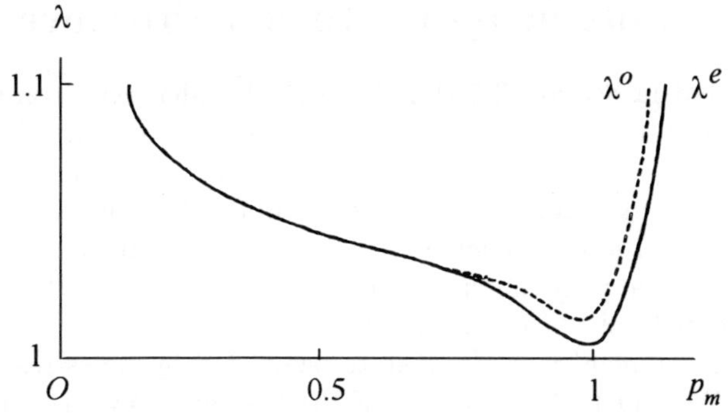

Figure 5.3: Load parameters for even and odd buckling modes of a cylindrical shell under non-homogeneous axial compression.

Table 5.2: Critical buckling loads and wave numbers for a cylindrical shell under combined loading

p_m	λ^{even}	λ^{odd}	λ^{even}	λ^{odd}	λ^{even}	λ^{odd}
1	2	3	4	5	6	7
0.50	1.0516	1.0519	1.0348	1.0348	–	–
0.90	1.0138	1.0165	1.0126	1.0126	1.0093	1.0104
0.95	1.0055	1.0111	1.0078	1.0078	1.0046	1.0093
1.00	1.0041	1.0143	∞	∞	1.0034	1.0136
1.05	1.0108	1.0240	1.0168	1.0203	1.0113	1.0222
1.15	1.0480	1.0660	1.0495	1.0640	–	–
1.30	1.1523	1.1756	1.1523	1.1726	–	–

odd buckling modes the numerical values of λ are given in columns 2 and 3. Columns 4 and 5 give the values for $p_m < 1$ found by formula (4) (they are the same for even and odd modes since the eigenvalues are asymptotically double) and the values for $p_m > 1$ obtained by formulae (6) and (9). For $p_m \approx 1$ formulae (4), (6) and (9) are not applicable, and that is why in the columns 6 and 7 the values of λ found by formula (10) are given.

The results in Table 5.2 illustrate the fact that the asymptotic formulae derived for $p_m < 1$, $p_m \approx 1$ and $p_m > 1$, cover all possible values of p_m and so the areas of their applicability intersect. The asymptotically double eigenvalues for $p_m < 1$ transforms for $p_m > 1$ to eigenvalues which differ by order μ (see (6) and (9)).

5.4 The Influence of Internal Pressure

It was noted in Section 3.5 that for a shell under axial compression, the tensile stress-resultant T_2^0, generated by the internal pressure does not affect the critical load and also that the buckling mode is axisymmetric. In the cases of non-homogeneous compression considered in this Section the buckling mode is not axially symmetric. Therefore, a more significant effect of the tensile stress T_2^0 on the value of the critical load should be expected. This question will be considered below (see also [97]).

Let the initial membrane stress state be defined by stress-resultants $T_1^0(y)$ and $T_2^0(y)$. We add to the left side of the first equation (1.4) the following term

$$-\lambda\,(-i\,\mu)^2 \frac{d}{dy}\left(t_2(y)\,\frac{dw_m}{dy}\right), \tag{5.4.1}$$

where t_2 is related to T_2^0 by an expression similar to (1.2). We assume that $t_2 < 0$.

Instead of (1.6), in this case function f has the form

$$f = \frac{d\,(p_m^2 + q^2)^4 + g\,k_2^2 p_m^4}{(t_1 p_m^2 + t_2 q^2)\,(p_m^2 + q^2)^2}. \tag{5.4.2}$$

The value of λ_0 may be found by the minimization of f by q, m and y. As in Section 5.1 we assume that relations (1.10) apply at the weakest generatrix.

For $p_m \geqslant 1$, the minimum of $\lambda_0 = \min f$ is reached for $q_0 = 0$ and it is equal to $\lambda_0 = \frac{1}{2}\,(p_m^2 + p_m^{-2})$.

For $p_m < 1$ the minimum is reached for $q_0 \neq 0$, and $\lambda_0 > 1$. To verify this, let us represent f in the form

$$f = f_1 f_2, \quad 2f_1 = z + \frac{1}{z}, \quad f_2 = \left(1 + \frac{t_2 q^2}{p_m^2}\right)^{-1}, \quad z = \frac{(p_m^2 + q^2)^2}{p_m^2}. \tag{5.4.3}$$

We have $f_1 \geqslant 1$, $f_2 \geqslant 1$, and $f_1 = 1$ for $q_1 = \sqrt{p_m - p_m^2}$, and $f_2 = 1$ for $q_2 = 0$. The inequality $\lambda_0 > 1$ is fulfilled since $q_1 \neq q_2$.

For different values of t_2 the dependence of λ_0 on p_m is shown in Figure 5.4.

Now consider a shell of moderate length. By virtue of (1.3) we may assume, that there exists a number m such that $p_m \simeq 1$. Unlike the case considered in Section 5.2, the case where $p_m = 1$ and $q_0 = 0$ is not special here for $t_2 < 0$. Indeed, from (2) we find that

$$f_{qq}^0 = -t_2^0, \quad f_{qy}^0 = 0, \quad f_{yy}^0 = \frac{d^2}{dy_0^2}\left(\frac{d + g\,k_2^2}{t_1}\right). \tag{5.4.4}$$

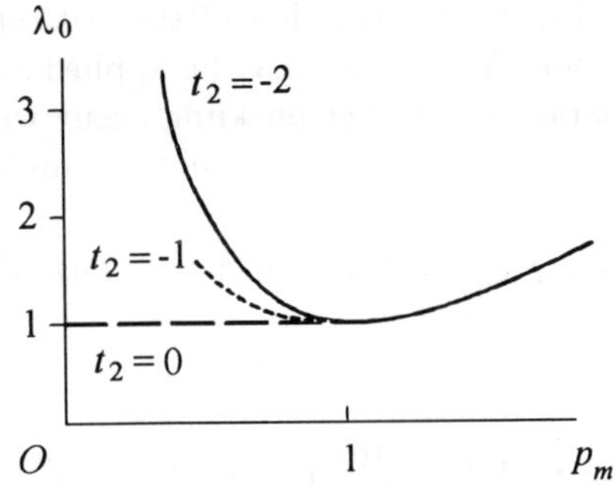

Figure 5.4: The load parameter for a cylindrical shell under non-homogeneous axial compression and an internal pressure.

Now the critical load and buckling mode can be found by formulae (4.2.28)

$$\begin{aligned} \lambda &= 1+\frac{\mu}{2}\left(f_{qq}^0 f_{yy}^0\right)^{1/2}+O\left(\mu^2\right), \quad c=\left(\frac{f_{qq}^0}{f_{yy}^0}\right)^{1/2}, \\ w &= \sin\frac{p_m x}{\mu}\exp\left\{-\frac{c}{2\mu}(y-y_0)^2\right\}\left(1+O\left(\mu^{1/2}\right)\right). \end{aligned} \tag{5.4.5}$$

The main difference between the present problem and a case without stress-resultant t_2 is that here, buckling necessarily occurs for $p_m \simeq 1$ while if $t_2 = 0$, buckling is also possible for $p_m < 1$ (see Figure 5.4).

It should be noted that $p_m \simeq 1$ corresponds to the deformation half-wave length L_x in the longitudinal direction, which is equal to

$$L_x = \pi R\mu = \pi\left(\frac{Rh}{\sqrt{12(1-\nu^2)}}\right)^{1/2}, \tag{5.4.6}$$

where R is the radius of the curvature of the directrix at its weakest point.

5.5 Buckling of a Non-Circular Cylindrical Shell

We will now examine the buckling of a non-circular cylindrical shell under homogeneous axial pressure. The shell parameters E, ν and h are assumed to be constant. As a typical size, R, we take the largest radius of the curvature of the directrix and let $y_0 = 0$ be the weakest generatrix. Then under the above assumptions

$$d = g = 1, \quad t_1 \equiv 2, \quad k_2 = k_2(y), \quad k_2(0) = 1. \tag{5.5.1}$$

Let $k_2''(0) > 0$, then for the construction of the critical load and buckling mode, the results of Sections 5.1 and 5.2 may be applied and we can write the formulae for the load parameter λ. Let buckling occur with m half-waves at the generatrix (see (1.3)). Then for $p_m < 1$, due to formula (1.17) we find

$$\lambda = 1 + 2\,\mu \left(\frac{1 - p_m}{p_m}\, k_2''(0) \right)^{1/2} + O(\mu^2), \tag{5.5.2}$$

For $p_m \simeq 1$ according to (2.9) we have

$$\begin{aligned} \lambda &= 1 + 2\,\mu^{4/3} \left(\frac{k_2''(0)}{4} \right)^{2/3} \Lambda_0(k) + O\,(\mu^2), \\ k &= \frac{p_m - 1}{\mu^{2/3}} \left(\frac{4}{k_2''(0)} \right)^{1/3}, \end{aligned} \tag{5.5.3}$$

where $\Lambda_0(k)$ is obtained from Figure 5.1. At last, for $p_m > 1$, formula (1.13) gives

$$\lambda = \frac{1}{2}(p_m^2 + p_m^{-2}) + \mu\,(2^{-1}(p_m^{-2} - p_m^{-6})\,k_2''(0))^{1/2} + O\,(\mu^2). \tag{5.5.4}$$

We consider for example, a shell with a cross-section in the form of an ellipse with semi-axes a and b $(a > b)$. Then, in (2)–(4) we should assume that

$$\begin{aligned} R &= \frac{b^2}{a}, \quad \mu^4 = \frac{a^2 h^2}{12\,(1 - \nu^2)\,b^4}, \quad k_2''(0) = 3\,\frac{b^2 - a^2}{a^2}, \\ p_m &= \frac{m\,\pi}{\sqrt[4]{12\,(1 - \nu^2)}}\,\frac{b}{L}\left(\frac{h}{a}\right)^{1/2}. \end{aligned} \tag{5.5.5}$$

By virtue of (1.2) the stress-resultant, T_1^0, is related to λ by the expression

$$T_1^0 = -\frac{E\,h^2 a\,\lambda}{b^2\sqrt{12\,(1 - \nu^2)}}. \tag{5.5.6}$$

To evaluate the critical value T_1^0 we should minimize λ from formulae (2)–(4) by m.

5.6 Cylindrical Shell with Curvature of Variable Sign

For $k_2 = 0$ in the case when $p_m \leqslant 1$, the solution constructed in Section 5.1 is not applicable. At the same time, the weakest generatrix is where $k_2 = 0$. It is assumed below that

$$k_2(0) = 0, \qquad k_2'(0) = k_{20}' > 0. \tag{5.6.1}$$

These conditions are fulfilled for a shell which has a generatrix with a point of inflection as shown in Figure 5.5. The weakest generatrix $y_0 = 0$ is marked with point A.

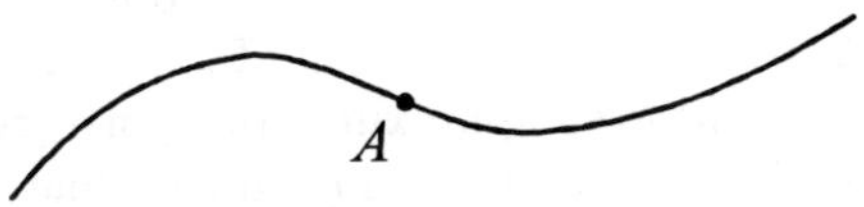

Figure 5.5: The generatrix of a cylindrical shell with the curvature of variable sign.

We now return to system of equations (1.4) and assume that $d = g = t_1 = 1$ for $y = 0$. As before, we seek solutions that exponentially decay as $|y|$ increases. Assuming that $k'_{20} \sim 1$, we find the minimum value of parameter λ, for which such a solution exists.

Let us consider a thin shell of moderate length. Then in (1.3) $l \sim 1$ and parameter p_m has values which are very close to each other. We minimize λ by m, assuming that parameter p_m is continuous.

We assume in system (1.4) that

$$y = \mu^t a^{-1}\eta, \quad p_m = \mu^\alpha a, \quad \lambda = \mu^\beta \lambda', \quad a, \lambda' \sim 1 \tag{5.6.2}$$

and equating the orders of the main terms we find that

$$t = \beta = 2/3, \qquad \alpha = 1/3. \tag{5.6.3}$$

Now, the solution of system (1.4) we seek in the form

$$\begin{gathered} w = \sum_{k=0}^{\infty} \varepsilon^k w_k(\eta), \quad \Phi = \sum_{k=0}^{\infty} \varepsilon^k \Phi_k(\eta), \\ \lambda = \varepsilon\lambda_0 + \varepsilon^3\lambda_1 + \cdots, \qquad \varepsilon = \mu^{2/3}, \end{gathered} \tag{5.6.4}$$

and we require that all functions $w_k(\eta)$, $\Phi_k(\eta)$ converge to zero as $\eta \to \pm\infty$. In the zeroth approximation we have the system of equations

$$\left(b^3\left(\frac{d^2}{d\,\eta^2} - 1\right)^2 - \Lambda\right) w_0 + \eta\,\Phi_0 = 0, \quad b^3\left(\frac{d^2}{d\,\eta^2} - 1\right)^2 \Phi_0 - \eta\, w_0 = 0, \tag{5.6.5}$$

where

$$b^3 = \frac{p_m^3}{\mu\, k'_{20}} = \frac{a^3}{k'_{20}}, \qquad \Lambda = \frac{\lambda_0 a}{k'_{20}}. \tag{5.6.6}$$

The next approximations for w_k and Φ_k satisfy non-homogeneous equations, the left sides of which coincide with the left sides of equations (5), and the right sides of which include functions which depend on the earlier approximations.

As in Section 4.2, due to the alternation of parity of functions $w_k(\eta)$, the solutions of equations in the even approximations converge to zero as $\eta \to \pm\infty$ always exist. The values λ_1, λ_2 and ... in (4) are evaluated from the existence condition for functions w_2, w_4, and ... which decrease to zero as $\eta \to \pm\infty$.

Consider the zeroth approximation in more detail. As in Section 5.2, after applying the Fourier transform (2.5) we arrive at the second order system of equations

$$\frac{d\,w^F}{d\,\omega} = b^3(\omega^2+1)^2\Phi^F, \quad \frac{d\,\Phi^F}{d\,\omega} = b^3(\omega^2+1)^2 w^F - \Lambda\, w^F. \tag{5.6.7}$$

By means of numerical integration we find the first eigenvalue, $\Lambda_0(b)$, for which a non-trivial solution of system (7) converging to zero for $\omega \to \pm\infty$ exists. We obtain from (6)

$$\lambda_0 = (k_{20}')^{2/3} b^{-1}\Lambda_0(b). \tag{5.6.8}$$

Figure 5.6 shows a plot of function $b^{-1}\Lambda_0(b)$. It corresponds to the buckling mode which is even by w and odd by Φ.

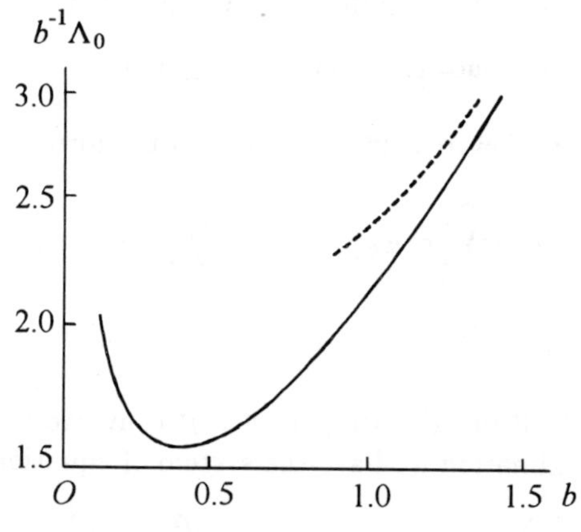

Figure 5.6: The function $b^{-1}\Lambda_0(b)$.

The minimum value of λ by p_m is reached for

$$b = 0.466, \qquad p_m = \mu^{1/3}(k_{20}')^{1/3} b \tag{5.6.9}$$

and it is equal to

$$\lambda = 1.577\,(\mu\,k'_{20})^{2/3} + O\,(\mu^2). \qquad (5.6.10)$$

For a discrete change in the parameter p_m, buckling occurs for one value of p_m determined by (1.3) which is closest to value (9).

Let us give the expression for the characteristic deformation half-wave length L_x in the direction of the x axis

$$L_x = \frac{\pi\,\mu\,R}{p_m} = \pi\,R\,b^{-1}\mu^{2/3}(k'_{20})^{-1/3} \qquad (5.6.11)$$

or in dimensional variables

$$L_x = \frac{\pi}{b^{1/3}}\left(\frac{h}{\sqrt{12\,(1-\nu^2)}}\right)^{1/3}\left(\frac{d}{d\,s}\left(\frac{1}{R_2}\right)\right)^{-1/3}. \qquad (5.6.12)$$

If the shell length is $L < L_x$, then it should be assumed that

$$m = 1, \qquad p_m = p_1 = \frac{\pi\,\mu}{l}. \qquad (5.6.13)$$

As before, the critical load is defined by formula (8), where $\Lambda_0(b)$ is taken from Figure 5.6, and for $b \gg 1$

$$\Lambda_0(b) = b^3 + \sqrt{2} + O\,(b^{-3}). \qquad (5.6.14)$$

The curve $b^2 + b^{-1}\sqrt{2}$ is shown in Figure 5.6 as a dotted line.

To derive formula (14) in system (7) we make the substitution $\omega = b^{-3/2}\omega_1$. Then in the zeroth approximation we come to the Weber equation [69]

$$\frac{d^2 w_0^F}{d\,\omega_1^2} + (\Lambda - b^3 - 2\,\omega_1^2)\;w_0^F = 0. \qquad (5.6.15)$$

For $\Lambda - b^3 = \sqrt{2}\,(1 + 2\,n)$, $n = 0, 1, 2\ldots$ this equation has solutions which converge to zero as $\omega_1 \to \pm\infty$. Formula (14) is obtained for $n = 0$.

Since in this problem it is inconvenient to use the radius of the curvature of the directrix as the characteristic shell size R, we can write the formula for the critical compressive stress-resultant T_1^0 in the dimensional variables. Accounting for (1.2) the main term of (4) gives

$$\begin{aligned} T_1^0 &= -\,E\,h\left(\frac{h^2}{12\,(1-\nu^2)}\frac{d}{ds}\left(\frac{1}{R_2}\right)_0\right)^{2/3} b^{-1}\Lambda_0(b), \\ b^3 &= b_m^3 = \frac{\pi^3 h\,m^3}{L^3\sqrt{12\,(1-\nu^2)}}\left[\frac{d}{ds}\left(\frac{1}{R_2}\right)_0\right]^{-1}. \end{aligned} \qquad (5.6.16)$$

The problem considered here is, to some degree, similar to the buckling problem of a simply supported plate under compression in one direction. The buckling of a short plate occurs with one half-wave in the longitudinal direction while the buckling of a long plate displays several half-waves [16].

5.7 Problems and Exercises

5.1 Examine the buckling of a non-circular cylindrical shell under homogeneous axial pressure, when $k_2(y) = y^{2n} + 1$. Find the critical pressure.

5.2 Find the critical compressive stress-resultant T_1^0 for a short cylindrical shell with curvature of variable sign where $b = b_1 \gg 1$.

Answer

$$T_1^0 = -\frac{E\,h^3\pi^2}{12\,(1-\nu^2)\,L^2}\left(1 + \frac{\sqrt{2}}{b_1^3}\right),$$

where the main term coincides with the Euler formula (3.4.8), while the second term in the parentheses takes into account the bending of the directrix.

5.3 Obtain critical load under axial compression of the non-circular cylindrical shell of moderate length and constant thickness h. The shell cross-section is an ellipse with the semi-axes a_0 and b_0 $(a_0 < b_0)$.
Hint Find the ellipse curvature and apply relation (2.11).

Answer

$$T_1 = \frac{Eh^2 a}{b^2\sqrt{12(1-\nu^2)}}\left(1 + 1.81\left(\frac{3(b^2-a^2)}{4a^2}\right)^{2/3}\mu^{4/3}\right),$$
$$\mu^4 = \frac{h^2 a^2}{12(1-\nu^2)b^4}.$$

To calculate the axial force multiply the stress-resultant T_1 by the length of the ellipse arc.

5.4 Obtain the critical load for the circular simply supported cylindrical shell of radius R, moderate length $L \sim R$ and constant thickness h under bending moment M_1.
Hint Use relation (2.11).

Answer

$$M_1 = 2\pi R^2 Eh\mu^2\left(1 + 0.72\mu^{4/3}\right), \qquad \mu^4 = \frac{h^2}{12(1-\nu^2)R^2}.$$

5.5 Obtain the critical load for the simply supported short ($L < \pi\mu R$) circular cylindrical shell of radius R and constant thickness h under bending moment M_1 .

Hint Use relation (1.13).

Answer

$$M_1 = 2\pi R^2 E h \mu^2 \left(\tfrac{1}{2}\left(p_1^2 + p_1^{-2}\right) + \mu \left(\tfrac{p_1^4 - 1}{2p_1^6}\right)^{1/2} \right),$$

$$\mu^4 = \frac{h^2}{12(1-\nu^2)R^2}, \quad p_1 = \frac{\mu\pi R}{L}.$$

5.6 Obtain the critical load for the cylindrical panel with the curvature of variable sign of the moderate length $L \sim \{a, b\}$, and constant thickness h under axial compression. The panel generatrix has a sine form $Y = a\sin(X/b)$, $-\pi b/2 \le X \le \pi b/2$. The panel is simply supported at the curvilinear edges $x = 0, L$ and clamped at the rectilinear edges $X = \pm\pi b/2$.

Hint Use relation (6.10).

Answer

$$T_1 = 1.577 E h \left(\frac{h^2 a}{12(1-\nu^2)b(a^2+b^2)} \right)^{2/3}.$$

Chapter 6

Buckling Modes Localized at a Point

In this Chapter we will consider the buckling of a thin shell for which the determining functions (see Sections 4.1 and 6.1) are variable in both directions under a membrane stress state. We seek the buckling modes which are localized in the neighbourhood of weak points which are assumed to be located far from the edges of the shell.

Under these assumptions we will construct the buckling modes of a convex shell [101, 160, 163] and a cylindrical shell under non-homogeneous axial compression [164].

6.1 Local Buckling of Convex Shells

Consider membrane stress state buckling that is defined by the initial stress-resultants T_i^0 and S^0. We assume that the determining functions (stress-resultants T_i^0 and S^0, metric coefficients A_1 and A_2, curvatures k_1 and k_2, thickness h and the elastic characteristics of the material E and ν) depend on the curvilinear coordinates α_1 and α_2 and are infinitely differentiable with respect to them. Here the values A, B, α and β are denoted by A_1, A_2, α_1 and α_2.

The buckling equations we take in the form of (2.3.8), (4.3.1)

$$\begin{aligned} \mu^2\Delta\,(d\,\Delta\,w) + \lambda\,\Delta_t w - \Delta_k\Phi &= 0, \\ \mu^2\Delta\,(g^{-1}\Delta\,\Phi) + \Delta_k w &= 0, \end{aligned} \tag{6.1.1}$$

where

$$\Delta\, w = \frac{1}{A_1A_2}\sum_i \frac{\partial}{\partial \alpha_i}\left(\frac{A_j}{A_i}\frac{\partial w}{\partial \alpha_i}\right),$$

$$\Delta_k w = \frac{1}{A_1A_2}\sum_i \frac{\partial}{\partial \alpha_i}\left(\frac{A_jk_j}{A_i}\frac{\partial w}{\partial \alpha_i}\right), \tag{6.1.2}$$

$$\Delta_t w = \frac{1}{A_1A_2}\sum_i \frac{\partial}{\partial \alpha_i}\left(\frac{A_jt_i}{A_i}\frac{\partial w}{\partial \alpha_i} + t_3\frac{\partial w}{\partial \alpha_j}\right), \quad i,j=1,2, \quad j\neq i.$$

Here, all of the linear variables are referred to the characteristic size of the neutral surface, R, and retain their original designations. The dimensionless stress-resultants t_i are related to the stress-resultants T_i^0 and S^0 by expressions (4.3.3)

$$\left(T_1^0, T_2^0, S^0\right) = -\lambda\, E_0h_0\mu^2\,(t_1, t_2, t_3)\,. \tag{6.1.3}$$

The small parameter μ and functions d and g are also defined by formulae (4.3.3).

We assume that $k_1k_2 > 0$. Let point $\alpha_i = \alpha_i^*$ be fixed and change the determining functions in system (1) by their values at this point. Then we come to system (3.1.1), with constant coefficients, the solution of which has been constructed in Section 3.1 in the form of (3.1.4)

$$(w, \Phi) = (w_0, \Phi_0)\, e^{i\,\mu^{-1}(p_1\alpha_1+p_2\alpha_2)}, \tag{6.1.4}$$

where λ satisfies the relation

$$\lambda = \frac{d\left(q_1^2+q_2^2\right)^4 + g\left(k_2q_1^2+k_1q_2^2\right)^2}{\left(q_1^2+q_2^2\right)^2\left(t_1q_1^2+2t_3q_1q_2+t_2q_2^2\right)}, \qquad q_i = \frac{p_i}{A_i}. \tag{6.1.5}$$

Now we vary the point $\alpha_i = \alpha_i^*$. The right side in (5) is a function of the four arguments p_i and α_i^* which we denote by $f\,(p_i, \alpha_i^*)$. Let

$$\lambda_0 = \min_{p_i,\alpha_i^*}\{f\} = f\left(p_i^0, \alpha_i^0\right), \tag{6.1.6}$$

where we seek the minimum for all α_i^* on the neutral surface and for real values of p_i. To evaluate this minimum it is convenient first to find the minimum by p_i for fixed α_i^* using the results of Section 3.1.

Let the minimum of (6) be attained at $p_i = p_i^0$, $\alpha_i^* = \alpha_i^0$ and let the weakest point α_i^0 be far from the edge of the shell. Then

$$\frac{\partial f}{\partial \alpha_i^0} = \frac{\partial f}{\partial p_i^0} = 0. \tag{6.1.7}$$

The solution of system (1) which decreases exponentially away from point α_i^0 in all directions, is constructed below. Here we will not consider the question of satisfying the boundary conditions but rather we will assume that the mentioned decay replaces the satisfaction of the boundary conditions.

The case when point α_i^0 coincides with a shell edge is more complex, since then we must satisfy the boundary conditions. This case is considered in Section 13.7.

The method of construction of the solution of system (1) is a generalization of the algorithm described in Section 4.2 for the case of two independent variables α_1 and α_2.

We also assume that the function $f(p_i, \alpha_i)$ has a strong minimum at point (p_i^0, α_i^0) and also that the quadratic form $d^2 f$ is positive definite

$$d^2 f = \sum_{i,j=1}^{4} \frac{\partial^2 f}{\partial x_i \, \partial x_j} dx_i dx_j > 0, \tag{6.1.8}$$

where $(x_1, x_2, x_3, x_4) = (p_1, p_2, \alpha_1, \alpha_2)$, and the derivatives are evaluated at point x_i^0.

We are looking for the formal asymptotic solutions of system (1) in the neighbourhood of point α_i^0 in the form [99]

$$(w, \Phi) = \left(\tilde{w}\,(\alpha_k, \mu),\, \tilde{\Phi}\,(\alpha_k, \mu)\right) e^{i\mu^{-1} S(\alpha_k)}, \tag{6.1.9}$$

and we require that

$$\Im S\,(\alpha_k^0) = 0, \quad \Im S\,(\alpha_k) > 0 \quad (\alpha_k \neq \alpha_k^0). \tag{6.1.10}$$

The last inequality leads to the fast damping of functions (9) and as a consequence provides the possibility of not satisfying the boundary conditions.

We can find functions $S(\alpha_k)$, $\tilde{w}(\alpha_k, \mu)$ and $\tilde{\Phi}(\alpha_k \mu)$ by substituting them into system (1). Let us make a substitution for the independent variables

$$\alpha_k = \alpha_k^0 + \mu^{1/2} \zeta_k, \qquad k = 1, 2, \tag{6.1.11}$$

and expand S into a series in ζ_k

$$S\,(\alpha_k) = \mu^{1/2} \sum_k \frac{\partial S}{\partial \alpha_k^0} \zeta_k + \frac{1}{2} \mu \sum_{k,l} \frac{\partial^2 S}{\partial \alpha_k^0 \, \partial \alpha_l^0} \zeta_k \zeta_l + S_R, \tag{6.1.12}$$

where $S_R = O\,(\mu^{3/2})$.

Substituting (12) into (9) we note that $\mu^{-1} S_R = O\,(\mu^{1/2})$ and this term may be included into $\tilde{w}$ and $\tilde{\Phi}$. Therefore, we seek $S(\alpha_k)$ in the form

$$S\,(\alpha_k) = \mu^{1/2}\,(p_1^0 \zeta_1 + p_2^0 \zeta_2) + \frac{\mu}{2} \sum_{k,l} s_{kl} \zeta_k \zeta_l, \tag{6.1.13}$$

where p_k^0 provides a minimum value of (6) and the symmetric matrix with constant elements $S_0 = \{s_{kl}\}$ is still unknown. Due to (10) $\Im S_0$ is a positive definite matrix. We seek functions $\tilde{w}$ and $\tilde{\Phi}$ in the form

$$\tilde{w} = \sum_{k=0}^{\infty} \mu^{k/2} w^k(\zeta_i), \quad \tilde{\Phi} = \sum_{k=0}^{\infty} \mu^{k/2} \Phi^k(\zeta_i), \tag{6.1.14}$$

where w^k and Φ^k are polynomials in ζ_i. Simultaneously we find the expansion of the load parameter from the existence condition for this solution

$$\lambda = \lambda_0 + \mu\,\lambda_1 + \mu^2\lambda_2 + \cdots \tag{6.1.15}$$

In Section 6.2 the algorithm for constructing the polynomials $w^k(\zeta_i)$ and $\Phi^k(\zeta_i)$ and unknown parameters λ_i and s_{kl} is given. Here we note that, as in Section 4.2 (case B) each eigenvalue (15) is proven to be asymptotically double. Due to (9) the buckling mode is given by the relation

$$w = (\Re\,\tilde{w}\,\cos z - \Im\,\tilde{w}\,\sin z)\,\exp\left\{-\frac{1}{2}\zeta^T S_2 \zeta\right\}, \tag{6.1.16}$$

where

$$\begin{aligned} z &= \mu^{-1}\left(\alpha_1 p_1^0 + \alpha_2 p_2^0\right) + \frac{1}{2}\zeta^T S_1 \zeta + \theta, \\ \zeta &= (\zeta_1, \zeta_2)^T, \quad S_0 = S_1 + i\,S_2. \end{aligned} \tag{6.1.17}$$

As in Section 4.2, the initial phase θ takes on one of two possible values ($0 \leqslant \theta_1,\ \theta_2 < 2\pi$), which cannot be determined by the proposed method. For the smallest eigenvalue corresponding to the critical value λ, in the zeroth approximation $\tilde{w} \simeq 1$ (see Section 6.2).

The size of the buckled area is of order $R\,h_*^{1/4}$ and is covered by a series of stretched pits the sizes of which are of order $R\,h_*^{1/4} \times R\,h_*^{1/2}$. These pits are inclined in the α_2 direction by angle φ_0 such that $\tan\varphi_0 = -p_2^0/p_1^0$ (see Figure 6.1).

The main difference between this mode and that described in Section 3.2 is the decay in the depth of the pits away from point α_i^0 (see Figure 6.1).

6.2 Construction of the Buckling Mode

We assume that $\tilde{w} = w_1$ and $\tilde{\Phi} = w_2$ and substitute (1.9) into (1.1). The result we write as

$$\sum_k H_{nk}\left(-i\mu\frac{\partial}{\partial\alpha_j},\,\alpha_j,\,\mu\right) w_k e^{i\mu^{-1}S(\alpha_j)} = 0, \quad n = 1, 2, \tag{6.2.1}$$

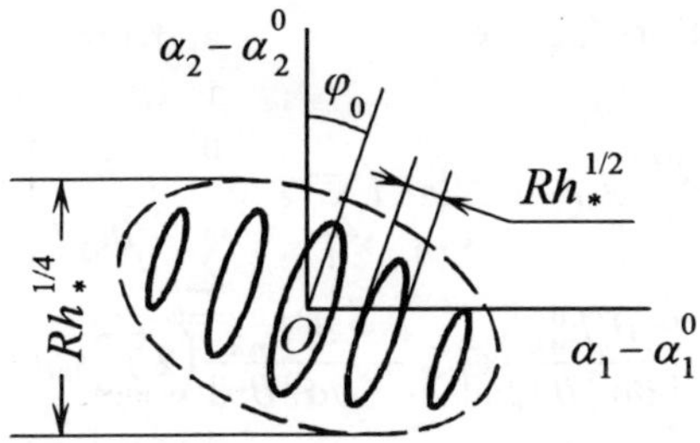

Figure 6.1: The buckling mode of a convex shell localized near the weakest point.

where $H(p_j, \alpha_j, \mu)$ are polynomials in p_j such that

$$H_{11}\left(-i\mu\,\frac{\partial}{\partial\alpha_j},\,\alpha_j,\,\mu\right)\,z = \mu^4\Delta\,(d\,\Delta\,z) + \mu^2\lambda\,\Delta_t z,$$
$$H_{21}z = -H_{12}z = \mu^2\Delta_k z, \quad H_{22}z = \mu^4\Delta\,(g^{-1}\Delta\,z). \tag{6.2.2}$$

We introduce the expansions

$$H_{nk}\,(p_j, \alpha_j, \mu) = H_{nk}^0(p_j, \alpha_j) - i\,\mu\,H_{nk}^1(p_j, \alpha_j) + \cdots, \tag{6.2.3}$$

then, by virtue of (1.2) and (1.15)

$$H_{11}^0 = d\,(q_1^2 + q_2^2)^2 - \lambda_0 H_t^0, \quad H_{12}^0 = -H_{21}^0 = k_2 q_1^2 + k_1 q_2^2,$$
$$H_{22}^0 = g^{-1}(q_1^2 + q_2^2)^2, \qquad H_t^0 = t_1 q_1^2 + 2t_3 q_1 q_2 + t_2 q_2^2, \tag{6.2.4}$$
$$H_{nk}^1 = \frac{1}{2A_1A_2}\sum_m \frac{\partial^2(A_1A_2H_{nk}^0)}{\partial\alpha_m\,\partial p_m} - i\,\lambda_1\delta_{n1}\delta_{k1}H_t^0, \quad q_m = \frac{p_m}{A_m},$$

where δ_{nk} is the Kronecker delta.

We transform (1) to the variables ζ_j using formulae (1.11) and taking into account (1.13) we obtain

$$\sum_k H_{nk}\left(p_j^0 + \mu^{1/2}\Big(\sum_l s_{jl}\zeta_l - i\,\frac{\partial}{\partial\zeta_j}\Big),\,\alpha_j^0 + \mu^{1/2}\zeta_j,\,\mu\right)\,w_k = 0. \tag{6.2.5}$$

If we expand the left side in a power series of $\mu^{1/2}$ we get

$$\sum_k \left(h_{nk}^0 + \mu^{1/2} l_{nk}^1 + \mu\, l_{nk}^2 + \cdots\right)\,w_k = 0, \tag{6.2.6}$$

where the first operators l_{nk}^{j} are

$$
\begin{aligned}
l_{nk}^{1} &= \sum_{j}\left[\frac{\partial h_{nk}^{0}}{\partial \alpha_{j}^{0}}\zeta_{j} + \frac{\partial h_{nk}^{0}}{\partial p_{j}^{0}}\left(\sum_{l} s_{jl}\zeta_{l} - i\frac{\partial}{\partial \zeta_{j}}\right)\right], \\
l_{nk}^{2} &= \frac{1}{2}\sum_{j,r}\left[\frac{\partial^{2} h_{nk}^{0}}{\partial \alpha_{j}^{0}\,\partial \alpha_{r}^{0}}\zeta_{j}\zeta_{r} + \frac{\partial^{2} h_{nk}^{0}}{\partial \alpha_{j}^{0}\,\partial p_{r}^{0}}\left(2\sum_{l} s_{rl}\zeta_{l}\zeta_{j} - \right.\right. \\
&\quad \left. - 2\,i\,\zeta_{j}\frac{\partial}{\partial \zeta_{r}} - i\,\delta_{jr}\right) + \frac{\partial^{2} h_{nk}^{0}}{\partial p_{j}^{0}\,\partial p_{r}^{0}}\left(\sum_{l,m} s_{jl}s_{rm}\zeta_{l}\zeta_{m} - \right. \\
&\quad \left.\left. - 2\,i\sum_{l} s_{jl}\zeta_{l}\frac{\partial}{\partial \zeta_{r}} - i\sum_{l} s_{ll} - \frac{\partial^{2}}{\partial \zeta_{j}\,\partial \zeta_{r}}\right)\right] - i\,h_{nk}^{1}.
\end{aligned}
\tag{6.2.7}
$$

Here h_{nk}^{j} are the values of functions H_{nk}^{j} at $\alpha_i = \alpha_i^0$, and $p_i = p_i^0$.

Substituting series (1.14) into (6) we get the sequence of systems of equations

$$\sum_{k} h_{nk}^{0} w_{k}^{0} = 0, \qquad n = 1, 2 \tag{6.2.8}$$

$$\sum_{k} \left(h_{nk}^{0} w_{k}^{1} + l_{nk}^{1} w_{k}^{0}\right) = 0, \tag{6.2.9}$$

$$\sum_{k} \left(h_{nk}^{0} w_{k}^{2} + l_{nk}^{1} w_{k}^{1} + l_{nk}^{2} w_{k}^{0}\right) = 0, \ldots \tag{6.2.10}$$

The determinant of system (8) we denote as

$$h^{0} = H^{0}\left(\alpha_i^0, p_i^0\right) = h_{11}^{0} h_{22}^{0} + \left(h_{12}^{0}\right)^{2} \tag{6.2.11}$$

and note that the equality $h^0 = 0$ corresponds to equality (1.6).

We next introduce $w_2^0 = h_{12}^0 (h_{22}^0)^{-1} w_1^0$. System (9) is compatible by virtue of (1.7) and (7). We can write

$$w_{2}^{1} = \left(h_{12}^{0} w_{1}^{1} - \sum_{k} l_{2k}^{1} w_{k}^{0}\right)\left(h_{22}^{0}\right)^{-1} \tag{6.2.12}$$

and the compatibility condition for system (10) is then

$$L_0 w_1^0 + \zeta^T D\, \zeta\, w_1^0 = 0, \tag{6.2.13}$$

where

$$L_0 Z = -\frac{1}{2}\,\mathrm{tr}\,(A\, Z_{\zeta\zeta}) - i\, Z_\zeta^T F\,\zeta - \left(\frac{i}{2}\,\mathrm{tr}\, F + \lambda_1 d_1\right) Z,$$

$$\zeta = (\zeta_1, \zeta_2)^T, \quad Z_\zeta = \left(\frac{\partial z}{\partial \zeta_1}, \frac{\partial z}{\partial \zeta_2}\right)^T, \quad Z_{\zeta\zeta} = \left\{\frac{\partial^2 z}{\partial \zeta_m\, \partial \zeta_n}\right\},$$

$$D = \frac{1}{2}\left[S_0 A\, S_0 + B^T S_0 + S_0 B + C\right], \quad F = A\, S_0 + B, \quad d_1 = h_t^0 h_{22}^0,$$

$$A = \left\{\frac{\partial^2 h^0}{\partial p_m^0\, \partial p_n^0}\right\}, \quad B = \left\{\frac{\partial^2 h^0}{\partial p_m^0\, \partial \alpha_n^0}\right\}, \quad C = \left\{\frac{\partial^2 h^0}{\partial \alpha_m^0\, \partial \alpha_n^0}\right\}. \tag{6.2.14}$$

Here ζ and Z_ζ are two-dimensional vectors, $Z_{\zeta\zeta}$, A, B and C are second order square matrices, $\mathrm{tr}\, X$ denotes the sum of the diagonal matrix elements X, and ^{T} is the matrix transpose operation.

Equation (13) has a solution in the form of a polynomial in ζ_1 and ζ_2 only if

$$2\, D = S_0 A\, S_0 + B^T S_0 + S_0 B + C = 0, \tag{6.2.15}$$

From this equation one can find S_0. To solve equation (15) we use the algorithm developed in [99]. We note that the fourth order matrix

$$\begin{pmatrix} A & B \\ B^T & C \end{pmatrix} \tag{6.2.16}$$

is positive definite due to inequality (1.8), since

$$\frac{\partial^2 h^0}{\partial x_m\, \partial x_n} = d_1\, \frac{\partial^2 f}{\partial x_m\, \partial x_n}. \tag{6.2.17}$$

Consider the system of vector equations

$$-B^T u - C\, v = \beta\, u, \qquad A\, u + B\, v = \beta\, v, \tag{6.2.18}$$

where u and v are two-dimensional vectors.

Let $\beta = \beta_j$ be the eigenvalues of β, for which system (16) has non-trivial solutions u_j and v_j, where $j = 1, 2, 3, 4$. Since matrix (17) is positive definite it follows that β_j are pure imaginary numbers.

Let

$$\beta_k = i\omega_k, \quad \beta_{k+2} = -i\omega_k, \quad \omega_k > 0, \quad k = 1, 2. \tag{6.2.19}$$

The following orthogonality condition is valid

$$u_m^T v_n - u_n^T v_m = 0, \qquad \beta_m + \beta_n \neq 0. \tag{6.2.20}$$

We introduce the second order matrices U and V the columns of which coincide with the vectors u_1, u_2 and v_1, v_2. Then

$$S_0 = U\,V^{-1}, \tag{6.2.21}$$

and S_0 does not depend on the normalization of u_k and v_k. We can verify directly the satisfaction of equation (15). The relation $u_1^T v_2 = u_2^T v_1$ leads to the symmetry of matrix S_0.

We will now prove that matrix S_2 is positive definite. Consider the quadratic form $T = \bar{x}^T S_0 x$. We must prove that $\Im T > 0$ for any $x \neq 0$. After substitution $y = V^{-1}x$ we get

$$T = \bar{y}^T \Gamma\, y, \qquad \Gamma = \{\gamma_{mn}\} = \bar{V}^T U = \{\bar{v}_m^T u_n\}. \tag{6.2.22}$$

From (18) we find

$$\bar{v}_k^T u_k = \frac{i}{2\omega_k}\left(u_k^T A\, u_k + 2\bar{u}_k^T B\, v_k + \bar{v}_k^T C\, v_k\right). \tag{6.2.23}$$

Due to the fact that matrix (16) is positive definite, the expression in the parentheses is also positive. Then we note that $\bar{u}_1$, $\bar{v}_1$ is the eigenvector that corresponds to $\beta_3 = -i\omega_1$ and use equality (20) $\bar{u}_1^T v_2 = \bar{v}_1^T u_2$. We then obtain $\gamma_{12} = \bar{\gamma}_{21}$ and it follows that $\Im T > 0$.

Now we come to the solution of equation (13). We make a substitution for the independent variables $\zeta = V\eta$. As a result, matrix F becomes

$$F' = V^{-1} F\, V = \operatorname{diag}(i\omega_1, i\omega_2). \tag{6.2.24}$$

In its turn, equation (13) may be written as

$$\begin{aligned} L_0' w_1^0 &= -\frac{1}{2}\sum_{k,m} a'_{km}\,\frac{\partial^2 w_1^0}{\partial\eta_k\,\partial\eta_m} + \\ &+ \sum_k \omega_k\eta_k\,\frac{\partial w_1^0}{\partial\eta_k} + \left(\frac{\omega_1+\omega_2}{2} - \lambda_1 d_1\right) w_1^0 = 0, \end{aligned} \tag{6.2.25}$$

where

$$A' = \{a'_{km}\} = V^{-1} A\,(V^{-1})^T.$$

We introduce arbitrary integers m, $n \geqslant 0$ and seek the solution of equation (25) in the form of a polynomial with the main term $\eta_1^m \eta_2^n$. Then for

$$\lambda_1^{(m,n)} = d_1^{-1}\left[\omega_1\left(m + \frac{1}{2}\right) + \omega_2\left(n + \frac{1}{2}\right)\right] \tag{6.2.26}$$

the unknown polynomial has the form

$$w_1^0 = P_{mn} = \eta_1^m \eta_2^n + \sum_{k<m,l<n} p_{kl} \eta_1^k \eta_2^l, \tag{6.2.27}$$

where the coefficients p_{kl} are uniquely defined and the sums of the indices $k+l$ at each term are odd if $m+n$ is odd and even if $m+n$ is even.

The critical load is determined by the expression

$$\lambda_1^{(0,0)} = \frac{\omega_1 + \omega_2}{2d_1}, \qquad w_1^0 \equiv 1. \tag{6.2.28}$$

We are not considering here the development of higher order approximations which may be made by methods similar to that presented in Section 4.2 (see also [160]).

6.3 Ellipsoid of Revolution Under Combined Load

We will now consider an example [101] that illustrates the material of the two previous Sections. We will study an prolate ellipsoid of revolution under a homogeneous external pressure q and a bending moment M_1. The vector $\boldsymbol{M}_1$ is assumed to be orthogonal to the axis of symmetry (see Figure 1.3). Parameters E, ν and h are constant.

Let us introduce a system of curvilinear coordinates θ, φ $(0 < \theta_0 \leqslant \theta \leqslant \pi - \theta_0,\ 0 \leqslant \varphi < 2\pi)$ on the neutral surface of the ellipsoid. Here we will use the same notation as in Section 4.4 (see Figure 4.4) and as the characteristic size, R, we will take the smaller semi-axis $R = a_0$.

Under only pressure q, buckling of the ellipsoid occurs in the neighbourhood of the weakest parallel (equator) $\theta_0 = \pi/2$ (see Section 4.4). Under the combined loading of the pressure and a moment M_1, the points of the equator are under different load conditions and we can expect that buckling will occur in the neighbourhood of the weakest point.

In the problem under consideration the initial stress-resultants are the following (1.4.6)

$$\begin{aligned} T_1^0 &= -\frac{1}{2} R_2 q - \frac{M_1 \cos\varphi}{\pi B^2 \sin\theta}, \qquad S^0 = -\frac{M_1 \cos\theta \sin\varphi}{\pi B^2 \sin\theta}, \\ T_2^0 &= \left(\frac{R_2^2}{2R_1} - R_2\right) q + \frac{M_1 R_2 \cos\varphi}{\pi B^2 R_1 \sin\theta}. \end{aligned} \tag{6.3.1}$$

Let

$$q = \frac{\lambda E h \mu^2}{a_0}, \qquad M_1 = \pi a_0^3 q m. \tag{6.3.2}$$

Then by virtue of (1.3) and (4.4.8) we get

$$t_1 = \frac{1}{2k_2} - \frac{m\,k_2^2\cos\varphi}{\sin^3\theta}, \quad t_2 = \frac{1}{k_2} - \frac{k_2}{2\delta^2} + \frac{m\,k_2^4\cos\varphi}{\delta^2\sin^3\theta},$$
$$t_3 = -\frac{m\,k_2^2\cos\theta\sin\varphi}{\sin^3\theta}, \quad k_2 = \left(\sin^2\theta + \delta^2\cos^2\theta\right)^{1/2}, \quad \delta = \frac{b_0}{a_0} > 1. \tag{6.3.3}$$

Let $m > 0$. If

$$m < \frac{(\delta^2-1)(4\delta^2-1)}{\delta^2+2}, \qquad \delta^2 > 1, \tag{6.3.4}$$

then $\theta^0 = \pi/2$, $\varphi^0 = 0$ is the weakest point. Assuming that conditions (4) are fulfilled, we can construct the buckling mode and find the critical load by means of the method discussed in Sections 6.1 and 6.2.

Under the above assumptions we have in (1.5)

$$d = g = 1, \quad A_1 = \frac{\delta^2}{k_2^3}, \quad A_2 = \frac{\sin\theta}{k_2}, \tag{6.3.5}$$

the values of t_i and k_2 are defined by formulae (3). In (1.6) the minimum

$$\lambda_0 = 2\left(m + \delta^2 - \frac{1}{2}\right)^{-1} \tag{6.3.6}$$

is attained at

$$p_1^0 = 0, \quad p_2^0 = \delta^{-1}, \quad \theta^0 = \pi/2, \quad \varphi^0 = 0. \tag{6.3.7}$$

By means of formulae (2.14), (2.17) and (1.5) we can find the elements of matrices A, B and C. In the case under consideration, matrices A and C are diagonal and B is a trivial matrix. The non-zero elements of matrices A and C are the following

$$\begin{aligned} a_{11} &= 4(\delta^2 + 2m - 1)\,\delta^{-8}(\delta^2 + m - 1/2)^{-1}, \\ a_{22} &= 8\,\delta^{-6}, \\ c_{11} &= 2\left[(\delta^2-1)(4\delta^2-1) - m(\delta^2+2)\right]\delta^{-8}(\delta^2 + m - 1/2)^{-1}, \\ c_{22} &= 2\,m\,\delta^{-8}(\delta^2 + m - 1/2)^{-1}. \end{aligned} \tag{6.3.8}$$

By directly solving equation (2.15) we find that

$$s_{11} = i\,c_{11}^{1/2}a_{11}^{-1/2}, \quad s_{22} = i\,c_{22}^{1/2}a_{22}^{-1/2}, \quad s_{12} = s_{21} = 0 \tag{6.3.9}$$

and the critical value of the load parameter is

$$
\begin{aligned}
\lambda &= \lambda_0 + \mu\,\lambda_1 + O\,(\mu^2), \\
\lambda_1 &= \frac{1}{2d_1}\left(\sqrt{a_{11}c_{11}} + \sqrt{a_{22}c_{22}}\right), \quad d_1 = \frac{1}{\delta^8}\left(\delta^2 + m - 1/2\right),
\end{aligned}
\tag{6.3.10}
$$

where λ_0 is given by formula (6). Formula (10) agrees with (2.28) and we can determine the other eigenvalues for the load parameter by means of (2.26)

$$
\lambda_1^{(n_1,n_2)} = \frac{1}{d_1}\left[\sqrt{a_{11}c_{11}}\left(n_1 + \frac{1}{2}\right) + \sqrt{a_{22}c_{22}}\left(n_2 + \frac{1}{2}\right)\right], \qquad n_1,\, n_2 = 0, 1, 2, \ldots
\tag{6.3.11}
$$

We note that for $m = 0$, (10) transforms to (4.4.9) which was obtained without the bending moment M_1.

The zeroth approximation for the buckling mode is the following

$$
\begin{aligned}
w(\theta, \varphi) = \exp&\left\{-\frac{1}{2\mu}\left[s_{11}\left(\theta - \frac{\pi}{2}\right)^2 + s_{22}\varphi^2\right]\right\}\cdot \\
&\left[\cos\left(\frac{\varphi}{\mu\,\delta} + \varphi_k\right) + O\,(\mu^{1/2})\right], \qquad k = 1, 2.
\end{aligned}
\tag{6.3.12}
$$

As before, the initial phases φ_1 and φ_2 cannot be obtained by this method. However, by taking into account symmetry we can find that $\varphi_1 = 0$ and $\varphi_2 = \pi/2$. The buckling pits are elongated in the meridional direction and their depths decrease away from point $\left(\theta = \dfrac{\pi}{2},\ \varphi = 0\right)$.

Let $0 < c < 1$. Then $|w| \leqslant c$, if

$$
s_{11}\left(\theta - \frac{\pi}{2}\right)^2 + s_{22}\varphi^2 + 2\,\mu\,\ln c \leqslant 0,
\tag{6.3.13}
$$

i.e. the pits cover an ellipse with its centre at the weakest point.

Finally, we will consider a numerical example. Let $\delta = 1.5$, $m = 1$, $a_0/h = 500$, $\nu = 0.3$ and $c = 0.1$. Using the above formulae we find that $\mu = 0.0246$, $\lambda_0 = 0.727$, $\lambda_1 = 1.145$ and $\lambda_0 + \mu\lambda_1 = 0.755$.

Ellipse (13) is elongated in the parallel direction and has semi-axes lengths 0.36 and 1.67 i.e. in the circumferential direction, the depth of the pits decreases sufficiently slowly. Each pit fills angle $\Delta\varphi = 0.116$ in the circumferential direction. For $\varphi_1 = 0$ the buckling mode is even by θ and by φ. In Figure 6.2 the contour lines for the buckling modes $w(\theta, \varphi)$ for $w = 0$ and $w = \pm 0.5$ are shown.

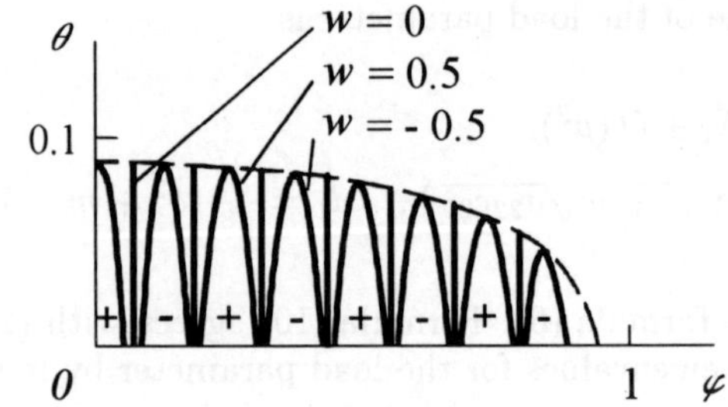

Figure 6.2: The buckling mode of a elliptic shell localized near the weakest point.

6.4 Cylindrical Shell Under Axial Compression

Let us now consider the buckling of a cylindrical shell under axial compression. The initial stress state is assumed to be membrane.

In the case of a circular cylindrical shell when the determining functions are constant and the shell edges are simply supported the waves due to buckling cover the entire neutral surface (see Section 3.4). If the compression is non-homogeneous in the circumferential direction, the buckling pits are localized in the neighbourhood of the weakest generatrix (see Chapter 5).

The general case of a non-circular cylindrical shell with variable determining functions is considered below. On the shell surface there may exist a weak point in the neighbourhood of which the buckling mode is localized. Assuming that this point exists and that it is far from the shell edges, approximate expressions for the critical load and the buckling mode are found.

We will use the system of equations (1.1) which in this case has the form

$$\begin{aligned} \mu^2\Delta\,(d\,\Delta\,W) + \lambda\,\frac{\partial}{\partial x}\left(t_1\,\frac{\partial W}{\partial x}\right) - k_2\,\frac{\partial^2\Phi}{\partial x^2} &= 0, \\ \mu^2\Delta\,(g^{-1}\Delta\,\Phi) + k_2\,\frac{\partial^2 W}{\partial x^2} &= 0, \end{aligned} \tag{6.4.1}$$

where

$$\begin{gathered} d = \frac{E\,h^3(1-\nu_0^2)}{E_0 h_0^3(1-\nu^2)}, \quad g = \frac{E\,h}{E_0 h_0}, \quad \mu^4 = \frac{h_0^2}{12\,(1-\nu_0^2)\,R^2}, \\ T_1 = -\lambda\,T\,t_1, \quad R_2 = R\,k_2^{-1}, \quad T = E_0 h_0 \mu^2, \quad \Delta = \frac{\partial^2}{\partial x^2} + \frac{\partial^2}{\partial y^2}. \end{gathered} \tag{6.4.2}$$

The functions $E(x,y)$, $h(x,y)$, $\nu(x,y)$, $t_1(x,y)$ and $k_2(y)$ are assumed to be infinitely differentiable.

Relation (1.5) may be rewritten as

$$\lambda = f(x, y, p, q) = f^0(x, y)\,\frac{\zeta + \zeta^{-1}}{2}, \tag{6.4.3}$$

where

$$f^0 = \frac{2\,k_2}{t_1}\,(dg)^{1/2}, \quad \zeta = \psi\, z^2, \quad \psi = \left(\frac{k_2^2 g}{d}\right)^{1/2}, \quad z = \frac{p}{p^2 + q^2}. \tag{6.4.4}$$

Evaluating the minimum of function f by all of its arguments we get

$$\lambda_0 = \min_{x,y} f^0 = f^0(x^0, y^0), \qquad \zeta = 1. \tag{6.4.5}$$

Let function f^0 attain its minimum at point $(x^0,\, y^0)$, the second differential

$$d^2 f^0 = \sum f^0_{kl}\, dx_k\, dx_l, \quad f^0_{kl} = \frac{\partial^2 f^0}{\partial x_k^0\, \partial x_l^0}, \quad (x, y) = (x_1, x_2) \tag{6.4.6}$$

at this point is the positive definite quadratic form and the point $(x^0,\, y^0)$ is situated far from the edges of the shell.

We choose R, h_0, E_0 and t_1 in such a way that

$$t_1 = 2, \qquad k_2 = d = g = 1 \tag{6.4.7}$$

at point (x^0, y^0). Then $\lambda_0 = 1$, which agrees with the Lorenz–Timoshenko formula (3.4.3).

As in Section 6.1 we seek a deflection shape that decreases away from the weakest point (x^0, y^0). However, since the values of p^0 and q^0 could not be uniquely determined from the minimum condition for f, we can not use the method employed in Section 6.1 for the present problem.

We apply Fourier transform

$$W(x, y) = \iint\limits_{-\infty}^{\infty} w(p, q)\, e^{iz_*}\, dp\, dq, \quad \Phi(x, y) = \iint\limits_{-\infty}^{\infty} \varphi(p, q)\, e^{iz_*}\, dp\, dq, \tag{6.4.8}$$

where $z_* = \mu^{-1}\left[p\,(x - x^0) + q\,(y - y^0)\right]$. Then we expand the coefficients of system (1) into a power series in $x - x^0$ and $y - y^0$. As a result, after the elimination of φ for $w(p, q)$ we get the equation

$$h\,w + i\,\mu \sum_k \frac{\partial}{\partial p_k}(h_k w) + \frac{1}{2}(i\,\mu)^2 \sum_{k,l} \frac{\partial^2 (h_{kl} w)}{\partial p_k\, \partial p_l} + \cdots = 0, \tag{6.4.9}$$

where

$$
\begin{aligned}
& h\,(p_k,\mu) = H\,(p_k, x_k^0, \mu), \quad h_k = \frac{\partial H}{\partial x_k^0}, \quad h_{kl} = \frac{\partial^2 H}{\partial x_k^0\,\partial x_l^0}, \;\dots \\
& (p_1, p_2) = (p, q), \qquad H = H^0 - i\,\mu\,H^1 + \cdots \\
& H^0 = d\,(p^2+q^2)^2 - \lambda\,t_1 p^2 + \frac{g\,k_2^2 p^4}{(p^2+q^2)^2}, \quad H^1 = \frac{1}{2}\left(\frac{\partial^2 H^0}{\partial x\,\partial p} + \frac{\partial^2 H^0}{\partial y\,\partial q}\right).
\end{aligned}
\tag{6.4.10}
$$

We transform equation (9) to the variables z and τ by means of the formulae

$$
z = \frac{p}{p^2+q^2}, \qquad \tau = \frac{q}{p^2+q^2} \tag{6.4.11}
$$

and we seek a solution which is localized at $z_0 = 1$, $\tau = \tau_0$ in the form ($z_0 = -1$, $\tau = -\tau_0$ leads to the complex conjugate solution)

$$
\begin{aligned}
& w\,(p,q,\mu) = \tilde{w}\,e^{-(a\zeta^2 + b\eta^2)/2}, \quad \tilde{w} = \sum_{k=0}^{\infty} \mu^{k/4} w_k(\zeta,\eta), \\
& z = 1 + \mu^{1/2}\zeta, \quad \tau = \tau_0 + \mu^{1/4}\eta, \quad \lambda = 1 + \mu\,\lambda_1 + \mu^{3/2}\lambda_2 + \cdots,
\end{aligned}
\tag{6.4.12}
$$

where w_k are unknown polynomials in ζ and η, a, b, τ_0 and λ_k are unknown constants ($\Re\,a > 0$, $\Re\,b > 0$).

Substitution (12) into equation (9) gives

$$
\left(L_0 + \mu^{1/4} L_1 + \mu^{1/2} L_2 + \cdots\right)\tilde{w} = 0, \tag{6.4.13}
$$

where

$$
\begin{aligned}
& L_0\tilde{w} = \left(l_0\zeta^2 - \lambda_1\right)\tilde{w} + i l_1\left(\zeta\frac{\partial}{\partial\zeta} - a\zeta^2 + \frac{1}{2}\right)\tilde{w} - \\
& \quad \frac{l_{11}}{2}\left(\frac{\partial}{\partial\zeta} - a\zeta\right)^2 \tilde{w}, \\
& L_1\tilde{w} = \eta\,L_0'\tilde{w} + \left[i\,l_2\zeta - l_{12}\left(\frac{\partial}{\partial\zeta} - a\,\zeta\right)\right]\left(\frac{\partial}{\partial\eta} - b\,\eta\right)\tilde{w}, \\
& L_2\tilde{w} = \frac{1}{2}\eta^2 L_0''\tilde{w} - \lambda_2\tilde{w} - \frac{l_{22}}{2}\left(\frac{\partial}{\partial\eta} - b\,\eta\right)^2\tilde{w} + L_*\tilde{w}, \\
& l_0 = 2, \quad l_j = 2\sum_k \frac{\partial\psi}{\partial x_k^0}\,\frac{\partial z}{\partial p_k^0}, \quad l_{ij} = a_{ij} + \frac{1}{4}\,l_i l_j, \\
& a_{ij} = \sum_{k,l} f_{kl}^0 \frac{\partial z_i}{\partial p_k^0}\,\frac{\partial z_j}{\partial p_l^0}, \quad (z_1, z_2) = (z, \tau), \quad i,j,k,l = 1,2.
\end{aligned}
\tag{6.4.14}
$$

Here f^0 and ψ are the same functions as in (4), l_j and l_{ij} are functions of τ_0, operators L_0' and L_0'' are obtained from L_0 after the substitutions instead of the coefficients l_1 and l_{11} their first and second derivatives by τ_0 (the first term in L_0 becomes zero). We denote the terms that do not influence $w_0(\zeta,\eta)$ and λ_2 due to (2.1) by $L_*\tilde{w}$.

Let us consider the equation $L_0 w_0 = 0$. This equation has a solution in the form of a polynomial in ζ if the coefficient of ζ^2 is equal to zero. This yields

$$a = (r - i\,l_1)\,l_{11}^{-1}, \qquad r = \left(4\,l_{11} - l_1^2\right)^{1/2} = (4\,a_{11})^{1/2} > 0. \tag{6.4.15}$$

For

$$\lambda_1 = \lambda_1^{(n)} = r\left(n + \frac{1}{2}\right), \qquad n = 0, 1, 2, \ldots \tag{6.4.16}$$

the equation $L_0 w_0 = 0$ has the solution

$$w_0(\zeta,\eta) = v_0(\eta)\,H_n(\theta), \quad \theta = \alpha\,\zeta, \quad \alpha = \left(\frac{2\,r}{l_{11}}\right)^{1/2}, \tag{6.4.17}$$

where $v_0(\eta)$ is still an unknown function, and $H_n(\theta)$ is the Hermite polynomial of the n-th order that satisfies the equation

$$M_n u \equiv \frac{d^2u}{d\theta^2} - 2\theta\,\frac{du}{d\theta} + 2n\,u = 0, \qquad H_0(\theta) \equiv 1. \tag{6.4.18}$$

The equation of the first approximation $L_0 w_1 + L_1 w_0 = 0$ after transition to the variable θ takes the form

$$M_n w_1 = F_1(\theta)\,\eta\,v_0(\eta) + F_2(\theta)\left(\frac{\partial}{\partial\eta} - b\,\eta\right)v_0 \equiv F(\theta,\eta), \tag{6.4.19}$$

where

$$\begin{aligned} F_1(\theta) &= -\frac{l_{11}'}{2\,r}\left(\alpha^2 H_n'' - 2\,a\,\theta\,H_n' - a\,H_n + \frac{a^2\theta^2}{\alpha^2}\,H_n\right) + \\ &\quad + i\,\frac{h_1'}{r}\left(\theta\,H_n' - \frac{a\,\theta^2}{\alpha}\,H_n + \frac{1}{2}\,H_n\right), \\ F_2(\theta) &= \frac{1}{r}\left(\frac{l_2\theta}{\alpha}\,H_n - l_{12}\Big(\alpha\,H_n' - \frac{a\,\theta}{\alpha}\,H_n\Big)\right) \end{aligned} \tag{6.4.20}$$

are known polynomials.

Equation (19) has a solution in the form of a polynomial if

$$\int_{-\infty}^{\infty} F(\theta,\eta)\,H_n(\theta)\,e^{-\frac{\theta^2}{2}}\,d\theta = 0. \tag{6.4.21}$$

Condition (21) gives the equation $dr/d\tau_0 = 0$ for the evaluation of τ_0 and equation (19) has the solution

$$w_1(\theta,\eta) = v_1(\eta)\,H_n(\theta) + \eta\,v_0(\eta)\,G_1(\theta) + \left(\frac{\partial v_0}{\partial \eta} - b\,\eta\,v_0\right) G_2(\theta), \qquad (6.4.22)$$

where $v_1(\eta)$ is an unknown function,

$$G_1 = A_1\theta\,H_n' + A_2 H_{n+2}, \quad G_2 = A_3 H_n' + A_4\theta\,H_n,$$

$$A_1 = -\,\bar{A}_2, \qquad A_2 = \left(4\,\alpha^2 r\right)^{-1}\left(a^2 l_{11}' + 2\,i\,a\,l_1'\right), \qquad (6.4.23)$$

$$A_3 = -\,\alpha\,r^{-1} l_{12} - 2\,A_4, \quad A_4 = -\,(\alpha\,r)^{-1}(a\,l_{12} + i\,l_2).$$

Now we come to the second approximation

$$L_0 w_2 + L_1 w_1 + L_2 w_0 = 0. \qquad (6.4.24)$$

Condition (21) of the existence of a solution of equation (24) in the form of a polynomial in ζ leads to

$$-\,Q\left(\frac{\partial}{\partial \eta} - b\,\eta\right)^2 v_0 + (R\eta^2 - \lambda_2)\,v_0 = 0, \qquad (6.4.25)$$

where

$$\begin{aligned} Q &= \frac{1}{2}\,l_{22} - r\,|A_4|^2 = \frac{1}{2}\,a_{11}^{-1}\left(a_{11}a_{22} - a_{12}^2\right) > 0, \\ R &= \frac{2n+1}{16\,r}\left[l_{11}''\,|a|^2 + i\,l_1''\left(\alpha^2 - 2a\right) - 4\,\left|a\,l_{11}' + i\,l_1'\right|^2\right] = \\ &= \frac{2n+1}{4}\,\frac{d^2 r}{d\tau_0^2} = \frac{1}{2}\,\frac{d^2\lambda_1}{d\tau_0^2}. \end{aligned} \qquad (6.4.26)$$

If

$$b = R^{1/2}Q^{-1/2} > 0, \quad \lambda_2 = \lambda_2^{(m,n)} = (2m+1)\,(Q\,R)^{1/2}, \quad m = 0, 1, \ldots \qquad (6.4.27)$$

then equation (25) has a solution in the form of the polynomial $v_0(\eta) = H_m\left(b^{-1/2}\eta\right)$.

The parameter τ in (12) could be found from the minimum condition for λ_1, thus $R > 0$ in (26).

We give the final expression for λ

$$\begin{aligned}\lambda = \lambda^{(m,n)} &= 1 + (2n+1)\,\mu\, a_{11}^{1/2} + \\ &+\mu^{3/2}(2m+1)\,(2n+1)^{1/2}(4\,a_{11})^{-3/4}\left[a_{11}''(a_{11}a_{22} - a_{12}^2)\right]^{1/2} + O\left(\mu^2\right) \\ & m, n = 0, 1, 2, \ldots\end{aligned} \tag{6.4.28}$$

As in Section 6.1 each eigenvalue $\lambda^{(m,n)}$ is asymptotically double. Buckling corresponds to the smallest eigenvalue, which is obtained from (28) for $m = n = 0$. The neighbouring eigenvalues allow us to conclude that the critical load is sensitive to any initial imperfections.

6.5 Construction of the Buckling Modes

Let us find the buckling mode which corresponds to the critical load obtained previously. In order to do this we substitute expression (4.12) into integral (4.8) assuming that $\tilde{w} = 1 + O\left(\mu^{1/4}\right)$. We get

$$\begin{aligned}W(x,y) &= \iint\limits_{-\infty}^{+\infty} \left(1 + O\left(\mu^{1/4}\right)\right) \exp\left\{\frac{i}{\mu}\left(p\,x' + q\,y'\right) - \right. \\ &\left. - \frac{1}{2}\,a\,\zeta^2 - \frac{1}{2}\,b\,\eta^2\right\} dp\,dq,\end{aligned} \tag{6.5.1}$$

where we denote $x' = x - x^0$ and $y' = y - y^0$. In (1) we come to the integrating variables ζ and η. By virtue of (4.11) and (4.12) we find

$$\begin{aligned}p &= p_0 + \mu^{1/2}p_2\zeta + \mu^{1/4}p_1\eta + \mu^{1/2}p_3\eta^2 + \cdots \\ q &= q_0 + \mu^{1/2}q_2\zeta + \mu^{1/4}q_1\eta + \mu^{1/2}q_3\eta^2 + \cdots\end{aligned} \tag{6.5.2}$$

where

$$\begin{aligned}&p_0 = \frac{1}{1+\tau_0^2}, \qquad q_0 = \tau_0 p_0, \qquad p_1 = q_2 = -2\,\tau_0 p_0^2, \\ &q_1 = -p_2 = \left(1 - \tau_0^2\right)p_0^2, \quad p_3 = \left(3\,\tau_0^2 - 1\right)p_0^3, \quad q_3 = \tau_0\left(\tau_0^2 - 3\right)p_0^3.\end{aligned} \tag{6.5.3}$$

Evaluating integral (1), we get

$$\begin{aligned}W &= C\,e^{-\xi}\exp\left\{\frac{i}{\mu}\left(p_0x' + q_0y'\right)\right\}\left(1 + O\left(\mu^{1/4}\right)\right), \\ \xi &= \frac{\left(p_1x' + q_1y'\right)^2}{2\,\mu^{3/2}\left(b - 2i\mu^{-1/2}\left(p_3x' + q_3y'\right)\right)} + \frac{\left(p_2x' + q_2y'\right)^2}{2\,\mu\,a}.\end{aligned} \tag{6.5.4}$$

It follows from (4) that buckling is accompanied by the formation of a series of elongated pits inclined by angle γ to the cylinder generatrix ($\tan\gamma = -\tau_0^{-1}$). The distance between the centres of the pits is $2\pi\mu\,(1+\tau_0^2)$. The factor $e^{-\xi}$ in (4) describes slowly varying pit amplitudes (depths) and phases.

We fix $c > 0$ and find the domain in which $\Re\,\xi < c$. Outside of this domain, the deflections are smaller than e^{-c}. The domain bound is given by the curve

$$\frac{(p_1x' + q_1y')^2\, b}{2\,\mu^{3/2}\left[b^2 + 4\mu^{-1}\,(p_3x' + q_3y')^2\right]} + \frac{(p_2x' + q_2y')^2\,\Re\, a}{2\,\mu\,|a|^2} = c, \qquad (6.5.5)$$

and we obtain the inequalities

$$\begin{aligned} |p_2x' + q_2y'| &\leqslant \left(\frac{2\,|a|^2 c\,\mu}{\Re\, a}\right)^{1/2} = \mu^{1/2}a_0, \\ |p_1x' + q_1y'| &\leqslant \mu^{3/4}b_0, \qquad b_0 = \left[\frac{2\,c}{b}\left(b^2 + \frac{8\,c\,|a|^2}{(1+\tau_0^2)\,\Re\, a}\right)\right]^{1/2}. \end{aligned} \qquad (6.5.6)$$

The domain obtained above is reminiscent of the form of an the ellipse with semi-axes

$$\mu^{1/2}a_0(1+\tau_0^2), \qquad \mu^{3/4}b_0(1+\tau_0^2).$$

The large semi-axis of the "ellipse" is inclined to the cylinder generatrix by the angle δ ($\tan\delta = 2\tau_0(1-\tau_0^2)^{-1}$).

The buckling mode is shown schematically in Figure 6.3. The series of strongly elongated pits inclined to axis x by angle γ covers an "ellipse" whose size is of order $\mu^{1/2} \times \mu^{3/4}$ that is inclined to axis x by angle δ.

Two linear combinations of the real and imaginary parts of (4) are the eigenfunctions. As was found in Section 6.1, we cannot find the coefficients of this linear combination by use of the present method.

Let us return to the evaluation of τ_0. By virtue of (4.14)

$$a_{11} = f_{11}^0(1-\tau_0^2)^2 - 4\,f_{12}^0\tau_0(1-\tau_0^2) + 4\,f_{22}^0\tau_0^2. \qquad (6.5.7)$$

The value of τ_0 may be found by minimizing the function a_{11}. According to the above assumption on $f^0(x,y)$, the value of τ_0 is unique for $f_{12}^0 \neq 0$ and

$$f_{12}^0\tau_0 > 0, \qquad |\tau_0| < \sqrt{1+\beta^2} - \beta, \qquad \beta = \left|\frac{f_{12}^0}{f_{11}^0}\right|. \qquad (6.5.8)$$

Let $f_{12}^0 = 0$. Then for $f_{11}^0 < 2\,f_{22}^0$ we get $\tau_0 = 0$.

We are not considering the case where $f_{11}^0 = 2\,f_{22}^0$ since $a_{11} = 2\,f_{22}^0(1+\tau_0^4)$ and $d^2a_{11}/d\,\tau_0^2 = 0$ for $\tau_0 = 0$. For $f_{11}^0 > 2\,f_{22}^0$ we get $\tau_0 = \pm\left(1 - f_{22}^0/f_{11}^0\right)^{1/2}$.

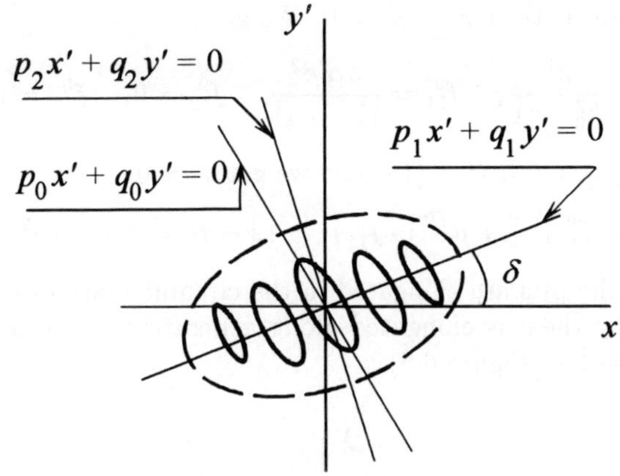

Figure 6.3: The localized buckling mode of a cylindrical shell near the weakest point.

In the last case, due to the existence of two values of τ_0, the critical load is asymptotically tetra-multiple and the buckling mode is a linear combination of the forms described above for $\pm\tau_0$. The expected buckling mode is shown schematically in Figure 6.4. The domain filled with pits is comprised of a combination of the two elongated "ellipses".

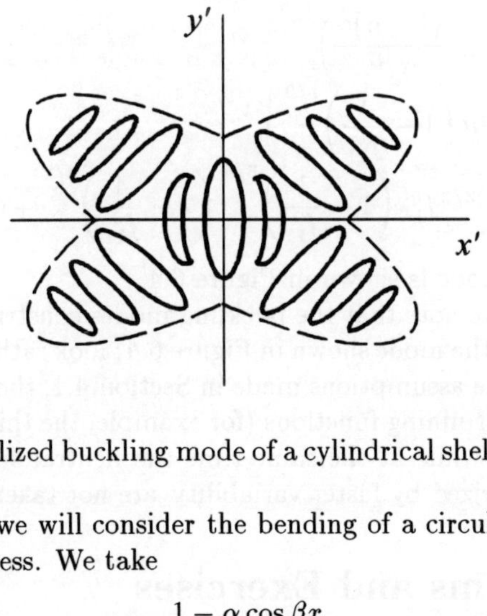

Figure 6.4: The localized buckling mode of a cylindrical shell at $\alpha\beta^2 > (1-\alpha)^2$.

As an example, we will consider the bending of a circular cylindrical shell with variable thickness. We take

$$t_1 = 2\cos y, \qquad h = h_0\,\frac{1-\alpha\cos\beta x}{1-\alpha}, \qquad 0 < \alpha < 1. \tag{6.5.9}$$

The weakest point is then $x^0 = y^0 = 0$. We get

$$f^0(x,y) = \frac{h^2}{h_0^2 \cos y}; \quad f_{11}^0 = \frac{2\,\alpha\,\beta^2}{(1-\alpha)^2}, \quad f_{12}^0 = 0, \quad f_{22}^0 = 1. \qquad (6.5.10)$$

For $f_{11}^0 < 2\,f_{22}^0$, i.e. $\alpha\beta^2 < (1-\alpha)^2$ we get $\tau_0 = 0$,

$$\lambda = 1 + \mu\,(f_{11}^0)^{1/2} + \mu^{3/2}(4\,f_{11}^0)^{-1/4}(2 - f_{11}^0)^{1/2} + O\,(\mu^2). \qquad (6.5.11)$$

In this case the pits are elongated in the circumferential direction and the "ellipse" filled by them is elongated along generatrix $y = 0$ in contrast with the previous case (see Figure 6.5).

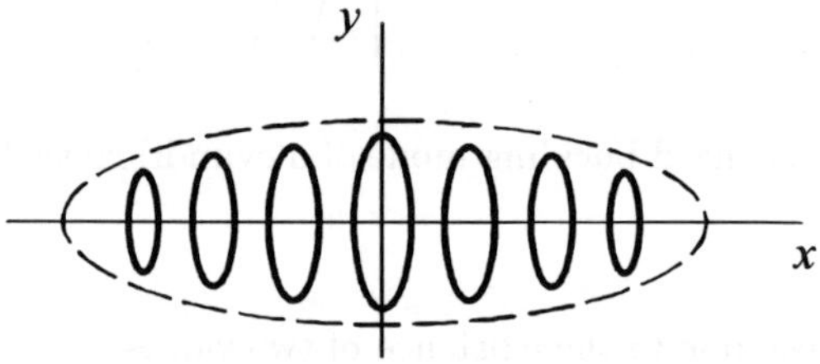

Figure 6.5: The localized buckling mode of a cylindrical shell at $\alpha\beta^2 < (1-\alpha)^2$.

For $\alpha\beta^2 > (1-\alpha)^2$ we find

$$\begin{aligned} \tau_0 &= \pm\left(1 - \frac{(1-\alpha)^2}{\alpha\,\beta^2}\right)^{1/2}, \\ \lambda &= 1 + 2\mu\left(1 - \frac{1}{f_{11}^0}\right)^{1/2} + \\ &\quad + \mu^{3/2} f_{11}^0\left(1 - \frac{1}{f_{11}^0}\right)^{5/4}\left(2 - \frac{4}{f_{11}^0}\right)^{1/2} + O\,(\mu^2), \end{aligned} \qquad (6.5.12)$$

and the buckling mode is shown in Figure 6.4.

In conclusion, we note that the buckling modes constructed in Sections 6.4 and 6.5 (especially the mode shown in Figure 6.4) look rather unusual. Perhaps this is related to the assumptions made in Section 4.1, the major one of which states that the determining functions (for example, the thickness) are variable, but change slowly, while at the same time the neutral surface imperfections, which are characterized by faster variability, are not taken into account.

6.6 Problems and Exercises

6.1. Prove that the parameter of the critical load λ has expansion (1.15) in integer powers of μ whereas mode (1.14) is expanded in the series in fractional powers of μ. In particular, prove that the term $\lambda_{3/2}\mu^{3/2}$ in (1.15) vanishes.

6.2. Consider a prolate truncated ellipsoid of revolution under the external pressure q and edge force P_1 (see Figure 1.3). For $P_1 = 0$ the equator is the weakest line. Find the position of the weakest point for $P_1 \neq 0$ and the critical load in the zeroth approximation.

Hint Use relations (1.4.6) to calculate the stress-resultants, formulae (4.4.8) for the parameters of ellipsoid and (1.5) to determine the weakest point and the critical load.

6.3. Consider prolate truncated ellipsoid of revolution under the external pressure q and edge force P_1 (see Figure 1.3). With the help of formulae (5.3) and (5.4) for $\mu = 0.02$, $\tau_0 = \pm 0.5$, $b = 1$ construct the level lines for four real buckling modes which are odd or even in x' and y' respectively.

Hint One of this modes is schematically plotted in Figure 6.2.

Chapter 7

Semi-momentless Buckling Modes

In this Chapter we will study the class of buckling problems for shells with zero Gaussian curvature under a membrane initial stress state. For this class of problems the buckling concavities are stretched along the asymptotic lines of the neutral surface and may be localized near one (the weakest) line. The additional stress state appearing under buckling is a semi-momentless state [119]. The method may be applied to cylindrical and conic shells of moderate length, the cross-section of which need not necessarily be circular and the edges of which are not plane curves. The two-dimensional problem may be reduced to a sequence of fourth order one-dimensional problems. In some particular cases for cylindrical shells, approximate solutions have been obtained.

In this Chapter we will use the results obtained in [100, 152, 161, 162, 163]. Buckling problems for non-circular cylindrical shells and shells with variable thickness are also studied in [6, 7, 8, 23, 34, 42, 83, 122, 149, 181] and others.

7.1 Basic Equations and Boundary Conditions

We will consider the buckling of a conic shell under a membrane stress state. In the neutral shell surface we introduce an orthogonal system of curvilinear coordinates s, φ, where $s = R^{-1}s'$, s' is the distance between the point of a surface and the cone vertex, R is the characteristic size of the neutral surface, φ is the coordinate in a directrix which is so chosen that the first quadratic surface form is equal to $d\sigma^2 = R^2(ds^2 + s^2 d\varphi^2)$. The radius of curvature, R_2, is equal to $R_2 = R\,s\,k^{-1}(\varphi)$. The shell is closed in the circumferential direction,

φ, and it has two edges

$$s_1(\varphi) \leq s \leq s_2(\varphi), \qquad 0 \leq \varphi \leq \varphi_1, \tag{7.1.1}$$

where φ_1 is the length of a curve which is formed by the intersection of a cone and a sphere of unit radius with its centre at the vertex of the cone. For example, for a straight circular cone with the vertex angle 2α we have $\varphi_1 = 2\pi \sin\alpha$.

We can rewrite system (6.1.1) in the form

$$\begin{aligned} \varepsilon^4 \Delta\,(d\,\Delta\, w) + \lambda\, \varepsilon^2 \Delta_t w - \Delta_k \Phi &= 0, \\ \varepsilon^4 \Delta\,(g^{-1} \Delta\, \Phi) + \Delta_k w &= 0, \end{aligned} \tag{7.1.2}$$

where

$$\begin{aligned} \Delta_t w &= \frac{1}{s}\left[\frac{1}{s}\frac{\partial}{\partial\varphi}\left(t_2 \frac{\partial w}{\partial \varphi}\right) + \frac{\partial}{\partial s}\left(t_3 \frac{\partial w}{\partial \varphi}\right) + \frac{\partial}{\partial \varphi}\left(t_3 \frac{\partial w}{\partial s}\right) + \right. \\ &\quad \left. + \frac{\partial}{\partial s}\left(s\, t_1 \frac{\partial w}{\partial s}\right)\right], \qquad \Delta\, w = \frac{1}{s^2}\frac{\partial^2 w}{\partial \varphi^2} + \frac{1}{s}\frac{\partial}{\partial s}\left(s\frac{\partial w}{\partial s}\right), \\ \Delta_k w &= \frac{k}{s}\frac{\partial^2 w}{\partial s^2}, \qquad (T_1^0, T_2^0, S^0) = -\lambda\, E_0 h_0 \varepsilon^6\,(t_1, t_2, t_3), \\ \varepsilon^8 &= \frac{h_0^2}{12\,(1-\nu_0^2)\,R^2}, \quad d = \frac{E\,h^3(1-\nu_0^2)}{E_0 h_0^3 (1-\nu^2)}, \quad g = \frac{E\,h}{E_0 h_0}. \end{aligned} \tag{7.1.3}$$

The functions $k(\varphi) > 0$, $s_i(\varphi)$, $t_i(s,\varphi)$, $d(s,\varphi)$, $g(s,\varphi)$ are assumed to be infinitely differentiable and the orders of their derivatives are not greater than the orders of the original functions.

Remark 7.1. The infinite differentiability of the coefficients of system (2) is necessary for the construction of all of the terms of the asymptotic series (2.3). To construct only the first few terms, the existence of only the first few derivatives is necessary.

The buckled stress state consists of the main stress state and the edge effect at the shell edges (1). Here the main state is semi-momentless [119] and in the circumferential direction it has an index of variation equal to 1/4 and $\dfrac{\partial^2 w}{\partial \varphi^2} \sim h_*^{-1/2} \dfrac{\partial^2 w}{\partial s^2}$. At each edge (1) four homogeneous boundary conditions are introduced but in constructing the main state we may satisfy only two boundary conditions at each edge. The problem of choosing these conditions is discussed in Section 8.4.

In Section 7.2 we will study only one case of simply supported edges (S_2 in Table 1.1)

$$T_m = v_t = w = M_m = 0 \qquad \text{at} \qquad s = s_1, \quad s = s_2, \tag{7.1.4}$$

where T_m and M_m are the stress-resultant and the stress-couple respectively in the direction orthogonal to the edge, and v_t is the displacement in the direction tangential to the edge (see (1.2.11) and (1.2.12)). To construct the main stress state with an error of order ε^2 we must satisfy the conditions

$$w = \Phi = 0 \qquad \text{at} \qquad s = s_1, \quad s = s_2. \tag{7.1.5}$$

If the shell edge is a plane curve supported by a diaphragm which is rigid in-plane but flexible out-of-plane, then we must reformulate boundary conditions (4) but conditions (5) are valid with an error of the same order ε^2. Some other variants of the boundary conditions may also be reduced to conditions (5) (see Section 8.4).

7.2 Buckling Modes for a Conic Shell

Let $t_1, t_2, t_3 = O(1)$. Then, due to formulae (3.6.11) and (1.3) we have estimates

$$\lambda \sim 1 \qquad \text{at} \qquad t_2 > 0, \tag{7.2.1}$$

$$\lambda \sim h_*^{-1/4} \qquad \text{at} \qquad t_2 = 0, \quad t_3 \neq 0, \tag{7.2.2}$$

and it is sufficient that conditions in estimates (1) and (2) be fulfilled, at least in some subdomain of the neutral surface G.

We start with the case when $t_2 > 0$ in some subdomain of G. Then, buckling occurs such that concavities are elongated in the longitudinal direction from one shell edge to the other. This is the situation where a conic or cylindrical shell is buckling under external pressure (see Section 3.5). In this Section we assume additionally that the shell generatrices are under different loading conditions (the meaning of this assumption will be clarified below) and the buckling mode does not spread over the entire neutral surface but rather is localized near the weakest generatrix.

Under these assumptions we can seek the solution of system (1.2) in the form

$$\begin{aligned} w(s, \varphi, \varepsilon) &= w_* \exp\left\{ i \left(\varepsilon^{-1/2} q \,\zeta + (1/2)\, a \,\zeta^2 \right) \right\}, \\ w_* &= \sum_{n=0}^{\infty} \varepsilon^{n/2} w_n(\zeta, s), \qquad \zeta = \varepsilon^{-1/2}(\varphi - \varphi_0), \\ \lambda &= \lambda_0 + \varepsilon \lambda_1 + \varepsilon^2 \lambda_2 + \dots, \qquad \Im a > 0, \end{aligned} \tag{7.2.3}$$

where $w_n(\zeta, s)$ are polynomials in ζ.

We seek the function Φ in the same form as (3). The value of parameter q which determines the buckling mode oscillation in the direction φ is real. The weakest generatrix is $\varphi = \varphi_0$ and parameter a characterizes the rate of decrease of the depth of the buckling concavities as we go away from the weakest generatrix. We note that mode (3) generalizes solution (4.2.6) for ordinary differential equation (4.2.1). Here solution (3) depends also on variable s.

To determine unknown functions w_n, Φ_n and the values of q, a, φ_0 and λ_n, we substitute solution (3) into system (1.2) and equalize the coefficients by the same powers of $\varepsilon^{1/2}$. We expand the coefficients of system (1.2) depending on φ in a power series of $\varphi - \varphi_0 = \varepsilon^{1/2}\zeta$.

First, it is convenient to express function Φ_* through w_* by virtue of second equation (1.2). We get

$$\begin{aligned}\Phi_* &= -\Delta_s\left[\frac{w_*}{q^4} - \frac{4\,\varepsilon^{1/2}}{q^5}\left(a\,\zeta\,w_* - i\,\frac{\partial w_*}{\partial \zeta}\right) + \right.\\ &\left.+\frac{10\varepsilon}{q^6}\left(a^2\zeta^2 w_* - 2\,i\,a\,\zeta\,\frac{\partial w_*}{\partial\zeta} - i\,a\,w_* - \frac{\partial^2 w_*}{\partial\zeta^2}\right)\right] + \\ &+\frac{2\,i\,\varepsilon}{g\,q^5}\frac{\partial g}{\partial\varphi}\Delta_s w_* + O\left(\varepsilon^{3/2}\right), \qquad \Delta_s = g s^3\frac{\partial^2}{\partial s^2}.\end{aligned} \tag{7.2.4}$$

Now, the first equation of (1.2) gives the sequence of equations for determining functions w_n, which may be written in the form

$$H_0 w_0 = 0, \quad H_0 w_1 + H_1 w_0 = 0, \quad H_0 w_2 + H_1 w_1 + H_2 w_0 = 0, \ \ldots \tag{7.2.5}$$

Here

$$\begin{aligned}H_0 z &= \frac{k^2}{g s^3 q^4}\Delta_s^2\, z + \frac{d q^4}{s^3}\, z - \lambda_0 N z, \quad N z = \frac{q^2 t_2}{s}\, z,\\ H_1 z &= \left(a\,\frac{\partial H_0}{\partial q} + \frac{\partial H_0}{\partial\varphi}\right)\zeta\, z - i\,\frac{\partial H_0}{\partial q}\frac{\partial z}{\partial\zeta},\\ H_2 z &= \frac{1}{2}\left(a^2\,\frac{\partial^2 H_0}{\partial q^2} + 2a\,\frac{\partial^2 H_0}{\partial q\,\partial\varphi} + \frac{\partial^2 H_0}{\partial\varphi^2}\right)\zeta^2 z -\\ &- i\left(a\,\frac{\partial^2 H_0}{\partial q^2} + \frac{\partial^2 H_0}{\partial q\,\partial\varphi}\right)\zeta\,\frac{\partial z}{\partial\zeta} - \frac{1}{2}\,\frac{\partial^2 H_0}{\partial q^2}\left(i\,z + \frac{\partial^2 z}{\partial\zeta^2}\right) -\\ &- \frac{i}{2}\,\frac{\partial^2 H_0}{\partial q\,\partial\varphi}\, z + H_* z - \lambda_1 N z,\\ H_* z &= i\,\lambda_0 q\left(\frac{\partial t_3}{\partial s}\, z + 2 t_3\,\frac{\partial z}{\partial s}\right).\end{aligned} \tag{7.2.6}$$

The substitution of solution (3) in the boundary condition $w = 0$ for $s = s_k(\varphi)$ gives

$$w_0(\zeta, s) = 0, \qquad w_1(\zeta, s) + \zeta\, s' \frac{\partial w_0}{\partial s} = 0,$$
$$w_2(\zeta, s) + \zeta\, s' \frac{\partial w_1}{\partial s} + \frac{\zeta^2}{2}\left((s')^2 \frac{\partial^2 w_0}{\partial s^2} + s'' \frac{\partial w_0}{\partial s}\right) = 0, \ldots \tag{7.2.7}$$

and from the condition $\Phi = 0$ we obtain, taking into account formula (4),

$$\frac{\partial^2 w_0}{\partial s^2} = 0, \qquad \frac{\partial^2 w_1}{\partial s^2} + \zeta\, s' \frac{\partial^3 w_0}{\partial s^3} = 0,$$
$$\frac{\partial^2 w_2}{\partial s^2} + \zeta\, s' \frac{\partial^3 w_1}{\partial s^3} + \frac{\zeta^2}{2}\left((s')^2 \frac{\partial^4 w_0}{\partial s^4} + s'' \frac{\partial^3 w_0}{\partial s^3}\right) - \frac{4\, i\, s'}{q} \frac{\partial^3 w_0}{\partial s^3} = 0. \tag{7.2.8}$$

Let us consider the boundary value problem in the zeroth approximation

$$H_0 w_0 \equiv \frac{k^2(\varphi_0)}{q^4} \frac{\partial^2}{\partial s^2}\left(g s^3 \frac{\partial^2 w_0}{\partial s^2}\right) + \frac{q^4 d}{s^3}\, w_0 - \lambda_0 N w_0 = 0,$$
$$w_0 = \frac{\partial^2 w_0}{\partial s^2} = 0 \quad \text{at} \quad s = s_1(\varphi_0), \quad s = s_2(\varphi_0). \tag{7.2.9}$$

In addition to the principal parameter λ_0 this problem also contains two parameters, q and φ_0. For fixed values of q and φ_0 the minimum eigenvalue λ_0 is a function of these parameters, i.e. $\lambda_0 = f(q, \varphi_0)$. Here, for problem (9) we consider only positive eigenvalues. We denote

$$\lambda_0^0 = \min_{q,\varphi_0} \{f(q, \varphi_0)\} = f(q_0, \varphi_0^0). \tag{7.2.10}$$

Below we will omit the superscript $^{(0)}$ on λ_0^0 and φ_0^0.

The conditions

$$\frac{\partial \lambda_0}{\partial q} = \frac{\partial \lambda_0}{\partial \varphi} = 0 \quad \text{at} \quad q = q_0, \quad \varphi = \varphi_0 \tag{7.2.11}$$

must be fulfilled.

As in the previous Sections (see formulae (4.2.11) and (6.1.8)) we assume that the second differential of function f at point (q_0, φ_0) is a positive definite quadratic form, i.e.

$$d^2 f = \lambda_{qq} dq^2 + 2\, \lambda_{q\varphi} dq\, d\varphi + \lambda_{\varphi\varphi} d\varphi^2 > 0, \tag{7.2.12}$$

where we denote the partial derivatives of f with respect to the corresponding variables for $q = q_0$, $\varphi = \varphi_0$ by λ_{qq}, $\lambda_{q\varphi}$, and $\lambda_{\varphi\varphi}$.

In order to evaluate the derivatives in (11) and (12) we may differentiate problem (9) by parameters q and φ_0. For example,

$$
\begin{gathered}
H_0 w_q + \frac{\partial H_0}{\partial q} w_0 - \frac{\partial \lambda_0}{\partial q} N w_0 = 0, \qquad w_q = \frac{\partial^2 w_q}{\partial s^2} = 0; \\
H_0 w_\varphi + \frac{\partial H_0}{\partial \varphi_0} w_0 - \frac{\partial \lambda_0}{\partial \varphi_0} N w_0 = 0, \\
w_\varphi + s' \frac{\partial w_0}{\partial s} = \frac{\partial^2 w_\varphi}{\partial s^2} + s' \frac{\partial^3 w_0}{\partial s^3} = 0, \qquad s = s_1, \quad s = s_2.
\end{gathered}
\tag{7.2.13}
$$

Problems (13) are non-homogeneous boundary value problems "on spectrum". The existence condition for a solution of a problem "on spectrum"

$$
H_0 z + G(s) = 0, \qquad z + g_{0k} = \frac{\partial^2 z}{\partial s^2} + g_{2k} = 0, \qquad s = s_k \tag{7.2.14}
$$

is the equality

$$
\int_{s_1}^{s_2} \bar{w}_0 G \, ds + \frac{k^2 s^3}{q^4} \left(g_{0k} \frac{\partial^3 \bar{w}_0}{\partial s^3} + g_{2k} \frac{\partial \bar{w}_0}{\partial s} \right) \Bigg|_{s=s_1}^{s=s_2} = 0, \tag{7.2.15}
$$

where the complex conjugate values are denoted by the bar symbol.

Now, equalities (11) transform to

$$
\begin{aligned}
\int_{s_1}^{s_2} \bar{w}_0 \frac{\partial H_0}{\partial q} w_0 ds &= 0, \\
\int_{s_1}^{s_2} \bar{w}_0 \frac{\partial H_0}{\partial \varphi_0} w_0 ds + \frac{k^2 s^3 s'}{q^4} \left(\frac{\partial w_0}{\partial s} \frac{\partial^3 \bar{w}_0}{\partial s^3} + \frac{\partial^3 w_0}{\partial s^3} \frac{\partial \bar{w}_0}{\partial s} \right) \Bigg|_{s_1}^{s_2} &= 0.
\end{aligned}
\tag{7.2.16}
$$

For derivatives in (12) we write only the expression for λ_{qq} which is

$$
-\lambda_{qq} \int_{s_1}^{s_2} \bar{w}_0 N w_0 ds + \int_{s_1}^{s_2} \bar{w}_0 \left(2 \frac{\partial H_0}{\partial q} w_q + \frac{\partial^2 H_0}{\partial q^2} w_0 \right) ds = 0. \tag{7.2.17}
$$

We come now to solution of the sequence of equations (5) taking into account boundary conditions (7) and (8). In the zeroth approximation we get

$$
w_0(\zeta, s) = P_0(\zeta)\, w_0^0(s), \tag{7.2.18}
$$

where the w_0^0 is an eigenfunction for problem (9) under conditions (11), and P_0 is an unknown function to be determined from the following approximations.

In the first approximation we get

$$H_0 w_1 + \left[\zeta\, P_0 \left(a\,\frac{\partial H_0}{\partial q} + \frac{\partial H_0}{\partial \varphi_0}\right) - i\, P_0' \frac{\partial H_0}{\partial q}\right] w_0^0 = 0,$$
$$w_1 + \zeta\, P_0 s' \frac{\partial w_0^0}{\partial s} = \frac{\partial^2 w_1}{\partial s^2} + \zeta\, P_0 s' \frac{\partial^3 w_0^0}{\partial s^3} = 0 \qquad \text{at} \quad s = s_k. \tag{7.2.19}$$

Due to formulae (13)–(16), the existence condition for the solution of problem (19) is analogous to equalities (11) from which the parameters q_0 and φ_0^0 have been found. In particular, it follows that a non-arbitrary generatrix may be introduced as the weakest one in formula (3). From (19) we then obtain

$$w_1(\zeta, s) = P_1(\zeta)\, w_0^0 + \zeta\, P_0(\zeta)(a w_q + w_\varphi) - i\, P_0'(\zeta) w_q, \tag{7.2.20}$$

where w_q and w_φ are solutions of problems (13) for $w_0 = w_0^0$ and both functions P_0 and P_1 are unknown.

In the second approximation we have the equation from which taking into account boundary conditions (7), (8) and due to (15), we get the existence condition for the solution w_2

$$L\, P_0 \equiv -\frac{1}{2}\, \lambda_{qq} P_0'' + b\, \zeta\, P_0' + \left(\eta - \lambda_1 + \frac{1}{2}\, b + c\, \zeta^2\right) P_0 = 0, \tag{7.2.21}$$

where

$$b = -\, i\,(a\,\lambda_{qq} + \lambda_{q\varphi}), \quad 2c = a^2 \lambda_{qq} + 2a\lambda_{q\varphi} + \lambda_{\varphi\varphi},$$
$$\eta = \frac{i}{2z} \left\{ \int_{s_1}^{s_2} \left(w_0^0 \frac{\partial \bar{H}_0}{\partial q}\, \bar{w}_\varphi - \bar{w}_0^0\, \frac{\partial H_0}{\partial q}\, w_\varphi \right) ds + \right.$$
$$\left. + \frac{4k^2 s^3 s'}{q^5} \left(\frac{d w_0^0}{ds} \frac{d^3 \bar{w}_0^0}{ds^3} - \frac{d \bar{w}_0^0}{ds} \frac{d^3 w_0^0}{ds^3} \right)\Bigg|_{s_1}^{s_2} - 2i \int_{s_1}^{s_2} \bar{w}_0^0 H_* w_0^0 ds \right\},$$
$$z = \int_{s_1}^{s_2} \bar{w}_0^0 N w_0^0 ds. \tag{7.2.22}$$

Condition $c = 0$ is necessary for the existence of a polynomial form solution of equation (21). From the square equation $c = 0$ according to (12) we find the unique value of a such that $\Im\, a > 0$.

For

$$c = 0, \quad \lambda_1 = \left(n + \frac{1}{2}\right) b + \eta, \quad n = 0, 1, 2, \ldots \tag{7.2.23}$$

equation (21) has the solution $P_0 = H_n(\zeta)$, where H_n are n-th degree Hermite polynomials. The most interesting case is when $n = 0$ and $H_n \equiv 1$ since in this case the value of λ_1 is a minimum. Indeed, due to (22) and (12) $b > 0$.

The following approximations may be constructed in a similar fashion. We note only that $w_k(\zeta, s)$ are either even or odd polynomials in ζ and the existence condition for w_{2k+2} gives

$$L\, P_{2k} + \lambda_k P_0 + F_{2k}(\zeta) = 0, \qquad k > 0, \tag{7.2.24}$$

where the L is the operator in the left side of equation (21) for $c = 0$.

The value of λ_k may be found from the existence condition for a polynomial form solution of (24). If the polynomials w_k, P_k and F_k are even, then the polynomials w_{k+1}, P_{k+1} and F_{k+1} are odd and vice-versa. That is why the equation $L\, P_{2k+1} + F_{2k+1} = 0$ always has a polynomial solution.

In fact, the values of λ_k $(k \geq 2)$ are not found below. First, because of the difficulty of the calculations and, second, since the value of λ_2 depends on the terms which were omitted in the derivation of system (1.2). That is why the above discussion is important only to justify the expansion by powers of ε (3) of the parameter λ. In [42] the correction of the second order λ_2 is found for a cylindrical shell with a slanted edge under uniform external pressure. For this aim instead of system (1.2) the more exact system (1.2.6) was used.

Separating the real and the imaginary parts in (3) we find that each eigenvalue λ is asymptotically double. The buckling mode has the form

$$\begin{aligned} w &= (\Re\, w_* \cos z - \Im\, w_* \sin z) \exp\left\{-\frac{1}{2}\zeta^2 \Im\, a\right\}, \\ z &= \varepsilon^{-1/2} q_0 \zeta + \frac{1}{2}\zeta^2 \Re\, a + \theta. \end{aligned} \tag{7.2.25}$$

As in Section 4.2 the method used here does not permit the determination of the initial phase $\theta = \text{const}$ which is equal to $0 \leq \theta_1, \theta_2 < 2\pi$, nor does it allow one to distinguish the corresponding eigenvalues.

Remark 7.2. It follows from the method of solution that the results obtained are also valid for a shell which is not closed in the circumferential direction. It is only necessary that the weakest generatrix be far from the shell rectilinear edges.

7.3 Effect of Initial Membrane Stress Resultants

In Section 7.2 we assumed in developing the solution that the initial stress-resultants T_1^0, T_2^0 and S^0 had the same orders and that the stress-resultant T_2^0 was compressive (see (2.1), and also (1.3), which represents the dimensional stress resultants T_1^0, T_2^0 and S^0 through the dimensionless ones t_1, t_2 and t_3).

Under this assumption, the terms $\lambda_0+\varepsilon\lambda_1$ in expansion (2.3) for the critical load depend only on the stress-resultant t_2 and not on t_1 and t_3.

Indeed, in this case the function $w_0^0(s)$ is real and $\eta = 0$ in (2.23). The value of λ_2 depends on t_1 and t_3 (see (2.3)) and to derive λ_2 it is necessary to assemble the next two approximations.

In order to estimate the effect of the stress-resultants t_1 and t_3 on λ we use the following method. We assume temporarily that the variables t_2, εt_1, and εt_3 are of the same order, i.e. the order of variables t_1 and t_3 are greater than the order of t_2. This assumption causes the operators N and H_* in (2.5) to change

$$\begin{aligned} N z &= \frac{q^2 t_2}{s}\, z - i\, q\, \varepsilon \left(\frac{\partial t_3}{\partial s}\, z + 2\, t_3 \frac{\partial z}{\partial s}\right) \equiv N_0 z + i\,\varepsilon\, N_1 z, \\ H_* z &= \varepsilon \left[\frac{\partial}{\partial s}\left(s\, t_1 \frac{\partial z}{\partial s}\right) - \frac{1}{2}\, \frac{\partial^2 t_3}{\partial s\, \partial \varphi_0}\, z\right]. \end{aligned} \tag{7.3.1}$$

The rest of the formulae obtained above are still valid. Taking Nz in the form of (1), we represent the boundary value problem (2.9) as following

$$H_0 w_0 = H_{00} w_0 - i\,\varepsilon\, \lambda_0 N_1 w_0 = 0 \tag{7.3.2}$$

and expand its solution in a power series in ε

$$w_0 = w_{00} + i\,\varepsilon\, w_{01} + \varepsilon^2 w_{02} + \ldots, \qquad \lambda_0 = \lambda_0^0 + \varepsilon^2 \lambda_{02} + \ldots \tag{7.3.3}$$

Here w_{00} and λ_0^0 are the same as in (2.9).

The function $w_{01}(s)$ is the solution of the non-homogeneous problem

$$H_{00} w_{01} - \lambda_0^0 N_1 w_{00} = 0, \quad w_{01} = \frac{\partial^2 w_{01}}{\partial s^2} = 0, \quad s = s_1, s_2, \tag{7.3.4}$$

which is always solvable due to (2.15) because

$$\int_{s_1}^{s_2} w_{00} N_1 w_{00}\, ds = 0. \tag{7.3.5}$$

The coefficient λ_{02} may be found from the solvability condition for the boundary value problem for w_{02} and it is equal to

$$\lambda_{02} = \lambda_0^0 z^{-1} \int_{s_1}^{s_2} w_{00} N_1 w_{01} ds, \quad z = \int_{s_1}^{s_2} w_{00} N_0 w_{00} ds. \tag{7.3.6}$$

For that $\lambda_{02} < 0$ since $z > 0$ and according to (4)

$$\int_{s_1}^{s_2} w_{00} N_1 w_{01} ds = -\int_{s_1}^{s_2} w_{01} N_1 w_{00} ds = -\left(\lambda_0^0\right)^{-1} \int_{s_1}^{s_2} w_{01} H_{00} w_{01} ds < 0. \tag{7.3.7}$$

By using (2.22) and (2.23) we can represent the critical value of the parameter λ in the form

$$\lambda = \lambda_0^0 + \varepsilon \lambda_1 + \varepsilon^2 \lambda_2' + \cdots \tag{7.3.8}$$

Here, $\lambda_1 = b/2$ and b are the same as in (2.23) and $\varepsilon^2 \lambda_2'$ are terms of order ε^2, depending on t_1 and t_3,

$$\begin{gathered} \lambda_2' = \frac{1}{z}\left[\int_{s_1}^{s_2}\left(w_{01}\frac{\partial H_{00}}{\partial q} w_{\varphi 0} + \lambda_0^0 w_{00}\frac{\partial N_1}{\partial q} w_{\varphi 0} - w_{00}\frac{\partial H_{00}}{\partial q} w_{\varphi 1} - \right.\right. \\ \left.\left. - w_{00} H_* w_{00} - \lambda_0^0 w_{00} N_1 w_{01}\right) ds + \frac{4k^2 s^3 s'}{q^5}\left(w_{01}' w_{00}''' - w_{00}' w_{01}'''\right)\Big|_{s_1}^{s_2}\right], \end{gathered} \tag{7.3.9}$$

where

$$w_{\varphi 0} = \frac{\partial w_{00}}{\partial \varphi}, \qquad w_{\varphi 1} = \frac{\partial w_{01}}{\partial \varphi}. \tag{7.3.10}$$

Substituting w_0 from (3) into (2.22) we may obtain (9).

In the particular case, when

$$\frac{\partial H_{00}}{\partial \varphi_0} = 0, \qquad \frac{\partial s_k}{\partial \varphi_0} = 0, \qquad k = 1, 2, \tag{7.3.11}$$

formula (9) is simplified and

$$\lambda_2' = z^{-1} \int_{s_1}^{s_2} \left(w_{00} H_* w_{00} + \lambda_0^0 w_{00} N_1 w_{01}\right) ds. \tag{7.3.12}$$

Now let us discuss the buckling modes described by function (2.3) when there are no compressive stress-resultants T_2^0 on the neutral surface but there are shear stress-resultants S^0.

Let $t_2 \leq 0$ and stress-resultant t_3 be such that εt_3 has order not less than the order of t_2. Note that we do not exclude the case when $t_2 \equiv 0$. Concerning the stress-resultant t_1, we assume that its order is not larger than the order of t_3.

The boundary value problem in the zeroth approximation has the form

$$\frac{k^2}{q^4}\frac{\partial^2}{\partial s^2}\left(s^3\frac{\partial^2 w_0}{\partial s^2}\right)+\frac{q^4 w_0}{s^3}+ \\ +\lambda_0\left(-\frac{q^2 t_2 w_0}{s}+i\,q\left(w_0\frac{\partial t_3'}{\partial s}+2t_3'\frac{\partial w_0}{\partial s}\right)\right)=0, \tag{7.3.13}$$

where $t_3' = \varepsilon t_3$ and the boundary conditions are of the same form as in (2.9). This problem has a real discrete spectrum but the eigenfunctions are complex.

Indeed, let w_0 be an eigenfunction. We multiply (13) by $\bar{w}_0$ and integrate it from s_1 to s_2. Then we get

$$\lambda_0=\frac{\int\limits_{s_1}^{s_2}\left(\frac{k^2 s^3}{q^4}\left|\frac{\partial^2 w_0}{\partial s^2}\right|^2+\frac{q^4}{s^3}\left|w_0\right|^2\right)ds}{\int\limits_{s_1}^{s_2}\left(\frac{q^2 t_2}{s}\left|w_0\right|^2+q\,t_3'\left(u\,\frac{\partial v}{\partial s}-v\,\frac{\partial u}{\partial s}\right)\right)ds}, \tag{7.3.14}$$

where we denote $w_0 = u + iv$. It is possible to show that there exists an infinite set of both positive and negative eigenvalues. As in Section 7.2 we denote the minimum positive eigenvalue λ_0 by $f(q, \varphi_0)$. We make the following evaluations in a manner similar to that in Section 7.2.

7.4 Semi-Momentless Buckling Modes of Cylindrical Shells

Let the shape and size of a cylindrical shell be determined by the following relation

$$s_1(\varphi) \leq s \leq s_2(\varphi), \quad 0 \leq \varphi \leq \varphi_2, \\ d\sigma^2 = R^2\left(ds^2 + d\varphi^2\right), \quad R_2 = \frac{R}{k(\varphi)}, \tag{7.4.1}$$

where the R is the characteristic radius of curvature. For a circular cylindrical shell, R is its radius and $\varphi_2 = 2\pi$, $k(\varphi) \equiv 1$.

The direct transition from a conic to a cylindrical shell is not simple since in (1) $s_1, s_2 \to \infty$ and $\varphi_2 \to 0$. But we still may use all of the relations derived in Sections 7.1-7.3 for the case of a cylindrical shell by replacing all of the factors s^n in them by unity.

For a cylindrical shell the boundary value problem in the zeroth approximation (2.9) has the form

$$\frac{k^2}{q^4}\frac{\partial^2}{\partial s^2}\left(g\,\frac{\partial^2 w_0}{\partial s^2}\right)+dq^4w_0-\lambda_0q^2t_2w_0=0,$$
$$w_0=\frac{\partial^2 w_0}{\partial s^2}=0 \quad \text{at} \quad s=s_1, \quad s=s_2. \tag{7.4.2}$$

All of the formulae in this Chapter are given for non-constant variables E, ν and h through which the variables d and g are expressed according to (1.3). If E, ν and h are constant then we assume that $d=g=1$.

If the variables g, d and t_2 do not depend on s then problem (2) has an explicit solution

$$w_0=\sin\frac{\pi\,(s-s_1)}{l}, \quad \lambda_0=\frac{dq^2}{t_2}+\frac{\pi^4k^2g}{t_2l^4q^6}, \quad l\,(\varphi)=s_2-s_1, \tag{7.4.3}$$

from which we find

$$\lambda_0=\min_{\varphi_0}\left\{\frac{4\pi}{l\,t_2}\left(\frac{d^3k^2g}{27}\right)^{1/4}\right\}, \qquad q_0^8=\frac{3\pi^4k^2g}{l^4d}\bigg|_{\varphi=\varphi_0}, \tag{7.4.4}$$

and the value of λ_1 in (2.3), by virtue of (2.22) and (2.23), is equal to

$$\lambda_1=1/2\left(\lambda_{qq}\lambda_{\varphi\varphi}-\lambda_{q\varphi}^2\right)^{1/2}, \tag{7.4.5}$$

where

$$\lambda_{qq}=\frac{16d}{t_2}, \qquad \lambda_{q\varphi}=8q_0\frac{\partial}{\partial\varphi_0}\left(\frac{d}{t_2}\right),$$
$$\lambda_{\varphi\varphi}=q_0^2\left[\frac{\partial^2}{\partial\varphi_0^2}\left(\frac{d}{t_2}\right)+\frac{l^4d}{3k^2g}\frac{\partial^2}{\partial\varphi_0^2}\left(\frac{k^2g}{l^4t_2}\right)\right], \tag{7.4.6}$$

Here, all of the functions are evaluated at the weakest generatrix $\varphi=\varphi_0$.

We note that the zeroth approximation for the critical load, λ_0, coincides with the value obtained by the Southwell–Papkovich formula for a circular cylindrical shell of constant length $L_1=Rl_0$ and radius Rk_0^{-1} (here $l_0=l\,(\varphi_0)$, $k_0=k\,(\varphi_0)$) (see (3.5.5)).

In the following examples we evaluate both the zeroth approximation and the first correction for a critical load. In Examples 1, 2 and 4 we assume that parameters E, ν and h are constant.

Example 7.1. Consider the buckling of a circular cylindrical shell of moderate length with a sloped edge under homogeneous external pressure p.

Let

$$s_1(\varphi) = 0, \qquad s_2(\varphi) = l\,(\varphi) = l_0 + (\cos\varphi - 1)\tan\beta, \tag{7.4.7}$$

where β is the edge slope angle.

In (3), (4) and (6) we assume that $t_2 = 1$, $k = 1$. Then

$$\begin{aligned} \varphi_0 &= 0, \qquad \lambda_0 = \frac{4\pi}{3^{3/4} l_0}, \qquad q_0^8 = \frac{3\pi^4}{l_0^4}, \\ \lambda_{qq} &= 16, \qquad \lambda_{q\varphi} = 0, \qquad \lambda_{\varphi\varphi} = \frac{4\pi\tan\beta}{3^{3/4} l_0^2} \end{aligned} \tag{7.4.8}$$

and the critical value of pressure is the following

$$p_0 = \frac{4\pi\, E\, h\, \varepsilon^6}{3^{3/4} R\, l_0}\, k_*, \quad k_* = 1 + 0.624 \left(\frac{h^2 \tan^4\beta}{(1-\nu^2)\, R^2}\right)^{1/8} + O\,(h_*^{1/2}). \tag{7.4.9}$$

For $k_* = 1$ we get the Southwell–Papkovich formula which corresponds to the maximum shell length. The second term in k_* takes into account the slope edge influence. For example, for $\beta = 45^0$, $\nu = 0.3$ and $R/h = 400$ the slope edge increases the critical load by 14%.

Example 7.2. Consider the buckling of a non-circular cylindrical shell of moderate length L_1 under homogeneous external pressure p. The weakest generatrix is φ_0 with minimum curvature $k\,(\varphi)$. As a characteristic linear size, R, we take the maximum radius of curvature of the generatrix.

Then $k\,(\varphi_0) = 1$ and we assume

$$t_2 = \frac{1}{k\,(\varphi)}, \qquad l = \frac{L_1}{R}, \qquad p = E\, h\, \varepsilon^6 \lambda\, R^{-1}. \tag{7.4.10}$$

Then due to (4)–(6) we get

$$\begin{aligned} \lambda_0 &= \frac{4\pi}{e^{3/4} l}, \quad \lambda_{qq} = 16, \quad \lambda_{q\varphi} = 0, \quad \lambda_{\varphi\varphi} = \frac{2\pi\, 3^{1/4}}{l} \frac{\partial^2 k}{\partial \varphi_0^2}, \\ p &= \frac{4\pi\, \varepsilon^6 E\, h}{3^{3/4} L_1}\, k_*, \quad k_* = 1 + \frac{3^{7/8}\varepsilon}{\sqrt{2\pi}} \left(l\, \frac{\partial^2 k}{\partial \varphi_0^2}\right)^{1/2} + O\,(\varepsilon^2). \end{aligned} \tag{7.4.11}$$

For $k_* = 1$, as in the previous example, we get the Southwell–Papkovich formula for a circular cylindrical shell of radius R.

Example 7.3. Now we will consider the buckling of a circular cylindrical shell of constant length and variable thickness in the circumferential direction under homogeneous external pressure p.

The weakest generatrix is that with the minimum thickness. In the neighbourhood of this generatrix ($\varphi_0 = 0$) let the shell thickness have the expansion

$$h(\varphi) = h_0 \left(1 + \frac{\alpha \varphi^2}{2}\right). \tag{7.4.12}$$

Suppose that the parameters E and ν are constant, we find that

$$d = \left(1 + \frac{\alpha \varphi^2}{2}\right)^3, \qquad g = \left(1 + \frac{\alpha \varphi^2}{2}\right)^{-1}. \tag{7.4.13}$$

Now (4)–(6) give

$$p = \frac{4\pi}{3^{3/4}} \frac{E h_0 \varepsilon^6}{L_1} k_*, \quad k_* = 1 + 3^{7/8} \left(\frac{2\alpha L_1}{3\pi R}\right)^{1/2} \varepsilon + O(\varepsilon^2). \tag{7.4.14}$$

Example 7.4. Let us look at the buckling of a cantilever circular cylindrical shell of constant length $L_1 = l\,R$ under non-homogeneous (in the circumferential direction) external pressure. Let the load parameter λ describe the pressure p by the expression $p = p_0 t_2$, where $p_0 = \lambda\, E\, h\, \varepsilon^6 R^{-1}$.

We introduce the dimensionless initial stress-resultants t_i as

$$\begin{aligned} t_1 &= -\frac{1}{2} s^2 (\alpha_1 \cos\varphi + 4\alpha_2 \cos 2\varphi), \\ t_2 &= 1 + \alpha_1 \cos\varphi + \alpha_2 \cos 2\varphi, \\ t_3 &= s (\alpha_1 \sin\varphi + 2\alpha_2 \sin 2\varphi), \quad \alpha_1, \alpha_2 \geq 0. \end{aligned} \tag{7.4.15}$$

This problem was solved in [61] by means of the Bubnov–Galerkin method using a large number of coordinate functions and below, we will compare our results with those obtained in [61].

Let $\alpha_1 + \alpha_2 > 0$. Then the weakest generatrix is $\varphi_0 = 0$. Keeping the two first terms in expansion (2.3) we get the critical value of parameter p_0 in the form

$$p = \frac{4\pi}{3^{3/4}} \frac{E\, h\, \varepsilon^6 k_*}{L_1 (1 + \alpha_1 + \alpha_2)}, \qquad k_* = 1 + \frac{\varepsilon \lambda_1}{\lambda_0}. \tag{7.4.16}$$

Using the relations derived in Section 7.2 yields

$$a = \frac{i}{4}\left[\lambda_0(\alpha_1 + 4\alpha_2)\right]^{1/2}, \ \lambda_0 = \frac{4\pi\, 3^{-3/4}}{l\,(1 + \alpha_1 + \alpha_2)}, \ \lambda_1 = \frac{-8\,a\,i}{1 + \alpha_1 + \alpha_2} \tag{7.4.17}$$

and the expression for the parameter k_* which takes into account the variable load is

$$k_* = 1 + 0.624\left(\frac{\alpha_1 + 4\alpha_2}{1+\alpha_1+\alpha_2}\right)^{1/2} y, \quad y = \left(\frac{h^2 L_1^4}{(1-\nu^2)\,R^6}\right)^{1/8}. \tag{7.4.18}$$

We introduce the length of a half-wave concavity in the circumferential direction $\Delta\varphi$ and parameter ρ which takes into account the decrease in depth of the concavity in the circumferential direction away from the weakest generatrix. We obtain

$$\Delta\varphi = \frac{\pi\,\varepsilon}{q_0} = 1.132\,y, \quad \rho = \frac{(\Delta\varphi)^2\,\Im a}{2\varepsilon} = 0.513\left(\frac{\alpha_1+4\alpha_2}{1+\alpha_1+\alpha_2}\right)^{1/2} y. \tag{7.4.19}$$

The depths of the concavities are proportional to $1, e^{-\rho}, e^{-4\rho}, \ldots$

We may use expansions (2.3) under the assumption that the number of waves in the circumferential direction is large. Taking into account the expressions for $\Delta\varphi$ and y we conclude that the proposed method is valid for shells of moderate length. As length L_1 increases, the value of $\Delta\varphi$ also increases and the accuracy of the method decreases.

From a qualitative point of view, formula (19) fully agrees with the results of [61]. Quantitatively, the error in formula (19) essentially depends on the problem parameters. The most unfavourable combination of parameters is $R/h = 100$ and $L_1/R = 10$ for which the error attains 15 % for some values of α_1 and α_2. Where h/R and L_1/R decrease, the error also decreases and for $R/h = 400$, $L_1/R = 2.5$ it is less than 5 %.

In Figure 17.4 of [61] the buckling mode for $R/h = 1000$, $L_1/R = 10$, $\alpha_1 = 0.5$ and $\alpha_2 = 0$ is shown. The localization of the concavities at the weakest generatrix $\varphi_0 = 0$ is clearly seen. In this case, formulae (20) give $\Delta\varphi = 36.4^0$, $\rho = 0.186$. This means that the depths of the concavities are proportional to 1, 0.83, 0.48, 0.19, 0.05. One can see in Figure 17.4 that $\Delta\varphi \approx 30^0$. We note that in this case the error in formula (19) is less than 1 %.

Formula (19) does not account for the initial stress-resultants T_1^0 and S^0. Their contribution may be estimated by use of (3.8) and (3.12). As a result one must add the term $0.181\,(\alpha_1+4\alpha_2)(1+\alpha_1+\alpha_2)^{-1}y^2$, to the value of k_*. This term partly takes into account terms of order ε^2.

In [61] the even functions in φ were used as the coordinate functions for the Bubnov–Galerkin method. That is why the buckling mode obtained, $w(s,\varphi)$, is also an even function in φ. For a load which is very close to the critical load, the odd buckling mode can also occur.

The method of this chapter is generalized in [40] for buckling of cylindrical shells (pipes) joined with a mitred joint under homogeneous external pressure and in [102] for wave propagation in a non-circular cylindrical shell.

7.5 Problems and Exercises

7.1. Consider the buckling of a circular cylindrical shell of moderate length with a sloped edge under homogeneous external pressure p. Plot the graph of the function of critical pressure vs edge slope angle when the other parameters are constant.

7.2. Consider the buckling of a non-circular cylindrical shell with the simple supported edges $s = 0, l$ of moderate length $L_1 = Rl$ under homogeneous external pressure p, when generatrix is an ellipse with semi-axes a_0 and b_0 $(a_0 > b_0)$. Find the critical pressure. Plot the graph for critical pressure vs. eccentricity.

Answer

The critical pressure may be found by formula (4.11) in which

$$\frac{\partial^2 k}{\partial \varphi_0^2} = 3\,(\delta_1^2 - 1),$$

where

$$R = \frac{b_0^2}{a_0}, \qquad k\,(\varphi) = \left(\cos^2\theta + \delta_1^2 \sin^2\theta\right)^{3/2},$$
$$\frac{d\theta}{d\varphi} = k_1, \qquad \delta_1 = \frac{a_0}{b_0} > 1.$$

For $R/h = 400$, $\nu = 0.3$, $l = 1$ and $\delta_1 = 2$ the critical pressure for an elliptic shell is 52 % larger than the critical pressure for a circular shell of radius R.

7.3. Consider the buckling of a non-circular cylindrical shell with the simple supported edges $s = 0, l$ of moderate length $L_1 = Rl$ under homogeneous external pressure p, when generatrix is an ellipse with semi-axes a_0 and b_0 $(a_0 > b_0)$. Note that for an elliptic shell the critical load obtained is asymptotically tetra-multiple since in the neighbourhood of each generatrices $\theta_0 = 0, \pi$ two buckling modes corresponding to nearly identical loads may exist. This problem permits the exact separation of the variables

$$\{w,\, \Phi\}(s, \varphi) = \{w,\, \Phi\}(\varphi) \sin\frac{\pi x}{l}$$

and may be reduced to the system of ordinary differential equations of the eights order.

Construct even and/or odd with respect to the ellipse diameters modes solving the problem numerically on the quarter of ellipse $0 \le \theta \le \pi/2$.

Similar calculations for the problem of shell free vibrations has been made in [81].

Hint. The evenness and oddness conditions have correspondingly the following forms

$$w' = w''' = \Phi' = \Phi''' = 0, \qquad w = w'' = \Phi = \Phi'' = 0.$$

7.4. Prove that the parameter of the critical load λ has expansion (2.3) in integer powers of ε whereas mode w_* is expanded in the series in fractional powers of ε. In particular, prove that the term $\lambda_{3/2}\varepsilon^{3/2}$ in (1.15) vanishes.

Chapter 8

Effect of Boundary Conditions on Semi-momentless Modes

In this chapter we will study the same problems as in Chapter 7 but we will present the solutions in another analytical form. Here we will seek the unknown functions as an integer power series of the small parameter ε.

Various boundary conditions in addition to simply supported are studied and the problem of the separation of the boundary conditions (i.e. of extracting two linear combinations of four boundary conditions), which must be satisfied under the development of the main stress state is solved.

The dependence of the critical load on the boundary conditions is also examined.

8.1 Construction Algorithm for Semi-Momentless Solutions

Consider system (7.1.2)

$$\begin{aligned} \varepsilon^4 \Delta\,(d\,\Delta\, w) + \lambda\,\varepsilon^2 \Delta_t w - \Delta_k \Phi = 0, \\ \varepsilon^4 \Delta\,(g^{-1} \Delta\,\Phi) + \Delta_k w = 0 \end{aligned} \tag{8.1.1}$$

under the same assumptions as in Section 7.1.

In Section 7.2 the solution of this system in the form of a power series (7.2.3) of $\varepsilon^{1/2}$ was developed. Here we will construct the solution in the form of a formal asymptotic integer powers series of ε.

$$w(s,\varphi,\varepsilon) = w^0 \exp\Big\{\frac{i}{\varepsilon}\int_{\varphi_0}^{\varphi} q(\varphi)\,d\varphi\Big\}, \qquad w^0 = \sum_{n=0}^{\infty}\varepsilon^n w_n^0(s,\varphi), \tag{8.1.2}$$
$$\lambda = \lambda_0 + \varepsilon\lambda_1 + \varepsilon\lambda_2 + \cdots$$

The function Φ is represented in the same form. As in the previous Sections we seek the solution which exponentially decreases away from the weakest generatrix $\varphi = \varphi_0$ that is provided by fulfilment of the conditions

$$\Im\, q(\varphi_0) = 0, \qquad \Im\, a > 0, \qquad a = \frac{dq}{d\varphi_0}. \tag{8.1.3}$$

(compare with the conditions in (7.2.3)).

The substitution of solutions (2) into equation (1) leads to a sequence of equations for the unknown functions $q(\varphi)$, $w_n(s,\varphi)$ and $\Phi_n(s,\varphi)$ and the values of λ_n. For simply supported edges (see conditions (7.1.4) and (7.1.5)) in the zeroth approximation we come to the boundary value problem

$$k\,\frac{\partial^2\Phi_0}{\partial s^2} - \frac{q^4 d}{s^3}\,w_0 + \lambda_0 N\,w_0 = 0, \quad k\,\frac{\partial^2 w_0}{\partial s^2} + \frac{q^4}{g s^3}\,\Phi_0 = 0,$$
$$N = \frac{q^2 t_2}{s}; \qquad w_0 = \Phi_0 = 0 \quad \text{at} \quad s = s_1(\varphi), \quad s = s_2(\varphi), \tag{8.1.4}$$

which after the elimination of Φ_0 is comparable to problem (7.2.9). In the solution of problem (7.2.9) we assumed that q and φ are the independent parameters and then the eigenvalue $\lambda_0 = f(q,\varphi)$ was found as a function of these parameters. Then, due to (7.2.10) we searched for the minimum value of λ_0 by q and φ and the corresponding values of λ_0^0, q_0 and φ_0.

Here we assume that the value of λ_0 is fixed and is equal to λ_0^0, the angle φ is the variable parameter and the function $q = q(\varphi)$ is unknown. Then the equality

$$\lambda_0^0 = f(q(\varphi),\varphi) \tag{8.1.5}$$

may be considered as an equation for evaluating $q(\varphi)$. If relations (7.2.11) and (7.2.12) are fulfilled then equation (5) has a unique solution in the neighbourhood of $\varphi = \varphi_0$ satisfying conditions (3).

It follows from the expansion of $f(q,\varphi)$ in formula (5) in a power series in $q - q_0$ and $\varphi - \varphi_0$ that

$$\lambda_{qq}(q-q_0)^2 + 2\lambda_{q\varphi}(q-q_0)(\varphi-\varphi_0) + \lambda_{\varphi\varphi}(\varphi-\varphi_0)^2 + \cdots = 0. \tag{8.1.6}$$

To find a we get the same quadratic equation $c = 0$ (see (7.2.22)).

By substituting solution (2) into equation (1) in the first approximation we obtain

$$
\begin{aligned}
k\,\frac{\partial^2 \Phi_1}{\partial s^2} - \frac{q^4 d}{s^3}\, w_1 + \lambda_0 N\, w_1 + \lambda_1 N\, w_0 + \\
+ \frac{i}{s^3}\left(4q^3 d\,\frac{\partial w_0}{\partial \varphi} + 2\,\frac{\partial}{\partial \varphi}\,(q^3 d)\, w_0\right) - \\
- \frac{i\,\lambda_0}{s}\left(2q\,t_2 \frac{\partial w_0}{\partial \varphi} + \frac{\partial}{\partial \varphi}\,(q\,t_2)\, w_0\right) - \\
- i\,\lambda_0 q\left(2t_3 \frac{\partial w_0}{\partial s} + \frac{\partial t_3}{\partial s}\, w_0\right) &= 0, \\
k\,\frac{\partial^2 w_1}{\partial s^2} + \frac{q^4}{g s^3}\,\Phi_1 - \frac{i}{s^3}\left(\frac{4q^3}{g}\,\frac{\partial \Phi_0}{\partial \varphi} + 2\,\frac{\partial}{\partial \varphi}\left(\frac{q^3}{g}\right)\Phi_0\right) &= 0
\end{aligned}
\tag{8.1.7}
$$

and $w_1 = \Phi_1 = 0$ at $s = s_k(\varphi)$, $k = 1, 2$.

Let $w_0(s,\varphi)$, $\Phi_0(s,\varphi)$ be a non-zero solution of problem (4). Then $V(\varphi)\, w_0$, $V(\varphi)\,\Phi_0$ is also its solution. This allows us to replace in formula (7) w_0 and Φ_0 by $V w_0$ and $V\Phi_0$ respectively. Problem (7) is a non-homogeneous boundary value problem "on spectrum". To obtain the compatibility condition for this problem we multiply the first equation (7) by $\bar{w}_0$, the second equation by $\bar{\Phi}_0$, add them and then integrate with respect to s.

We then obtain the equation for the function $V(\varphi)$

$$
\begin{gathered}
i\,\frac{\partial \lambda_0}{\partial q}\,\frac{dV}{d\varphi} + \frac{i}{2z}\,\frac{d}{d\varphi}\left(\frac{\partial \lambda_0}{\partial q}\, z\right) V + \lambda_1 V = 0, \\
z = \int_{s_1}^{s_2} \bar{w}_0 N\, w_0 ds, \qquad \frac{d}{d\varphi} = \frac{\partial}{\partial \varphi} + \frac{dq}{d\varphi}\,\frac{\partial}{\partial q}.
\end{gathered}
\tag{8.1.8}
$$

Due to (7.2.11) we obtain $\partial\lambda_0/\partial q = 0$ at $\varphi = \varphi_0$ and that is why the solution of equation (8) in the form

$$
V(\varphi) = (\varphi - \varphi_0)^n \Big[v_0 + v_1(\varphi - \varphi_0) + \cdots\Big], \qquad n = 0, 1, \ldots \tag{8.1.9}
$$

exists for

$$
\lambda_1 = -i\Big(a\,\lambda_{qq} + \lambda_{q\varphi}\Big)\Big(n + \frac{1}{2}\Big). \tag{8.1.10}
$$

This result coincides with (7.2.23). In particular, for $n = 0$ we obtain formula (7.4.5)

$$
\lambda_1 = \frac{1}{2}\Big(\lambda_{qq}\lambda_{\varphi\varphi} - \lambda_{q\varphi}^2\Big)^{1/2}. \tag{8.1.11}
$$

We will not consider higher order approximations because these formulae are very complicated and also because system (1) is not sufficiently accurate since it does not contain some terms which effect the second approximation.

8.2 Semi-Momentless Solutions

In Section 8.1 we developed only the functions w and Φ and studied simply supported boundary conditions. In order to examine other types of boundary conditions it is necessary to have expressions for the other functions which describe the stress-strain state of the shell.

In this Section we will find the displacements ($u_1 = u$, $u_2 = v$), the normal rotation angles (γ_i), the tangential stress-resultants (T_i, S), the stress-couples (M_i, H) and the shear stress-resultants (Q_i).

Let us introduce the dimensionless values (marked with prime symbols) with the relations

$$\begin{aligned} (u_i', w') &= \frac{1}{R}(u_i, w), \qquad (T_i', S', Q_i') = \frac{1}{E_0 h_0}(T_i, S, Q_i), \\ (M_i', H') &= \frac{1}{E_0 h_0 R}(M_i, H). \end{aligned} \tag{8.2.1}$$

Later, we will omit the prime symbols for simplicity.

Any of mentioned functions we denote by x and seek them in the form similar to (1.2),

$$x(s, \varphi, \varepsilon) = x^0 \exp\left\{\frac{i}{\varepsilon}\int_{\varphi_0}^{\varphi} q(\varphi)\, d\varphi\right\}, \quad x^0 = \varepsilon^{\alpha_0(x)} \sum_{n=0}^{\infty} \varepsilon^n x_n^0(s, \varphi), \tag{8.2.2}$$

where the α_0 is the index of intensity depending on the choice of the function x and $\alpha_0(w) = \alpha_0(\Phi) = 0$. As a rule, we will find two first terms of series (2).

The second equation in (1.1) gives the relation between the functions Φ^0 and w^0

$$\begin{aligned} \Phi^0 = &-\frac{s^3 k\, g}{q^4}\frac{\partial^2 w^0}{\partial s^2} - \frac{4 i\, \varepsilon\, g\, s^3}{q^5}\frac{\partial}{\partial \varphi}\left(k\,\frac{\partial^2 w^0}{\partial s^2}\right) - \\ &- 2 i\, \varepsilon\, s^3 \frac{\partial}{\partial \varphi}\left(\frac{g}{q^5}\right) k\, \frac{\partial^2 w^0}{\partial s^2}. \end{aligned} \tag{8.2.3}$$

Due to formulae (1.5.6) we get

$$T_1 = \varepsilon^4\left(\frac{1}{s^2}\frac{\partial^2 \Phi}{\partial \varphi^2} + \frac{1}{s}\frac{\partial \Phi}{\partial s}\right), \quad S = -\varepsilon^4 \frac{\partial}{\partial s}\left(\frac{1}{s}\frac{\partial \Phi}{\partial \varphi}\right), \quad T_2 = \varepsilon^4 \frac{\partial^2 \Phi}{\partial s^2}, \tag{8.2.4}$$

and after the substitution of (2) and (3) we obtain

$$\begin{aligned} T_1^0 &= -\varepsilon^2 \frac{q^2}{s^2}\Phi^0 + \frac{i\varepsilon^3}{s^2}\left(2q\frac{\partial\Phi^0}{\partial\varphi} + q'\Phi^0\right) + \\ &\quad \varepsilon^4\left(\frac{1}{s^2}\frac{\partial^2\Phi^0}{\partial\varphi^2} + \frac{1}{s}\frac{\partial\Phi^0}{\partial s}\right), \\ S^0 &= -i\,q\,\varepsilon^3\frac{\partial}{\partial s}\left(\frac{\Phi^0}{s}\right) - \varepsilon^4\frac{\partial}{\partial s}\left(\frac{1}{s}\frac{\partial\Phi^0}{\partial\varphi}\right), \\ T_2^0 &= \varepsilon^4\frac{\partial^2\Phi^0}{\partial s^2}. \end{aligned} \tag{8.2.5}$$

To evaluate the displacements we use formulae

$$\begin{aligned} \frac{1}{s}\left(\frac{\partial v}{\partial\varphi} - kw + u\right) &= \varepsilon_2 = \frac{1}{g}\left(T_2 - \nu T_1\right), \\ \frac{1}{s}\frac{\partial u}{\partial\varphi} + s\frac{\partial}{\partial s}\left(\frac{v}{s}\right) &= \omega = \frac{2(1+\nu)}{g}S \end{aligned} \tag{8.2.6}$$

and according to formula (2) we find

$$\begin{aligned} u^0 &= \frac{\varepsilon^2 s^2 k}{q^2}\frac{\partial}{\partial s}\left(\frac{w^0}{s}\right) + \\ & i\,\varepsilon^3 s^2\frac{\partial}{\partial s}\left(\frac{1}{q^2 s}\frac{\partial}{\partial\varphi}\left(\frac{kw^0}{q}\right) + \frac{1}{qs}\frac{\partial}{\partial\varphi}\left(\frac{kw^0}{q^2}\right)\right) + \\ & \varepsilon^4\left(s^2\frac{\partial}{\partial s}\left(\frac{\nu\Phi^0}{s^2 g}\right) - \frac{2(1+\nu)s}{g}\frac{\partial}{\partial s}\left(\frac{\Phi^0}{s}\right) - \frac{ks^2}{q^4}\frac{\partial}{\partial s}\left(s\frac{\partial}{\partial s}\left(\frac{w^0}{s}\right)\right) - \right. \\ & - s^2\frac{\partial}{\partial s}\left(\frac{1}{sq^2}\frac{\partial}{\partial\varphi}\left(\frac{1}{q}\frac{\partial}{\partial\varphi}\left(\frac{kw^0}{q}\right)\right)\right) + \\ & \left. + \frac{1}{qs}\frac{\partial}{\partial\varphi}\left(\frac{1}{q^2}\frac{\partial}{\partial\varphi}\left(\frac{kw^0}{q}\right) + \frac{1}{q}\frac{\partial}{\partial\varphi}\left(\frac{kw^0}{q^2}\right)\right)\right), \\ v^0 &= -i\,\varepsilon\frac{kw^0}{q} + \frac{\varepsilon^2}{q}\frac{\partial}{\partial\varphi}\left(\frac{kw^0}{q}\right) + \\ & + i\,\varepsilon^3\left(\frac{1}{q}\frac{\partial}{\partial\varphi}\left(\frac{1}{q}\frac{\partial}{\partial\varphi}\left(\frac{kw^0}{q}\right)\right) + \frac{ks^2}{q^3}\frac{\partial}{\partial s}\left(\frac{w^0}{s}\right) - \frac{\nu q\Phi^0}{gs}\right). \end{aligned} \tag{8.2.7}$$

Then, by using formulae

$$
\begin{aligned}
\gamma_1 &= -\frac{\partial w}{\partial s}, \qquad \gamma_2 = -\frac{1}{s}\frac{\partial w}{\partial \varphi}, \\
\varkappa_1 &= \frac{\partial^2 w}{\partial s^2}, \qquad \varkappa_2 = \frac{1}{s^2}\frac{\partial^2 w}{\partial \varphi^2}, \\
\tau &= \frac{\partial}{\partial s}\left(\frac{1}{s}\frac{\partial w}{\partial \varphi}\right), \\
M_1 &= \varepsilon^8 \nu\, d\varkappa_1, \qquad M_2 = \varepsilon^8 d\varkappa_2, \\
H &= \varepsilon^8(1-\nu)\, d\tau, \\
Q_{1*} &= -\frac{1}{s}\frac{\partial}{\partial s}(s\, M_1) + \frac{M_2}{s} - \frac{1}{s}\frac{\partial H}{\partial \varphi} + \lambda\,(t_1\gamma_1 + t_3\gamma_2)\,\varepsilon^6, \\
Q_2 &= -\frac{1}{s}\frac{\partial M_2}{\partial \varphi} + \lambda\,(t_3\gamma_1 + t_2\gamma_2)\,\varepsilon^6,
\end{aligned}
\tag{8.2.8}
$$

we find

$$
\begin{aligned}
\gamma_1^0 &= -\frac{\partial w^0}{\partial s}, \qquad \gamma_2^0 = -\frac{iq}{\varepsilon s}\, w^0 - \frac{1}{s}\frac{\partial w^0}{\partial \varphi}, \\
M_1^0 &= \nu\, M_2^0, \quad M_2^0 = -\frac{\varepsilon^6 d\, q^2}{s^2}\, w^0 + \frac{i\,\varepsilon^7 d}{s^2}\left(2q\,\frac{\partial w^0}{\partial \varphi} + q' w^0\right), \\
H^0 &= i\,\varepsilon^7(1-\nu)\, dq\, \frac{\partial}{\partial s}\left(\frac{w^0}{s}\right) + \varepsilon^8(1-\nu)\, d\,\frac{\partial}{\partial s}\left(\frac{1}{s}\frac{\partial w^0}{\partial \varphi}\right), \\
Q_{1*}^0 &= \varepsilon^6 q^2\left(-\frac{dw^0}{s^3} + \frac{1}{s}\frac{\partial}{\partial s}\left(\frac{\nu\, dw^0}{s}\right) + \frac{(1-\nu)\, d}{s}\frac{\partial}{\partial s}\left(\frac{w^0}{s}\right)\right) + \\
&\quad + \lambda\,\varepsilon^6\left(t_1\gamma_1 + t_3\gamma_2\right), \\
Q_2^0 &= \frac{i\,\varepsilon^5 dq^3}{s^3}\, w^0 + \frac{\varepsilon^6}{qs^3}\left(3\, dq^3\,\frac{\partial w^0}{\partial \varphi} + \frac{\partial}{\partial \varphi}\left(dq^3\right) w^0\right).
\end{aligned}
\tag{8.2.9}
$$

The expressions for the generalized displacements u_m, v_t, w and γ_m and for the corresponding generalized stresses T_m, S_t, Q_m and M_m at edge $s = s_0(\varphi)$ which does not coincide with the line of curvature we find according to formulae

(1.2.11)

$$
\begin{aligned}
u_m^0 &= u^0 \cos\chi + v^0 \sin\chi, \qquad v_t = -u^0 \sin\chi + v^0 \cos\chi, \\
\gamma_m^0 &= \gamma_1^0 \cos\chi + \gamma_2^0 \sin\chi, \qquad \gamma_t^0 = -\gamma_1^0 \sin\chi + \gamma_2^0 \cos\chi, \\
T_m^0 &= T_1^0 \cos^2\chi + 2S^0 \sin\chi\cos\chi + T_2^0 \sin^2\chi, \\
S_t^0 &= (T_2^0 - T_1^0)\sin\chi\cos\chi + S^0 \cos 2\chi, \\
Q_{m*}^0 &= \left(Q_1^0 - \frac{1}{s}\left(\frac{i\,q}{\varepsilon}\,H_{mt}^0 + \frac{\partial H_{mt}^0}{\partial\varphi}\right)\right)\cos\chi + \\
&\quad + \left(Q_2^0 - \frac{\partial H_{mt}^0}{\partial s}\right)\sin\chi + \lambda\,\varepsilon^6\left(t_m\gamma_m^0 + t_\tau\gamma_t^0\right), \qquad (8.2.10) \\
H_{mt}^0 &= (M_2^0 - M_1^0)\sin\chi\cos\chi + H^0\cos 2\chi, \\
Q_1^0 &= \varepsilon^6 q^2\left(-\frac{dw^0}{s^3} + \frac{1}{s}\frac{\partial}{\partial s}\left(\frac{\nu\, dw^0}{s}\right)\right), \\
t_m &= t_1\cos^2\chi + 2t_3\sin\chi\cos\chi + t_2\sin^2\chi, \\
t_\tau &= (t_2 - t_1)\sin\chi\cos\chi + t_3\cos 2\chi, \\
M_m^0 &= M_1^0\cos^2\chi + 2H^0\sin\chi\cos\chi + M_2^0\sin^2\chi.
\end{aligned}
$$

Here, $\boldsymbol{m}$ and $\boldsymbol{t}$ are the geodesic normal and the tangent to curve $s = s_0(\varphi)$, the angle χ is shown in Figure 8.1. The values of Q_{1^0} and Q_{m*}^0 are given for the case when the edge load maintains its original direction.

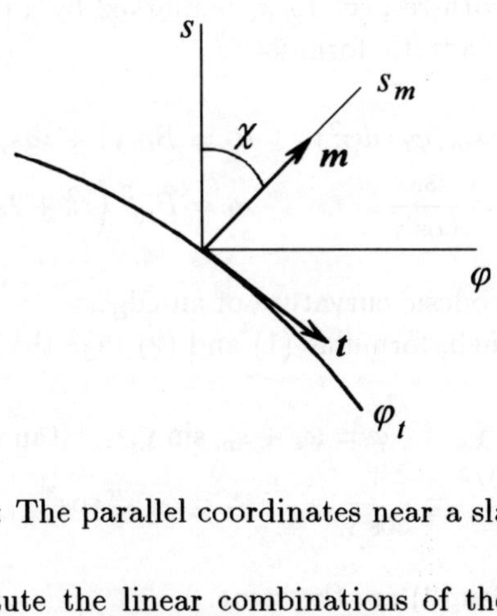

Figure 8.1: The parallel coordinates near a slanted edge.

One must substitute the linear combinations of these solutions and the edge effect solutions into the boundary conditions. The edge effect solutions are developed in the next section.

8.3 Edge Effect Solutions

Consider a conic shell where the edge does not coincide with the line of curvature and is given by the equation $s = s_0(\varphi)$ in the coordinate system s, φ introduced in Section 7.1.

In the neighbourhood of an edge we introduce the orthogonal system of parallel coordinates s_m, φ_t ([24]), where the curve $s_m = 0$ coincides with the shell edge, the lines $\varphi_t = \text{const}$ are the geodesic normals to the edge, φ_t is the value of φ at the edge and $R\, s_m$ is the distance between the point and the edge as measured along the geodesic normal (see Figure 8.1). Let $|\chi| < \frac{\pi}{2}$, i.e. the generatrices are nowhere tangent to the edge.

The coordinates s, φ and the coordinates s_m, φ_t are related by formulae

$$
\begin{aligned}
s &= s\,(s_m, \varphi_t) = \left(s_0^2 + 2\, s_0 s_m \cos\chi + s_m^2\right)^{1/2} = \\
&= s_0 + s_m \cos\chi + \frac{s_m^2}{2 s_0} \sin^2\chi + O\left(s_m^3\right), \\
\varphi &= \varphi\,(s_m, \varphi_t) = \varphi_t + \arcsin\left(\frac{s_m \sin\chi}{s}\right) = \\
&= \varphi_t + \frac{s_m}{s_0} \sin\chi - \frac{s_m^2}{s_0^2} \sin\chi\cos\chi + O\left(s_m^3\right), \\
s_0 &= s_0(\varphi_t), \qquad \chi = \chi\,(\varphi_t), \qquad \tan\chi = -\frac{s_0'}{s_0},
\end{aligned}
\tag{8.3.1}
$$

where the derivative with respect to φ_t is marked by a prime symbol.

The first surface quadratic form is

$$
\begin{aligned}
&d\sigma^2 = R^2\left(ds_m^2 + B_t^2(s_m, \varphi_t)\, d\varphi_t^2\right), \quad && B_t = B_{t0}\left(1 + \varkappa_t s_m\right) + O\left(s_m^3\right), \\
&B_{t0} = \left(s_0^2 + {s_0'}^2\right)^{1/2} = \frac{s_0}{\cos\chi}, && \varkappa_t = B_{t0}^{-3}\left(s_0^2 + 2{s_0'}^2 - s_0 s_0''\right),
\end{aligned}
\tag{8.3.2}
$$

where $R^{-1}\varkappa_t$ is the geodesic curvature of an edge.

For a cylindrical shell, formulae (1) and (2) take the form

$$
\begin{aligned}
&s = s_0 + s_m \cos\chi, \qquad \varphi = \varphi_t + s_m \sin\chi, \qquad \tan\chi = -\, s_0', \\
&B_{t0} = \left(1 + {s_0'}^2\right)^{1/2} = \frac{1}{\cos\chi}, \qquad \varkappa_t = -\, s_0'' \cos^3\chi,
\end{aligned}
\tag{8.3.3}
$$

and the rest of formulae (2) are the same.

The edge effect solutions in the case when the curvilinear coordinates do not coincide with the lines of curvature are constructed in [24, 52].

Let y be one of the functions describing the stress-strain state of a shell. We seek these solutions as the formal series

$$\begin{gathered} y\left(s_m, \varphi_t, \varepsilon\right)=y^e e^{m r\left(\varphi_t\right)} \exp \left\{\frac{i}{\varepsilon} \int_{\varphi_0}^{\varphi} q \, d \varphi\right\}, \\ y^e=\varepsilon^{\alpha_e(y)} \sum_{n=0}^{\infty} \varepsilon^n y_n^e\left(m, \varphi_t\right), \quad m=\frac{s_m}{\mu}, \quad \mu=\varepsilon^2, \end{gathered} \tag{8.3.4}$$

where y_n^e are polynomials in m.

The variability of solutions (4) with respect to s_m is greater than with respect to φ_t (the indices of variation are equal to 1/2 and 1/4 respectively). The functions w_0^e and w_1^e do not depend on m and they are arbitrary functions in φ_t. The functions w_n^e, $(n \geqslant 2)$ are polynomials in m and they contain arbitrary terms depending on φ_t. The number $\alpha_e(y)$ describes the edge effect intensity for function y.

The operators included in equations (1.1) in the coordinate system s_m, φ_t have the form

$$\begin{aligned} \Delta w &= \frac{1}{B_t}\frac{\partial}{\partial s_m}\left(B_t\frac{\partial w}{\partial s_m}\right)+\frac{1}{B_t}\frac{\partial}{\partial \varphi_t}\left(\frac{1}{B_t}\frac{\partial w}{\partial \varphi_t}\right) \simeq \frac{\partial^2 w}{\partial s_m^2}, \\ \Delta_k w &= \frac{1}{B_t}\left[\frac{\partial}{\partial s_m}\left(\frac{B_t}{R_t}\frac{\partial w}{\partial s_m}\right)+\frac{\partial}{\partial s_m}\left(\frac{1}{R_{mt}}\frac{\partial w}{\partial \varphi_t}\right)+\right. \\ &\quad \left. +\frac{\partial}{\partial \varphi_t}\left(\frac{1}{R_{mt}}\frac{\partial w}{\partial s_m}\right)+\frac{\partial}{\partial \varphi_t}\left(\frac{1}{R_m B_t}\frac{\partial w}{\partial \varphi_t}\right)\right] \simeq \\ &\simeq \frac{1}{R_t}\left(\frac{\partial}{\partial s_m}+\frac{R_t}{B_t R_{mt}}\frac{\partial}{\partial \varphi_t}\right)^2 w, \end{aligned} \tag{8.3.5}$$

where

$$\frac{1}{R_t}=\frac{k\cos^2\chi}{s_0}, \qquad \frac{1}{R_{mt}}=-\frac{k\sin\chi\cos\chi}{s_0}, \qquad \frac{1}{R_m}=\frac{k\sin^2\chi}{s_0}. \tag{8.3.6}$$

The approximate expressions (5) are given with an error of order ε^2. All of the solutions found later in this section will be determined with the same accuracy.

After the substitution of (4) into (1.1) we get the equation for r

$$r^4+r_0^4\left(1+\frac{\varepsilon b}{r}\right)^4=0, \quad r_0^4=\frac{g k^2\cos^4\chi}{d s_0^2}, \quad b=\frac{i q s_0'}{s_0 B_{t0}}, \tag{8.3.7}$$

from which we get r

$$r=r_j=r_0\beta_j+\varepsilon b+O\left(\varepsilon^2\right), \qquad \beta_j^4+1=0, \qquad j=1,2,3,4 \tag{8.3.8}$$

and the relations

$$\Phi^e = -\frac{(g\,d)^{1/2}}{\beta_j^2}\,w^e = -\frac{g\,k\,\cos^2\chi}{s_0}\left(\frac{1}{r_j^2}+\frac{2\varepsilon\,b}{r_j^3}\right)w^e,$$
$$w^e = w_0^e + \varepsilon\,w_1^e, \qquad \Phi^e = \Phi_0^e + \varepsilon\,\Phi_1^e. \tag{8.3.9}$$

We find the tangential stress-resultants T_m and S_t by formulae (see (1.5.6))

$$T_m = \varepsilon^4\left[\frac{1}{B_t}\frac{\partial}{\partial\varphi_t}\left(\frac{1}{B_t}\frac{\partial\Phi}{\partial\varphi_t}\right)+\frac{1}{B_t}\frac{\partial B_t}{\partial s_m}\frac{\partial\Phi}{\partial s_m}\right],$$
$$S_t = -\,\varepsilon^4\frac{\partial}{\partial s_m}\left(\frac{1}{B_t}\frac{\partial\Phi}{\partial\varphi_t}\right), \tag{8.3.10}$$

from which we obtain

$$T_m^e = \varepsilon^2\left(r\,\varkappa_t - \frac{q^2}{B_{t0}^2}\right)\Phi^e + \frac{i\,\varepsilon^3}{B_{t0}^2}\left(2q\,\frac{\partial\Phi^e}{\partial\varphi_t} + B_{t0}\frac{\partial}{\partial\varphi_t}\left(\frac{q}{B_{t0}}\right)\Phi^e\right),$$
$$S_t^e = -\,\varepsilon\,\frac{i\,q\,r}{B_{t0}}\,\Phi^e - \varepsilon^2\frac{r}{B_{t0}}\frac{\partial\Phi^e}{\partial\varphi_t}. \tag{8.3.11}$$

To develop the displacements u_m, v_t we use the relations

$$\frac{(1-\nu^2)\,T_m}{g} = \varepsilon_m + \nu\,\varepsilon_t = \frac{\partial u_m}{\partial s_m} - \frac{w}{R_m} +$$
$$+\,\nu\left(\frac{1}{B_t}\frac{\partial v_t}{\partial\varphi_t} + \frac{1}{B_t}\frac{\partial B_t}{\partial s_m}\,u_m - \frac{w}{R_t}\right) \simeq 0, \tag{8.3.12}$$
$$\frac{2\,(1+\nu)\,S_t}{g} = \omega_{mt} = \frac{1}{B_t}\frac{\partial u_m}{\partial\varphi_t} + B_t\,\frac{\partial}{\partial s_m}\left(\frac{v_t}{B_t}\right) + \frac{2w}{R_{mt}}$$

and finally we obtain

$$u_m^e = \varepsilon^2\,\frac{k\,(\sin^2\chi + \nu\,\cos^2\chi)}{r\,s_0}\,w^e - \varepsilon^3\,\frac{2i\,k\,\nu\,q\,\sin\chi\cos^2\chi}{r^2 s_0^2}\,w^e,$$
$$v_t^e = \varepsilon^2\,\frac{2k\,\sin\chi\cos\chi}{r\,s_0}\,w^e +$$
$$+\,\varepsilon^3\left(\frac{2i\,(1+\nu)\,q}{g\,B_{t0}}\,\Phi^e - \frac{i\,q\,k\,(\sin^2\chi + \nu\,\cos^2\chi)}{r^2 s_0\,B_{t0}}\,w^e\right). \tag{8.3.13}$$

The expressions for the values of γ_m, M_m and Q_{m*}, which may be contained in some boundary conditions are the following

$$\gamma_m^e = \varepsilon^{-2} r\,w^e, \qquad M_m^e = \varepsilon^4 d\,r^2 w^e,$$
$$Q_{m*} = Q_m - \frac{1}{B_t}\frac{\partial H_{mt}}{\partial\varphi_t}, \qquad Q_{m*}^e = -\,\varepsilon^2 d\,r^3 w^e. \tag{8.3.14}$$

We note that in the case when $\tan\chi = s_0' = 0$ (i.e. the edge coincides with the line of curvature or is tangent to it) the order of the value v_t^e decreases.

8.4 Separation of Boundary Conditions

In this Section we will discuss the problem of the determination of the two main boundary conditions from four boundary conditions (1.2.16)

$$\begin{aligned} u_m &= 0 \quad \text{or} \quad T_m = 0, \\ v_t &= 0 \quad \text{or} \quad S_t = 0, \\ w &= 0 \quad \text{or} \quad Q_{m*} = 0, \\ \gamma_m &= 0 \quad \text{or} \quad M_m = 0 \quad \text{at} \quad s = s_0, \end{aligned} \tag{8.4.1}$$

which must be satisfied in developing the semi-momentless solutions. Here, we will consider only the simple case when the generalized displacement or generalized force corresponding to it is equal to zero at the edge. In the other words, it is assumed that the boundary stiffness in each direction coinciding with the axes of the trihedron $(\bar{m}, \bar{t}, \bar{n})$, is equal either to infinity or to zero.

Note that elastic support of the type $u_m + \beta T_m = 0$ is not considered.

In Table 8.1 the intensity indices of the semi-momentless solutions α_0 which are found in Section 8.2 and of the edge effect solutions α_e (see Section 8.3) are given for the functions included in conditions (1).

We make a difference between the case of $\sin\chi = 0$ and $\sin\chi \neq 0$. The first case takes place if a shell edge coincides with the line of curvature. In particular, if a shell of revolution is limited by a parallel. We have the same case when $\sin\chi = 0$ only at the point of the intersection of the edge and the weakest parallel $\varphi = \varphi_0$.

The determination of the main boundary conditions for shells of revolution is discussed in [57]. For the case of a free edge, the main boundary conditions are found in [24, 25, 123].

The relative intensity of semi-momentless solutions (compared with edge effect solutions) is characterized by the value of $\Delta\alpha = \alpha_0 - \alpha_e$ (see Table 8.1). For the main boundary conditions, the value of $\Delta\alpha$ must be strictly less than for the additional boundary conditions. For that to be true the following *condition, A*, must be fulfilled (see [57]):

in the zeroth approximation both the semi-momentless solutions included in the main boundary conditions and the edge effect solutions included in the additional boundary conditions must be independent.

In order to satisfy this condition, as a rule, it is necessary to use linear combinations of boundary conditions. Let $x_1 = x_2 = 0$ be the independent

Table 8.1: The intensity indices

I	II	III	IV	V	VI
x	α_0	$x_0^0 + \varepsilon\, x_1^0$	α_e	$x_0^e + \varepsilon\, x_k^e$	$\Delta\alpha$
		$s_0' = 0, \qquad \sin\chi = 0$			
u	2	$\dfrac{\partial}{\partial s}\left(\dfrac{w^0}{s}\right)$	2	$r^{-1}w^e$	0
v	1	w^0	3	r^2w^e	-2
w	0	w^0	0	w^e	0
γ_1	0	$\dfrac{\partial w^0}{\partial s}$	-2	$r\,w^e$	2
T_1	2	Φ^0	2	$\left(r^{-1} + b_1 r^{-2}\right) w^e$	0
S	3	$\dfrac{\partial}{\partial s}\left(\dfrac{\Phi^0}{s}\right)$	1	$r^{-1}w^e$	2
Q_{1*}	6	$w^0 + a_1\dfrac{\partial w^0}{\partial s}$	2	r^3w^e	4
M_1	6	w^0	4	r^2w^e	2
		$s_0' \neq 0, \qquad \sin\chi \neq 0$			
u_m	1	$w^0 + \varepsilon\, a_2\dfrac{\partial w^0}{\partial s}$	2	$\left(r^{-1} + \varepsilon\, b_2 r^{-2}\right) w^e$	-1
v_t	1	w^0	2	$\left(r^{-1} + \varepsilon\, b_3 r^{-2}\right) w^e$	-1
w	0	w^0	0	w^e	0
γ_m	-1	$w^0 + \varepsilon\, a_3\dfrac{\partial w^0}{\partial s}$	-2	$r\,w^e$	1
T_m	2	Φ^0	2	$\left(r^{-1} + \varepsilon\, b_4 r^{-2}\right) w^e$	0
S_t	2	$\Phi^0 + \varepsilon\, a_4\dfrac{\partial \Phi^0}{\partial s}$	1	$\left(r^{-1} + \varepsilon\, b_5 r^{-2}\right) w^e$	1
Q_{m*}	5	$w^0 + \varepsilon\, a_5\dfrac{\partial w^0}{\partial s}$	2	r^3w^e	3
M_m	6	$w^0 + \varepsilon\, a_6\dfrac{\partial w^0}{\partial s}$	4	r^2w^e	2
u	2	$\dfrac{\partial}{\partial s}\left(\dfrac{w^0}{s}\right) + \varepsilon\, a_7\dfrac{\partial^2}{\partial s^2}\left(\dfrac{w^0}{s}\right)$	2	$\left(r^{-1} + \varepsilon\, b_6 r^{-2}\right) w^e$	0
v	1	$w^0 + \varepsilon\, a_8\dfrac{\partial w^0}{\partial s}$	2	$\left(r^{-1} + \varepsilon\, b_7 r^{-2}\right) w^e$	-1
S_*	3	$\dfrac{\partial}{\partial s}\left(\dfrac{\Phi^0}{s}\right)$	1	$\left(r^{-1} + \varepsilon\, b_8 r^{-2}\right) w^e$	2
		Remark. If $t_3 \neq 0$ for Q_{1*} we assume that $\alpha_0 = 5, \qquad \Delta\alpha = 3.$			

linear combinations forming the main boundary conditions and $x_3 = x_4 = 0$ provide the additional boundary conditions. We consider the value of

$$\delta = \min\{\Delta\alpha_3, \Delta\alpha_4\} - \max\{\Delta\alpha_1, \Delta\alpha_2\}, \quad \Delta\alpha_i = \Delta\alpha\,(x_i). \tag{8.4.2}$$

We conclude that the problem of separation of the boundary conditions is solved if $\delta \geqslant 2$.

In order determine whether or not the solutions are linearly dependent in Table 8.1, we represent the functions which, with an error of order ε^2, are proportional to the semi-momentless solutions and to the edge effect solutions respectively. The coefficients of proportionality and the coefficients a_i and b_i may be found from the formulae given in Sections 8.2 and 8.3.

In calculating the entries of column *III* for the semi-momentless solutions, the terms containing derivatives of unknown functions with respect to φ have been eliminated. For example, consider the value of v^0 at $s_0' \neq 0$. Retaining the first two terms in relation (2.7) we obtain

$$v^0 = C_1 w^0 + C_2 \frac{\partial w^0}{\partial \varphi}, \quad C_i = C_i(\varphi, \varepsilon), \quad C_2 = O\,(\varepsilon\, C_1), \tag{8.4.3}$$

If the boundary condition $v^0 = 0$ is introduced at edge $s = s^0$ then $w^0(s_0(\varphi), \varphi) = O\,(\varepsilon)$. Differentiating this relation with respect to φ we obtain

$$\frac{\partial w^0}{\partial \varphi} + s_0' \frac{\partial w^0}{\partial s} = O\,(\varepsilon), \tag{8.4.4}$$

and then

$$v^0 = C_1 w^0 - C_2 s_0' \frac{\partial w^0}{\partial s} + O\left(\varepsilon^2 C_1 w^0\right). \tag{8.4.5}$$

It follows from Table 8.1 that the semi-momentless solutions v, v_t, w and M_1 are proportional to each other with an error of order ε^2 and solutions u_m, v_t, w, γ_m, Q_{m*} and M_m are proportional to each other with an error of order ε. The edge effect solutions are proportional if they contain the multiplier r (see Column *V*) in equal powers or in powers which differ by four.

Below, the determination of the main boundary conditions is made for all 16 variations of boundary conditions that follow from (1), though not all of these variants are equally interesting for the practical problems.

First, let us study a simple case when the edge tangent at the weakest point is orthogonal to the generatrix ($s_0' = 0$). In this case due to (3.7) and (3.8), $b = 0$ and the values r in (3.4) are proportional to the roots of equation (3.8) $\beta_j^4 + 1 = 0$.

Assuming at the beginning that the edge under consideration $s = s_0(\varphi)$ is $s = s_2(\varphi)$ $(s_2 > s_1)$. Then we can write

$$\beta_{1,2} = \frac{1}{\sqrt{2}} \pm \frac{i}{\sqrt{2}}. \tag{8.4.6}$$

The edge effect integrals which decreases away from the edge $s = s_2$ correspond to these roots.

Let us examine the case of excluding the edge effect solutions from the main boundary condition

$$x^0 + x^e\left(C_1 r_1^\beta + C_2 r_2^\beta\right) = 0, \qquad r_j = r_0 \beta_j, \quad j = 1, 2 \tag{8.4.7}$$

by the use of two additional boundary conditions

$$a_i^0 + a_i^e\left(C_1 r_1^{\alpha_i} + C_2 r_2^{\alpha_i}\right) = 0, \qquad i = 1, 2. \tag{8.4.8}$$

Here, two cases are possible. In the first β is equal to one of the values α_i or differs from it by 4. For example, let $\beta = \alpha_1$. Then the linear combination

$$x^0 - \frac{x^e a_1^0}{a_1^e} = 0 \tag{8.4.9}$$

does not contain any edge effect solutions.

In the second case $\beta \neq \alpha_1, \alpha_2$ and excluding the edge effect solutions we obtain the formula

$$x^0 = x^e \sum_{\substack{i=1,2\\ i\neq j}} (-1)^{n_i} \operatorname{sgn}(\beta_{ji} - \beta_i)\, 2^{c_i} \frac{a_i^0}{a_i^e}\, r_0^{\beta - \alpha_i}, \tag{8.4.10}$$

where

$$\begin{aligned} &\beta - \alpha_i = 4\, n_i + \alpha_i^0, && \alpha_j - \alpha_i = 4\, n_{ji} + \alpha_{ji}, \\ &c_i = \frac{1}{2}\left|\alpha_i^0 - 2\right|(-1)^{\alpha_j^0}, && \{\alpha_i^0, \alpha_{ij}\} = \{1, 2, 3\}, \end{aligned} \tag{8.4.11}$$

and n_i and n_{ji} are integers.

For example, at $\beta = -1$, $\alpha_1 = 0$, $\alpha_2 = 1$ formula (10) has the form

$$x^0 = x^e\left(\frac{\sqrt{2}}{r_0}\frac{a_1^0}{a_1^e} - \frac{1}{r_0^2}\frac{a_2^0}{a_2^e}\right). \tag{8.4.12}$$

Naturally, by means of this method we exclude only the main terms of the edge effect solutions. If we wish to exclude the higher order terms it is necessary to repeat the procedure.

We also neglect the mutual effect of the edge effect solutions at edges s_1 and s_2. That is why below we will only discuss the conditions at one of the edges without dependence on the boundary conditions at the other edge. Below it is assumed by default that $s_0 = s_2 > s_1$.

Remark 8.1. If we consider edge $s_0 = s_1 < s_2$ then in (10) we must change the value r_0 by $-r_0$. If $s_0' \neq 0$ then in (3.7) $b \neq 0$ and formula (10) must be corrected. In the case when $s_0 = s_2$, the value of r_0 must be changed by $r_0 + \varepsilon\, b\sqrt{2}$ and in the case when $s_0 = s_1$ it must be changed by $-r_0 + \varepsilon\, b\sqrt{2}$.

The 16 boundary condition variants considered here may be split into four groups depending on which boundary conditions are to be satisfied in the integration of system (1.4)

1) the **clamped support** group consists of 6 variants (1111, 1110, 1101, 1100, 1011, 1010) and in the zeroth approximation leads to the conditions

$$w_0 = \frac{\partial w_0}{\partial s} = 0; \tag{8.4.13}$$

2) the **simple support** group also consists of 6 variants (0111, 0110, 0101, 0100, 0011, 0010)

$$w_0 = \Phi_0 = 0; \tag{8.4.14}$$

3) the **weak support** group contains 2 variants (1001, 1000)

$$\frac{\partial}{\partial s}\left(\frac{w_0}{s}\right) = \frac{\partial}{\partial s}\left(\frac{\Phi_0}{s}\right) = 0; \tag{8.4.15}$$

4) and the **free edge** group also contains 2 variants (0001, 0000)

$$\Phi_0 = \frac{\partial \Phi_0}{\partial s} = 0. \tag{8.4.16}$$

We can write (1.2) for the loading parameter

$$\lambda = \lambda_0 + \varepsilon\,\lambda_1 + \varepsilon^2\lambda_2 + \cdots$$

The groups differ from each other by the main terms λ_0. But it occurs that within one group the values of λ_0 and λ_1 coincide and in all cases the value of λ_1 is determined by (1.11).

The difference between the boundary conditions within one group are revealed in the terms $\varepsilon^2\lambda_2$. Below, for the clamped support and simple support groups, we can write the main boundary conditions more accurately by keeping the terms of order ε^2 which allow us to evaluate the value of λ_2. For the weak support and free edge support groups this cannot be done since the accuracy of the edge effect solutions developed in Section 8.3 is not sufficient. Also, such boundary conditions are quite rare in real buckling problems.

It should also be noted that in most cases we specify the group by the tangential boundary conditions. The variants 1011, 1010, 0011 and 0010, in which the tangential condition $S = 0$ is the additional boundary condition are clearly exceptions.

We will now study the clamped support group in detail. For variants 1111, 1110, 1101 and 1100 the main boundary conditions are the tangential conditions

$$u^0 = v^0 = 0, \tag{8.4.17}$$

which in the zeroth approximation by virtue of (2.7) lead to conditions (13). According to (2.7) we get

$$w^0 = O\left(\varepsilon^2\right), \qquad \frac{\partial w^0}{\partial s} = O\left(\varepsilon^2\right) \quad \text{at} \quad s = s_0. \tag{8.4.18}$$

From this point onward we will normalize the solutions assuming that away from the edges the deflection w^0 is of order unity.

We begin with the case of clamped support 1111 (in other words $u = v = w = \gamma_1 = 0$). According to estimate (18), the edge effect intensity is ε^2 times less than what is shown in Table 8.1 and

$$\Delta\,\alpha\,(u) = 0, \quad \Delta\,\alpha\,(v) = -2, \quad \Delta\,\alpha\,(w) = 2, \quad \Delta\,\alpha\,(\gamma_1) = 4, \tag{8.4.19}$$

i.e. due to (2) $\delta = 2$ for conditions (17). In order to make conditions (17) more accurate we eliminate the edge effect in condition $u = 0$ by means of the relations

$$\begin{gathered} u^0 + \varepsilon^2\,\frac{k\,\nu}{s_0}\left(r_1^{-1}C_1 + r_2^{-1}C_2\right) = 0, \\ w^0 + C_1 + C_2 = 0, \qquad r_1 C_1 + r_2 C_2 = 0. \end{gathered} \tag{8.4.20}$$

According to formula (12) we can more precisely determine the main boundary conditions which take into account terms of order ε^2

$$u^0 - \varepsilon^2\,\frac{k\,\nu\,\sqrt{2}}{s_0 r_0}\,w^0 = 0, \qquad v^0 = 0. \tag{8.4.21}$$

It follows from Table 8.1 that the edge effect intensity of function v is two orders of magnitude smaller (with respect to ε) than the intensity of function u and the condition $v^0 = 0$ does not change.

For the variant 1110 (ie. $u = v = w = M_1 = 0$) similar calculations give

$$u^0 - \varepsilon^2 \frac{k\,\nu}{s_0 r_0 \sqrt{2}}\, w^0 = 0, \qquad u^0 = 0. \tag{8.4.22}$$

For the remaining two variants, 1101 and 1100, (i.e. $u = v = Q_{1*} = 0$, $\gamma_1 = 0$ or $M_1 = 0$), conditions (17) may be used with an error of order ε^5. To estimate the error, we take into account the previously noted decay of the intensity of the edge effect by virtue of (18) and note that the edge effect solutions for the functions u and Q_{1*} are proportional to each other. The parameters λ differ in cases 1101 and 1100 by the same order of ε^5

In cases 1011 and 1010 (ie. $u = S_1 = w = \gamma_1 = 0$ or $M_1 = 0$) the conditions

$$u^0 = w^0 = 0, \tag{8.4.23}$$

are main ones and $\delta = 2$ due to (2) and Table 8.1.

In case 1011, conditions (23) after being made more precise are the following

$$u^0 + \varepsilon \frac{i\,\nu\, s_0}{g\, q} S^0 = 0, \quad w^0 + \frac{i\, s_0^2 r}{\sqrt{2}\,\varepsilon\, g\, q\, k} S^0 = 0, \tag{8.4.24}$$

and in case 1010, the second term in the left hand side of the second condition (24) increases by a factor of two.

Now let us study the simple support group and write the modified main boundary conditions for all 6 variants

$$\begin{aligned}
&T_1^0 - \varepsilon^2 \frac{g\,k}{s_0^2 r_0}\left(\frac{q^2}{s_0 r_0} - \sqrt{2}\,\varkappa_t s_0\right) w^0 - \\
&\quad -\varepsilon^4 \frac{g\,k}{s_0^2 r_0^2}\left(\varkappa_t s_0 - \frac{\sqrt{2}\,q^2}{s_0 r_0}\right)\frac{\partial w^0}{\partial s} = 0, \quad v^0 = 0 \quad (0111);\\
&T_1^0 + \varepsilon^2 \frac{g\,k\,\varkappa_t}{s_0 r_0 \sqrt{2}}\, w^0 = 0, \qquad v^0 = 0 \quad (0110);\\
&T_1^0 + \varepsilon^4 \frac{g\,k\,q^2}{s_0^3 r_0^3 \sqrt{2}}\frac{\partial w^0}{\partial s} = 0, \qquad v^0 = 0 \quad (0101);\\
&T_1^0 = 0, \qquad v^0 = 0 \quad (0100);\\
&T_1^0 + \varepsilon^4 \frac{g\,k\,q^2}{s_0^3 r_0^3 \sqrt{2}}\frac{\partial w^0}{\partial s} - \varepsilon \frac{i\, s_0}{q}\left(\varkappa_t - \frac{q^2}{s_0^2 r_0 \sqrt{2}}\right) S^0 = 0,\\
&w^0 - \varepsilon^2 \frac{1}{r_0\sqrt{2}}\frac{\partial w^0}{\partial s} + \frac{i\, s_0^2 r_0}{\varepsilon\, g\, k\, q\, \sqrt{2}} S^0 = 0 \qquad (0011);\\
&T_1^0 - \varepsilon \frac{i\, s_0 \varkappa_t}{q} S^0 = 0, \quad w^0 + \frac{i\, s_0^2 r_0 \sqrt{2}}{\varepsilon\, g\, q\, k} S^0 = 0 \qquad (0010).
\end{aligned} \tag{8.4.25}$$

It follows from these formulae that for the four first variants, the tangential boundary conditions are the main ones and for the remaining two variants that the non-tangential condition, $w^0 = 0$, becomes the most important. The correction terms in conditions (25) are of order ε^2 compared to the main ones. Conditions $T_1^0 = v^0 = 0$ in case 0100 provide an error of order ε^4.

Now we will briefly discuss the problem of separation of the boundary conditions at a slanting edge ($s_0' \neq 0$). As in the case when $s_0' = 0$, the boundary conditions split into four groups

1) the clamped support group (1111, 1110, 1101, 1100),
2) the special group (1011, 1010),
3) the simple support group (0111, 0110, 0101, 0100, 0011, 0010, 1001,1000),
4) the free edge group (0001, 0000).

In the zeroth approximation for the clamped support group, the simple support group and the free edge group, the main boundary conditions (13), (14) and (16) are the same as in the case when $s_0' = 0$, but the redistribution of the boundary conditions between the groups must be noted. Variants 1011 and 1010, which previously belonged to the clamped support group, now form a special group for which the main boundary conditions in the zeroth approximation have the form

$$w^0 = \frac{\partial w^0}{\partial s} - a_i\,\Phi^0 = 0. \tag{8.4.26}$$

The variants 1001 and 1000 which earlier formed the weak support group are now contained in the simple support group. Here we keep only the correction terms in the main boundary conditions that are of order ε compared to the main terms, unlike the case $s_0' = 0$ when we keep the corrections terms up to order ε^2.

Let us first study the clamped support group. It follows from Table 8.1 at $s_0' \neq 0$ that the main terms of solutions u_m^0 and v_t^0 are linearly dependent. That is why we introduced their linear combinations u^0 and v^0 (see Table 8.1) for which we can find the edge effect solutions by means of formulae

$$u^e = u_m^e \cos\chi - v_t^e \sin\chi, \quad v^e = u_m^e \sin\chi + v_t^e \cos\chi. \tag{8.4.27}$$

In all of the variants the conditions $u^0 = v^0 = 0$ dominate and according to (2.7) they give with an error of order ε^2

$$w^0 = \frac{\partial w^0}{\partial s} - \frac{2i\,\varepsilon\, s_0'}{q}\,\frac{\partial^2 w^0}{\partial s^2} = 0. \tag{8.4.28}$$

In order to derive the main boundary conditions (26) for the special group it is necessary to exclude the edge effect solutions. This may be done by means of formula (10). As a result we can write for variant 1011

$$u_m^0 = S_t^0 - \frac{\varepsilon\, i \sqrt{2}\, k\, g\, q\, \cos^3\chi}{r_0 s_0^3}\, w^0 = 0 \quad (1011) \tag{8.4.29}$$

from which we get conditions (26) with

$$a_1 = \frac{q^2 r_0 \sin^2\chi}{\sqrt{2}\, k\, g\, s_0 \cos\chi}. \tag{8.4.30}$$

In a similar way we obtain $a_2 = 2\, a_1$ in formula (26) for variant 1010. Here we are neglecting terms of order ε.

For the simple support group we find that, with an error of order ε^2, for variants 0111, 0110, 0101, 0100, 0011, and 0010, the conditions $T_m^0 = v_t^0 = 0$ (or $w^0 = 0$) are dominant and that leads to conditions (14). For variants 1001 and 1000 the main conditions are $u_m^0 = S_t^0 = 0$ which due to (2.10), (2.7) and (2.5) take on the form

$$w^0 + \frac{i\,\varepsilon\, s_0}{q\, \sin\chi \cos\chi}\, \frac{\partial w^0}{\partial s} = \Phi^0 - \frac{i\,\varepsilon\, s_0}{q\, \sin\chi \cos\chi}\, \frac{\partial \Phi^0}{\partial s} = 0. \tag{8.4.31}$$

Finally we study the free edge conditions 0000 (i.e. $T_m^0 = S_t^0 = Q_{m*}^0 = M_m^0 = 0$). According to formulae (2.5) and (2.10) solutions T_m^0 and S_t^0 are linearly dependent (the main terms) and that is why instead of S_t^0 we use the linear combination

$$S_*^0 = S^0 \cos\chi + T_2^0 \sin\chi. \tag{8.4.32}$$

According to Table 8.1, the difference between the orders of the semi-momentless solutions and the orders of the edge effect solution are equal to each other

$$\Delta\,\alpha\,(T_m) = 0, \quad \Delta\,\alpha\,(S_*) = 2, \quad \Delta\,\alpha\,(Q_m*) = 3, \quad \Delta\,\alpha\,(M_m) = 2, \tag{8.4.33}$$

from which it follows that the condition $T_m^0 = 0$ is the dominant one.

To derive the second main condition we eliminate from function S_* the edge effect solutions (in the zeroth and first approximations) by using the rest of the boundary conditions. As a result we obtain the linear combination

$$S_*^0 + \frac{i\, q}{\varepsilon\, k}\, Q_{m*}^0 + \frac{q^2 \sin\chi}{\varepsilon^2 k\, s_0}\, M_m^0 = 0, \tag{8.4.34}$$

for which $\Delta\alpha = 0$. After the transformations, the main boundary conditions, with an error of order ε^2, take on the form

$$\Phi^0 = \frac{\partial \Phi^0}{\partial s} - \frac{2i\,\varepsilon\, d\, q^3 s_0'}{k\, s_0^3}\, w^0 - \frac{i\,\varepsilon\, q\, \lambda_0}{k}\left(t_3 - \frac{t_2 s_0'}{s_0}\right)\frac{\partial w^0}{\partial s} = 0. \tag{8.4.35}$$

We note that, under the construction of conditions (34) it is necessary to use the refined edge effect solutions (see Remark 8.1).

8.5 The Effect of Boundary Conditions on the Critical Load

In Section 8.4 some variants of the main boundary conditions are introduced. We can write them in the general form

$$l_{ik}^0 + \varepsilon\, l_{ik}^1 + \cdots = 0 \quad \text{at} \quad s = s_k(\varphi), \tag{8.5.1}$$

where

$$l_{ik}^j = l_{ik}^j(w^0, \Phi^0) = A_{ik}^j w^0 + B_{ik}^j \frac{\partial w^0}{\partial s} + C_{ik}^j \Phi^0 + D_{ik}^j \frac{\partial \Phi^0}{\partial s}, \qquad i,k = 1,2, \tag{8.5.2}$$

and the coefficients A_{ik}^j, B_{ik}^j, ... may depend on φ and q.

Following the transformations of Section 8.1 we find that the boundary value problem in the zeroth approximation consists of equations (1.4) and the boundary conditions

$$l_{ik}^0(w_0, \Phi_0) = 0 \quad \text{at} \quad s = s_k(\varphi). \tag{8.5.3}$$

Conditions (3) may be of one of the types (4.13)–(4.16) and (4.25) and in all the cases the boundary value problem in the zeroth approximation is self-adjoint.

In the first approximation we come to the boundary value problem consisting of equations (1.7) and the boundary conditions

$$l_{ik}^0(w_1, \Phi_1) + l_{ik}^1(w_0, \Phi_0) = 0 \quad \text{at} \quad s = s_k(\varphi). \tag{8.5.4}$$

As in Section 8.1 we introduce the function $V(\varphi)$ and write the compatibility condition for the first approximation

$$i\,\frac{\partial \lambda_0}{\partial q}\frac{dV}{d\varphi} + \frac{i}{2z}\frac{d}{d\varphi}\left(z\,\frac{\partial \lambda_0}{\partial q}\right) V + (\lambda_1 - \eta)\, V = 0. \tag{8.5.5}$$

The value of $\partial\lambda_0/\partial q = 0$ at $\varphi = \varphi_0$. The solution of this equation of type (1.9) which is regular at $\varphi = \varphi_0$ exists if

$$\lambda_1 = \eta + (1/2 + n)\left(\lambda_{qq}\lambda_{\varphi\varphi} - \lambda_{q\varphi}^2\right)^{1/2}. \tag{8.5.6}$$

Here, the value of η depends on the type of boundary conditions in the zeroth approximation. For boundary conditions (4.13)–(4.16) it is equal to

$$\begin{aligned} \eta \;=\; & -\frac{1}{z}\left\{\frac{2s'q^3 i}{s^3}\left(\frac{|\Phi_0|^2}{g} - d\,|w_0|^2\right) + i\,q\,\lambda_0 |w_0|^2\left(\frac{t_2 s'}{s} - t_3\right) + \right. \\ & \left.\left. + k\left(\frac{\partial \Phi_1}{\partial s}\bar{w}_0 - \Phi_1\frac{\partial \bar{w}_0}{\partial s} + \frac{\partial w_1}{\partial s}\bar{\Phi}_0 - w_1\frac{\partial \bar{\Phi}_0}{\partial s}\right)\right\}\right|_{s_1}^{s_2}. \end{aligned} \tag{8.5.7}$$

In this Section we will not study boundary conditions (4.25). From the physical sense of the problem η is real. Direct calculations show that in all of the cases considered above (except for the special group 1011 and 1010 at $s' \neq 0$ which have not been considered) we get

$$\eta = 0, \tag{8.5.8}$$

i.e. the critical load parameter may be found by the formula

$$\lambda = \lambda_0 + \varepsilon\,\lambda_1 + O\left(\varepsilon^2\right), \quad \lambda_1 = \frac{1}{2}\left(\lambda_{qq}\lambda_{\varphi\varphi} - \lambda_{q\varphi}^2\right)^{1/2}. \tag{8.5.9}$$

If we take the boundary conditions contained in one group both for $s' = 0$ and for $s' \neq 0$ and evaluate λ, then the differences in the results is of order ε^2.

The complete development of the term $\varepsilon^2\lambda_2$ in the expansion for λ in ε is a complex problem the solution of which one must start by using a more precise initial system than system (1.1). We will not solve this problem here and the discussion will be limited to the development of the $\varepsilon^2\lambda_2'$ portion of this term which inside one group of boundary conditions depends on the particular choice of boundary conditions. Here we will assume that $s' = 0$, i.e. the edge is straight.

We can write the regularity condition for the second approximation (which is similar to conditions (6) and (7)) and keep only the terms which depend on the boundary conditions. Then we get

$$\begin{aligned} \varepsilon^2\lambda_2' &= \varepsilon^2(\eta_2 - \eta_1), \\ \eta_{2,1} &= -\frac{k}{z}\left(\frac{\partial\Phi_2}{\partial s}\,\bar{w}_0 - \Phi_2\,\frac{\partial\bar{w}_0}{\partial s} + \frac{\partial w_2}{\partial s}\,\bar{\Phi}_0 - w_2\,\frac{\partial\bar{\Phi}_0}{\partial s}\right)\Bigg|_{s=s_{2,1}}, \end{aligned} \tag{8.5.10}$$

where η_2 and η_1, which are independent of each other, take into account the boundary condition effect at $s = s_2$ and at $s = s_1$. The formulae for η_2 are given below.

Remark 8.2. For specific boundary condition we represent the loading parameter in the form

$$\lambda = \lambda_0 + \varepsilon\,\lambda_1 + \varepsilon^2(\lambda_2' + \lambda_2'') + O(\varepsilon^3),$$

where the term λ_2' is determined above and the term λ_2'' depending on the error of system (1.1) is unknown. Since the terms λ_0, λ_1, and λ_2'' are the same for the boundary conditions from one group the differences $\lambda_{2i}' - \lambda_{2j}'$ may be used to analyze the effect of the boundary conditions within one group. Here indexes i and j are the numbers of boundary conditions within the group. As it follows from relation (10) effects λ_{2j}' are to be summarized for the both edges.

If we do not know λ_2'' then the term λ_2' may be omitted in the expression for λ. For a cylindrical shell the value λ_2'' is found in [42]. Numerical results for the various shell parameters and for various boundary conditions (identical for the both edges) are ptrsented in [22]

Consider the clamped support group (4.13). The modified main boundary conditions for all 6 variants contained in the group have the form of (4.17), (4.21), (4.22), (4.24) or generally

$$w^0 = \varepsilon^2 C_1, \qquad \frac{\partial w^0}{\partial s} = \frac{w^0}{s} + \varepsilon^2 C_2, \tag{8.5.11}$$

where C_1 and C_2 linearly depend on Φ_0, $\dfrac{\partial \Phi_0}{\partial s}$. According to formulae (10) we find that

$$\eta_2 = -\frac{k}{z}\left(C_2\bar{\Phi}_0 - C_1\, s\,\frac{\partial}{\partial s}\left(\frac{\bar{\Phi}_0}{s}\right)\right). \tag{8.5.12}$$

By means of formulae (2.7) and (2.5) we can obtain the expressions for C_1 and C_2 for all 6 variants

$$\begin{aligned}
&C_1 = a_1, && C_2 = a_2 + a_3 && (1111);\\
&C_1 = a_1, && C_2 = a_2 + \frac{1}{2}a_3 && (1110);\\
&C_1 = a_1, && C_2 = a_2 && (1101,\ 1100);\\
&C_1 = a_4, && C_2 = a_2 + a_5 && (1011);\\
&C_1 = 2a_4, && C_2 = a_2 + a_5 && (1010),
\end{aligned} \tag{8.5.13}$$

where

$$\begin{aligned}
a_1 &= -\frac{\nu\, q^2}{g\, s\, k}\,\Phi_0, & a_3 &= -\frac{\nu^2 q^4\sqrt{2}}{g\, s^3 k\, r_0}\,\Phi_0,\\
a_4 &= -\frac{s^2 r_0}{g\, k\sqrt{2}}\,\frac{\partial}{\partial s}\left(\frac{\Phi_0}{s}\right), & a_5 &= -\frac{\nu\, q^2}{g\, k}\,\frac{\partial}{\partial s}\left(\frac{\Phi_0}{s}\right).
\end{aligned} \tag{8.5.14}$$

In order to simplify (12) and (13) we use Remark 8.2 and change (13) by

$$\begin{aligned}
&C_1 = 0, && C_2 = a_3 && (1111);\\
&C_1 = 0, && C_2 = \frac{1}{2}a_3 && (1110);\\
&C_1 = 0, && C_2 = 0 && (1101,\ 1100);\\
&C_1 = a_4 - a_1, && C_2 = a_5 && (1011);\\
&C_1 = 2a_4 - a_1, && C_2 = a_5 && (1010),
\end{aligned} \tag{8.5.15}$$

which leads to the following expressions for η_2

$$\eta_2^{(1)} = \frac{\nu^2 q^4 \sqrt{2}}{g\, s^3 r_0 z} \left|\Phi_0\right|^2 \quad (1111);$$

$$\eta_2^{(2)} = 1/2\, \eta_2^{(1)} \qquad (1110); \qquad \eta_2^{(3)} = 0 \quad (1101,\ 1100);$$

$$\eta_2^{(4)} = \frac{\nu\, q^2}{g\, z}\left(\bar{\Phi}_0 \frac{\partial \Phi_0}{\partial s} + \Phi_0 \frac{\partial \bar{\Phi}_0}{\partial s}\right) - \frac{s^3 r_0}{g\, z\sqrt{2}} \left|\frac{\partial}{\partial s}\left(\frac{\Phi_0}{s}\right)\right|^2 \quad (1011);$$

$$\eta_2^{(5)} = \frac{\nu\, q^2}{g\, z}\left(\bar{\Phi}_0 \frac{\partial \Phi_0}{\partial s} + \Phi_0 \frac{\partial \bar{\Phi}_0}{\partial s}\right) - \frac{s^3 r_0 \sqrt{2}}{g\, z} \left|\frac{\partial}{\partial s}\left(\frac{\Phi_0}{s}\right)\right|^2 \quad (1010). \tag{8.5.16}$$

Here r_0 and z are evaluated by (3.7) and (1.8). By calculating η_1 (for edge $s = s_1$), the value of r_0 must be changed by $-r_0$.

It is simple to verify that the larger value of η_2 corresponds to the more stiff support namely

$$\eta_2^{(1)} > \eta_2^{(2)} > \eta_2^{(3)}, \quad \eta_2^{(1)} > \eta_2^{(4)} > \eta_2^{(5)}, \quad \eta_2^{(2)} > \eta_2^{(5)}. \tag{8.5.17}$$

We will now study the simple support group. By differentiating the equality $w(s(\varphi), \varphi) = 0$ at $s' = 0$ twice with respect to φ we obtain

$$\frac{\partial^2 w}{\partial \varphi^2} = -\, s'' \frac{\partial w}{\partial s}, \tag{8.5.18}$$

and then according to (2.7), condition $v^0 = 0$ takes on the form

$$w^0 = \varepsilon^2 \frac{\varkappa_t s^2}{q^2} \frac{\partial w^0}{\partial s}. \tag{8.5.19}$$

By using relations (2.5) and (19) we write conditions (4.25) in the form

$$\Phi^0 = \varepsilon^2 \left(a_1^{(i)} \frac{\partial w^0}{\partial s} + a_2^{(i)} \frac{\partial \Phi^0}{\partial s} \right), \quad w^0 = \varepsilon^2 \left(a_3^{(i)} \frac{\partial w^0}{\partial s} + a_4^{(i)} \frac{\partial \Phi^0}{\partial s} \right), \tag{8.5.20}$$

and then due to (10) we obtain

$$\eta_2^{(i)} = \frac{k}{z}\left(a_1^{(i)} \left|\frac{\partial w_0}{\partial s}\right|^2 + a_2^{(i)} \frac{\partial \Phi_0}{\partial s}\frac{\partial \bar{w}_0}{\partial s} + a_3^{(i)} \frac{\partial w_0}{\partial s}\frac{\partial \bar{\Phi}_0}{\partial s} + a_4^{(i)} \left|\frac{\partial \Phi_0}{\partial s}\right|^2 \right). \tag{8.5.21}$$

Here the index i ($i = 1, 2, \dots, 6$) means the number of the boundary condition variant (using the same sequence of variants as in (4.25)). For simplicity we use Remark 8.2 and give only the non-zero coefficients $a_k^{(i)}$

$$
\begin{aligned}
a_1^{(1)} &= \frac{g\,k\,s}{q^2 r_0^2}\left(\frac{r_0 \varkappa_t^2 s^2 \sqrt{2}}{q^2} - 2\varkappa_t + \frac{q^2\sqrt{2}}{s^2 r_0}\right) \qquad (0111);\\
a_1^{(2)} &= \frac{g\,k\,\varkappa_t^2 s^3}{q^4 r_0 \sqrt{2}} \quad (0110); \qquad a_1^{(3)} = \frac{g\,k}{s\,r_0^3\sqrt{2}} \quad (0101);\\
a_1^{(5)} &= a_1^{(3)}, \qquad a_2^{(5)} = a_3^{(5)} = \frac{1}{r_0\sqrt{2}} - \frac{\varkappa_t s^2}{q^2},\\
a_4^{(5)} &= -\frac{s\,r_0}{g\,k\sqrt{2}} \quad (0011);\\
a_2^{(6)} &= a_3^{(6)} = -\frac{s^2\varkappa_t}{q^2}, \qquad a_4^{(6)} = 2a_4^{(5)} \qquad (0010).
\end{aligned}
\tag{8.5.22}
$$

As with the clamped support group, the inequalities

$$
\eta_2^{(1)} > \eta_2^{(2)} > \eta_2^{(4)}, \quad \eta_2^{(1)} > \eta_2^{(3)} > \eta_2^{(4)}, \quad \eta_2^{(1)} > \eta_2^{(5)} > \eta_2^{(6)} \tag{8.5.23}
$$

are fulfilled for arbitrary values of $\dfrac{\partial w_0}{\partial s}$ and $\dfrac{\partial \Phi_0}{\partial s}$.

formulae (10), (16) and (21) may be used to study of the effect of boundary conditions on the critical load. The main disadvantage of this approach is that the initial stress state is assumed to be membranous up to the edge while, in reality, the initial stress state is not truly membrane at the edge and the magnitude of the initial bending stresses (strains) depends on the particular type of boundary condition. This question will be discussed in more detail in Chapter 14.

The results obtained in this Chapter are valid if the shell has an initial tensile strain such that under the external load the stress state becomes truly membrane.

Remark 8.3. Here we obtain the formula for the critical load in the form $\lambda = \lambda_0 + \varepsilon\lambda_1 + \varepsilon^2\lambda_2' + \cdots$, where the main term, λ_0, is found from the solution of the boundary value problem in the zeroth approximation. The term $\varepsilon\lambda_1$ takes into account the localization of the buckling mode at the weakest generatrix.

For boundary conditions which are contained in one group, the first two terms coincide and the third summand, $\varepsilon^2\lambda_2'$, takes into account the effect of the boundary condition within the group and does not depend on the localization effect. These results may be used also for axisymmetrically loaded cylindrical and conic shells of revolution. In these cases $\lambda = \lambda_0 + \varepsilon^2\lambda_2'$, the buckling pits occupy the entire shell surface and the term $\varepsilon^2\lambda_2'$ (as in the general case) takes into account the boundary condition effect.

8.6 Boundary Conditions and Buckling of a Cylindrical Shell

As it was noted above in Section 7.4, the formulae for a cylindrical shell may be obtained from the formulae for a conic shell given above by the formal substitution of 1 instead of s^α. In this case the neutral surface metric is defined by (7.4.1).

The boundary value problem in the zeroth approximation consists of the equations

$$k\,\frac{\partial^2 w_0}{\partial s^2} + \frac{q^4}{g}\,\Phi_0 = 0, \quad k\,\frac{\partial^2 \Phi_0}{\partial s} - q^4 d w_0 + \lambda q^2 t_2 w_0 = 0 \tag{8.6.1}$$

and the boundary conditions at edges $s = s_1(\varphi)$ and $s = s_2(\varphi)$ which, according to the type of support, may have the form of (4.13)–(4.16) and condition (4.15) is to be replaced by $\dfrac{\partial w_0}{\partial s} = \dfrac{\partial \Phi_0}{\partial s} = 0$.

In cases when the values of d, g and t_2 do not depend on s, problem (1) may be resolved in the elementary functions. We will consider three variants of the main boundary conditions at two edges, namely

(i) clamped support—clamped support,
(ii) clamped support—simple support,
(iii) simple support—simple support.

In the following formulae the value of α corresponds to the minimum eigenvalues.

(i) If at both edges, clamped support conditions (4.13) are introduced then

$$w_0 = \frac{\cos\alpha\zeta}{\cos\alpha/2} - \frac{\cosh\,\alpha\zeta}{\cosh\,\alpha/2}, \quad \Phi_0 = \sqrt{\frac{dg}{3}}\left(\frac{\cos\alpha\zeta}{\cos\alpha/2} + \frac{\cosh\,\alpha\zeta}{\cosh\,\alpha/2}\right), \tag{8.6.2}$$
$$\alpha = 4.73, \qquad \zeta = \frac{s - s_1}{l} - \frac{1}{2}.$$

(ii) If at edge $s = s_2$ clamped support conditions (4.13) and at edge $s = s_1$ simple support conditions (4.14) are introduced then

$$w_0 = \frac{\sin\alpha\zeta}{\sin\alpha} - \frac{\sinh\,\alpha\zeta}{\sinh\,\alpha}, \quad \Phi_0 = \sqrt{\frac{dg}{3}}\left(\frac{\sin\alpha\zeta}{\sin\alpha} + \frac{\sinh\,\alpha\zeta}{\sinh\,\alpha}\right), \tag{8.6.3}$$
$$\alpha = 3.9266, \qquad \zeta = \frac{s - s_1}{l}.$$

(iii) If at both edges simple support conditions (4.13) apply then

$$w_0 = \sqrt{2}\sin\alpha\zeta, \quad \Phi_0 = \sqrt{\frac{2dg}{3}}\sin\alpha\zeta, \quad \alpha = \pi, \quad \zeta = \frac{s - s_1}{l}. \tag{8.6.4}$$

To simplify the following formulae we will assume that the normalization in (7.1.3) is such that $d = g = k = t_2 = 1$ at the weakest generatrix $\varphi = \varphi_0$. Then, due to (3.7) $r_0 = 1$.

As in Section 7.4 we find

$$\lambda_0 = \frac{4\alpha}{3^{3/4}l}, \quad q_0^2 = \frac{3^{1/4}\alpha}{l}, \quad l(\varphi) = s_2 - s_1, \quad z = 3^{1/4}\alpha, \tag{8.6.5}$$

where z is the same as in (1.8) and (5.10).

We assume that edge $s = s_2$ is considered and $s_2'(\varphi_0) = 0$. In order to evaluate the value of η_2, which takes into account the effect of the support (see (5.10)), we use formulae (5.16), (5.21) and (5.22). We can then write the values of functions w_0 and Φ_0 and their derivatives which are contained in these formulae

$$\begin{aligned}
&\Phi_0 = 1.155, && \frac{\partial\Phi_0}{\partial s} = \frac{5.366}{l} && \text{(clamped—clamped)},\\
&\Phi_0 = 1.155, && \frac{\partial\Phi_0}{\partial s} = \frac{4.53}{l} && \text{(simple support—clamped)},\\
&\frac{\partial w_0}{\partial s} = \frac{5.71}{l}, && \frac{\partial\Phi_0}{\partial s} = \frac{3.12}{l} && \text{(clamped—simple support)},\\
&\frac{\partial w_0}{\partial s} = -\frac{4.44}{l}, && \frac{\partial\Phi_0}{\partial s} = -\frac{2.565}{l} && \text{(simple support—simple support)}.
\end{aligned} \tag{8.6.6}$$

The effect of the support at edge $s = s_2$ on the critical load λ can be described by the parameter η_* introduced by

$$\lambda \simeq \lambda_0 + \varepsilon^2\eta_2 = \lambda_0\left(1 + \frac{\varepsilon^2}{l}\eta_*\right), \quad \eta_* = \frac{l\,\eta_2}{\lambda_0}. \tag{8.6.7}$$

The term $\varepsilon\lambda_1$ is not included in (7) since it does not depend on the boundary condition variant within a given group.

If we consider the buckling of an axisymmetrically loaded circular cylindrical shell with straight edges and with constant values of E, ν and h then $\lambda_1 = 0$ (see Remark 8.3). Formula (7) takes into account only the effect of the support on edge $s = s_2$. For edge $s = s_1$, a similar term must be introduced. If edges $s = s_1$ and $s = s_2$ are supported identically then the effect of the type of support is doubled.

Table 8.2: Parameter η_* corresponding to the boundary conditions

	clamped—clamped		simple support—clamped	
I	*II*	*III*	*IV*	*V*
1111	$1.414\,\nu^2$	0.29	$1.414\,\nu^2$	0.29
1110	$0.707\,\nu^2$	0.14	$0.707\,\nu^2$	0.14
1101, 1100	0	0	0	0
1011	$1.494\,\nu - 0.788$	−0.77	$1.524\,\nu - 0.816$	−0.81
1010	$1.494\,\nu - 1.576$	−2.56	$1.524\,\nu - 1.631$	−2.66

Table 8.3: The simply supported edge

	clamped—simple support		simple support—simple support	
I	*II*	*III*	*IV*	*V*
0111	$1.295 - 0.354\,\varkappa_* + 0.070\,\varkappa_*^2$	2.27	$1.223 - 0.419\,\varkappa_* + 0.071\,\varkappa_*^2$	1.55
0110	$0.024\,\varkappa_*^2$	0.87	$0.036\,\varkappa_*^2$	1.31
0101	0.647	1.47	0.612	1.39
0100	0	0	0	0
0011	$1.161 - 0.139\,\varkappa_*$	0.88	$1.125 - 0.241\,\varkappa_*$	0.36
0010	$-0.386 - 0.193\,\varkappa_*$	−2.63	$-0.408 - 0.241\,\varkappa_*$	−3.11

In columns *II* and *IV* of Tables 8.2 and 8.3, the values of η_* corresponding to the boundary conditions of column *I* are given. Table 8.2 is for the clamped edge, $s = s_2$, and Table 8.3 corresponds to the simply supported edge, $s = s_2$. Columns *II* in both Tables correspond to the clamped edge $s = s_1$ and columns *IV* are for the simply supported edge, $s = s_1$.

It follows from (7) and Tables 8.2 and 8.3 that the boundary condition effect within one group is determined by the value of $\varepsilon^2 l^{-1} \eta_*$ and depends on

Table 8.4: Main boundary conditions

	$s = s_1$	$s = s_2$	$\lambda_0 l$
1	clamped	clamped	8.30
2	clamped	simple support	6.89
3	simple support	simple support	5.51
4	clamped	weak support	4.15
5	clamped	free	3.29
6	simple support	weak support	2.75
7	simple support	free	0
8	weak support	weak support	0
9	weak support	free	0
10	free	free	0

three dimensionless parameters

$$\frac{\varepsilon^2}{l} = \frac{\sqrt{Rh}}{L}\left(12\,(1-\nu^2)\right)^{-1/4}, \quad \nu, \quad \varkappa_* = l\,\varkappa_t = -\frac{L_0 s''}{R}, \tag{8.6.8}$$

and for the straight edge $\varkappa_* = 0$.

As an example, consider the same shell as in Example 7.1. We take $R/h = 400$, $L_0/R = 4$, $\beta = 45^0$, and $\nu = 0.3$. Then $\varkappa_* = 4$, and $\varepsilon^2 l^{-1} = 0.0227$.

In columns III and V of Tables 8.2 and 8.3 the value $\varepsilon^2 l^{-1}\eta_*$ for the above mentioned parameters is given (in percentages).

One can see that, within the clamped support group, the boundary condition effect at edge $s = s_2$ does not exceed 3 %, and for the simple support group it does not exceed 5 %. This influence increases (as $\sqrt{h_*}$) with the shell thickness.

In general the boundary condition effect within a given group is small. The effect of the boundary conditions on the parameter λ when going from one boundary condition group to another is stronger. The reader may compare the results.

Table 8.4 includes all possible combinations of main boundary conditions in the zeroth approximation and the corresponding values of λ_0. For the first three lines in Table 8.4, the dependence of the parameter λ on the full boundary conditions was discussed above. The method of studying buckling proposed in Chapters 7 and 8 is not valid for the variants shown in the last four lines in Table 8.4 since in these cases the buckling mode is not localized. The cases of shells of revolution are discussed in Chapter 12.

The boundary condition effect on the critical normal pressure for the buckling of a circular cylindrical shell has been extensively studied in [22, 61, 113, 136, 143, 149] and others. In [143] the numerical results are given as a function of the parameter $z_m = L\,(Rh)^{-1/2}(1-\nu^2)^{1/4}$ which is related to our parameter $\varepsilon^2 l^{-1}$ (see (7) and (8)). The effect of the boundary conditions on the shell parallel on the upper critical load of the shell reinforced with the ring is examined in [41, 42].

In [39] buckling of a mitred pipe joint consisting of two connected circular cylindrical thin shells with slope edges under homogeneous external pressure is analyzed.

8.7 Conic Shells Under External Pressure

Consider a conic shell with constant parameters E, ν and h under a constant external pressure p. We assume that

$$t_2 = \frac{s}{k\,(\varphi)}, \qquad p = \frac{\lambda\, E\, h\, \varepsilon^6}{R} = \lambda\, E \left(\frac{h}{R}\right)^{5/2} \left(12\,(1-\nu^2)\right)^{-3/4} \tag{8.7.1}$$

and write the zeroth approximation system in the form [103]

$$\frac{\partial^2 w_0}{\partial y^2} + \frac{a\,\Phi_0}{(1-\varkappa\, y)^3} = 0, \qquad \frac{\partial^2 \Phi_0}{\partial y^2} - \frac{a\, w_0}{(1-\varkappa\, y)^3} + a^{1/2} b\, w_0 = 0, \tag{8.7.2}$$

which is convenient for comparison with a cylindrical shell.

Here introduce the designations

$$\begin{aligned} s &= (1-\varkappa\, y)\, s_2, \qquad l = s_2 - s_1, \quad \varkappa = l\, s_2^{-1}, \\ q^4 &= a\,\psi, \quad \psi = k\, l^{-2} s_2^3, \quad \lambda_0 = b\,\eta, \qquad \eta = k^{3/2} s_2^{-3/4} l^{-1}. \end{aligned} \tag{8.7.3}$$

Shell edges $s_1(\varphi)$ and $s_2(\varphi)$ due to (3) are transformed into $y = 1$ and $y = 0$ at which the two main boundary conditions must be introduced. Here we will only consider shells with clamped (4.13) or simply supported (4.14) edges.

For $\varkappa = 0$ we obtain a cylindrical shell and the case of $\varkappa = 1$ corresponds to a conic shell closed in a vertex. In the last case of system (2) for $y = 1$ there is a singular point and the boundary conditions at $y = 1$ are replaced by the limiting conditions for the solution.

System (2) is not integrable analytically in the known functions. By numerical solution of boundary value problem (2) with corresponding boundary conditions at edges $y = 0, 1$ for the fixed values of a and $\varkappa$ one can obtain function $b = b\,(a, \varkappa)$ and then find its minimum with respect to a and $\varkappa$.

Conditions (7.2.11) determining λ_0^0 and the values of q_0, and φ_0 may be written in the form

$$\frac{\partial \lambda_0}{\partial q} = \frac{\partial b}{\partial a} = 0, \qquad \frac{\partial \lambda_0}{\partial \varphi} = b\,\frac{\partial \eta}{\partial \varphi} + \eta\,\frac{\partial b}{\partial \varkappa}\,\frac{d\varkappa}{d\varphi} = 0. \tag{8.7.4}$$

The first of these equations determines the functions

$$a = a\,(\varkappa), \qquad b^0(\varkappa) = b\big(a\,(\varkappa),\,\varkappa\big), \tag{8.7.5}$$

And from the second of equations (4) we find the weakest generatrix $\varphi = \varphi_0$ and the corresponding values of

$$\lambda_0^0 = \eta_0 b_0^0, \qquad q_0^4 = a_0 \psi_0, \tag{8.7.6}$$

where

$$\begin{array}{lll} a_0 = a\,(\varkappa_0), & b_0^0 = b^0(\varkappa_0), & \\ \varkappa_0 = \varkappa\,(\varphi_0), & \psi_0 = \psi\,(\varphi_0), & \eta_0 = \eta\,(\varphi_0). \end{array} \tag{8.7.7}$$

To use formula (1.11) to evaluate the correction $\varepsilon\lambda_1$ of the value λ_0^0 one must find the values λ_{qq}, $\lambda_{q\varphi}$, and $\lambda_{\varphi\varphi}$. Due to (3), (4), and (5) we get

$$\begin{aligned} \lambda_{qq} &= \eta_0 b_{aa} a_q^2, \\ \lambda_{q\varphi} &= \eta_0 b_{aa} a_q \left(a_\varphi - a_\varkappa \varkappa_\varphi\right), \\ \lambda_{\varphi\varphi} &= b\,\eta_{\varphi\varphi} + b_\varkappa \left(\eta_0 \varkappa_{\varphi\varphi} + 2\eta_\varphi \varkappa_\varphi\right) + \\ &\qquad + \eta_0 b_{aa} a_\varphi \left(a_\varphi - 2a_\varkappa \varkappa_\varphi\right) + \eta_0 b_{\varkappa\varkappa} \varkappa_\varphi^2, \end{aligned} \tag{8.7.8}$$

where

$$a_q = \frac{4\,q_0^3}{\psi_0}, \qquad a_\varphi = -\frac{a_0 \psi_\varphi}{\psi_0}. \tag{8.7.9}$$

With subscripts we denote the derivatives with respect to the corresponding variables calculated at $q = q_0$ and $\varphi = \varphi_0$. To derive (8) we use the relation $b_{a\varkappa} = -b_{aa} a_\varkappa$ obtained by means of differentiating the first of equations (4) by $\varkappa$.

In order to use formulae (8) for calculations it is necessary to know the functions $b^0(\varkappa)$, $a(\varkappa)$, $b_\varkappa(\varkappa)$, $a_\varkappa(\varkappa)$, $b_{aa}(\varkappa)$, and $b_{\varkappa\varkappa}(\varkappa)$, which are determined in the numerical solution of boundary value problem (2).

Figure 8.2 shows plots of the functions $b^0(\varkappa)$, $a_0(\varkappa)$, $b_{aa}(\varkappa)$, and $b_{\varkappa\varkappa}(\varkappa)$ for four boundary condition variants: simple support—simple support (SS), simple support—clamped (SC), clamped—simple support (CS), and clamped—clamped (CC). In each case, the wider shell edge is mentioned first.

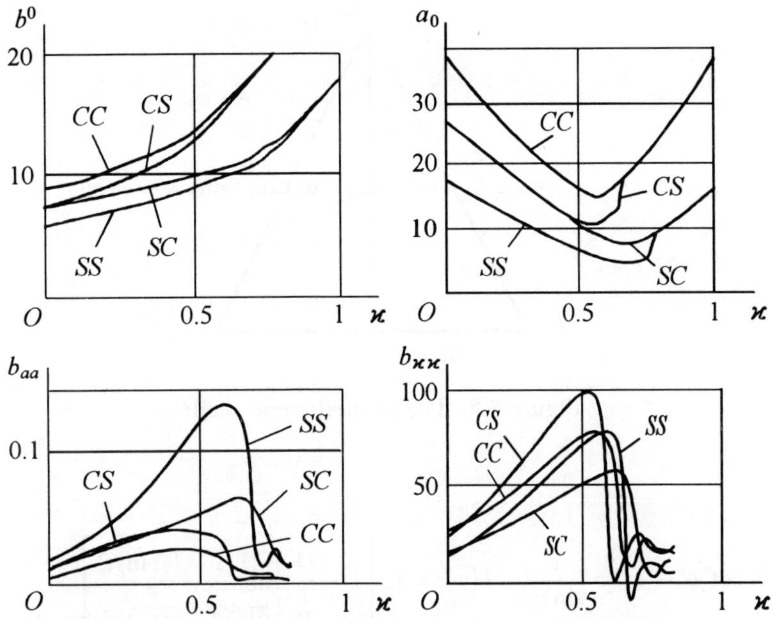

Figure 8.2: The functions $b^0(\varkappa)$, $a_0(\varkappa)$, $b_{aa}(\varkappa)$, and $b_{\varkappa\varkappa}(\varkappa)$.

The functions $b_\varkappa$ and $a_\varkappa$ may be found approximately by means of numerically differentiating curves b^0 and a which are shown in Figure 8.2.

As it is expected for small values of $\varkappa$, the curves SC and CS for functions $b^0(\varkappa)$ and $a_0(\varkappa)$ in Figure 8.2 converge since for shells which are nearly cylindrical it is not important which of the edges is clamped or simply supported. On the contrary, for $\varkappa \sim 1$ the pair of curves labelled SS, SC and CS, CC converge since then the shell stiffness increases as the shell edge approaches the cone vertex and the type of support for the narrow edge becomes unimportant.

System (2) at $y = \varkappa^{-1}$ has an essential singular point [142, 180]. If $\varkappa$ is close to 1, then the singular point is close to edge point $y = 1$ that influences the behaviour of the curves as shown in Figure 8.2. It is especially notable for functions $b_{aa}(\varkappa)$ and $b_{\varkappa\varkappa}(\varkappa)$ since, after differentiating, the oscillation amplitude increases.

Under these assumptions the existence of the weakest generatrix $\varphi = \varphi_0$ is connected with the variability of functions $\eta(\varphi)$ and $\varkappa(\varphi)$ (see (3)) which is determined by the dependence of functions s_1, s_2, and k on φ.

In particular, we are interested in a circular conic shell with a slanted edge and vertex angle 2α (see Figure 8.3). The upper edge is inclined by angle β to the lower edge. As a characteristic size, R, we take the lower edge radius.

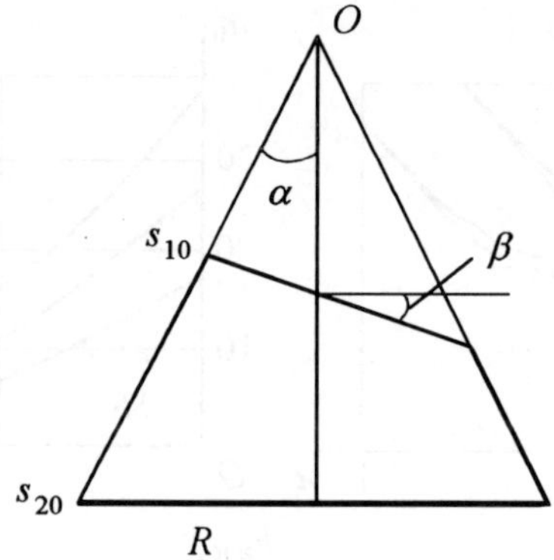

Figure 8.3: The slanted conic shell.

Then

$$
\begin{aligned}
&k = \cot\,\alpha, \quad s_2 = \frac{1}{\sin\alpha}, \quad l\,(\varphi) = s_2\left[1 - \frac{(1+c)\,(1-l_0\sin\alpha)}{1 + c\,\cos\dfrac{\varphi}{\sin\alpha}}\right], \\
&0 \leqslant \varphi \leqslant 2\pi\sin\alpha, \quad \varkappa = l\sin\alpha, \quad c = \tan\,\alpha\,\tan\,\beta < 1, \\
&b^*(\varkappa) = \varkappa^{-1}b^0(\varkappa), \quad \psi = \frac{\cos\alpha}{l^2\sin^4\alpha}, \quad \eta = \frac{1}{l}\,(\cos\alpha)^{3/2}, \\
&\lambda_0 = \sin\alpha\,(\cos\alpha)^{3/2}b^*(\varkappa),
\end{aligned}
\tag{8.7.10}
$$

where the $l_0 = l\,(0)$ corresponds to the longest generatrix.

From the last formula it follows due to (4) that function b^* attains its minimum value at the weakest generatrix. Function $b^*(\varkappa)$ is shown in Figure 8.4.

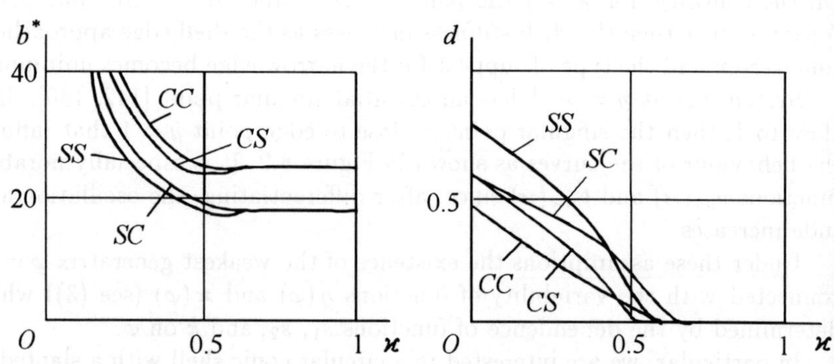

Figure 8.4: The functions $b^*(\varkappa)$ and $d(\varkappa)$.

In the buckling problem of a circular cylindrical shell with a slanted edge under a homogeneous external pressure, the longest generatrix is also the weakest (see Example 7.1). It then follows from (10) and Figure 8.4 for a conic shell with $\varkappa_m \lesssim 0.5$ that the longest generatrix also is the weakest one. Here, we denote the value of $\varkappa$ corresponding to the longest generatrix by $\varkappa_m$.

If $\varkappa > 0.5$ the dependence of b^* on $\varkappa$ is more complex.

If the narrow shell edge is clamped (see the curves CC and SC in Figure 8.4), then function $b^*(\varkappa)$ is monotonic and at $\varkappa \to 1$ (the full cone) it converges to $b^*(1) = 26.75$ for clamped support of the wide edge (CC) and to $b^*(1) = 18.30$ for simple support of the wide edge (SC).

The difference between $b^*(\varkappa)$ and $b^*(1)$ is less than 1 % for $\varkappa > \varkappa_1$, where $\varkappa_1 = 0.55$ in case CC and $\varkappa_1 = 0.63$ in case SC. That is why, if $\varkappa_m > \varkappa_1$, we recommend the use of $\lambda = \lambda_0$ at $b^* = b^*(1)$ in the analysis. The correction term $\varepsilon\lambda_1$ (see (1.2)) may be neglected since it takes into account the local character of the buckling mode, but in our case $b^*(\varkappa)$ is close to constant. This contradicts the assumption about the local character of the buckling mode.

If the narrow shell edge is simply supported then the function $b^*(\varkappa)$ has an infinite number of minimums at $\varkappa \to 1$, only one of which at $\varkappa = \varkappa_0$ can be seen in Figure 8.4 (see curves CS and SS).

We have $\varkappa_0 = 0.53$, $b^*(\varkappa_0) = 26.27$ in case CS and $\varkappa_0 = 0.59$, $b^*(\varkappa_0) = 17.85$ in case SS. For $\varkappa_m > \varkappa_0$ we recommend that one use in the analysis $\lambda = \lambda_0$ at $b^* = b^*(\varkappa_0)$ and that the term $\varepsilon\lambda_1$ be neglected for the same reason as before.

Let us now return to the case when $\varkappa_m < 0.5$ and by means of (1), (1.2), (1.11) and (8) represent the critical pressure in the form

$$p = E\left(\frac{h}{R}\right)^{5/2}\left(12\left(1-\nu^2\right)\right)^{-3/4}\lambda_0\left[1+\varepsilon\left(\frac{d\left(\varkappa_0\right)\tan\beta}{(1+c)\left(\cos\alpha\right)^{3/2}}\right)^{1/2}+O\left(\varepsilon^2\right)\right], \tag{8.7.11}$$

where

$$\begin{aligned} \varkappa_0 &= l_0 \sin\alpha, \qquad \lambda_0 = \frac{b^0}{l_0}(\cos\alpha)^{3/2}, \\ d &= \frac{4a_0^{3/2}b_{aa}(1-\varkappa_0)}{b^0}\left(1-\frac{\varkappa_0 b_\varkappa}{b^0}\right). \end{aligned} \tag{8.7.12}$$

For the boundary condition variants considered, the function $d(\varkappa)$ is shown in Figure 8.4. For $\varkappa > 0.5$ this function approaches zero which agrees with the earlier proposal about neglecting the term $\varepsilon\lambda_1$. For $\alpha = \varkappa_0 = 0$, (11) is valid for a cylindrical shell with a slanted edge and in the case when both edges are simply supported it coincides with (7.4.9).

For a circular conic shell with straight edges we assume that $d = 0$ in (11), and that the number of waves n in the circumferential direction is equal to

$$n = \frac{q \sin \alpha}{\varepsilon} = \left(\frac{a_0^2(\varkappa)\, R^2 h^2 \cos^2 \alpha}{12\,(1-\nu^2)\, L^4} \right)^{1/8}, \tag{8.7.13}$$

where $a_0(\varkappa)$ is shown in Figure 8.2.

8.8 Problems and Exercises

8.1 Derive formula (4.10).

8.2. Solve numerically the buckling problem for circular cylindrical shell under external normal pressure that is a linear function in coordinate s for different main boundary conditions at edges $s = 0, 1$. The results represent in the form of a table similar to Table 8.4.

Hint. Integrate system (6.1) for $k = g = q = 1$, $t_2 = s/l$.

8.3 Find the weakest generatrix of the truncated non-straight conic shell under external homogeneous pressure. The shell edges are the circles lying in the parallel planes and one of the generatrices is orthogonal to these planes. Consider various boundary conditions.

Hint. Use the results of Section 8.7. The required generatrix may be found from the equation $\partial\eta/\partial\varphi = 0$.

Chapter 9

Torsion and Bending of Cylindrical and Conic Shells

In this Chapter we will consider the buckling of shells of zero Gaussian curvature with the same modes stretched along the generatrices as in Chapters 7 and 8. It is assumed that the initial stress-resultant T_2^0 equals to zero or small and that buckling occurs due to the shear stress-resultant S^0. Such a stress state may appear in a shell which has loads applied to its edge. The internal pressure that leads to the appearance of hoop, T_2^0, and axial, T_1^0, tensile stress-resultants on reinforced shells are also considered.

When writing this Chapter we used the results obtained in [61, 97, 98, 103, 152, 159].

9.1 Torsion of Cylindrical Shells

Let us consider the buckling of a cylindrical shell of moderate length under combined loading assuming that the shear stresses are dominant.

Instead of formulae (7.1.3) we introduce new notation

$$\left(T_1^0, T_2^0, S^0\right) = \lambda\, E_0 h_0 \varepsilon^5 \left(\varepsilon^{-1} t_1^0, \varepsilon\, t_2^0, t_3^0\right), \tag{9.1.1}$$

which means that the critical value of the shear stress-resultant S^0 has order greater than the critical value of the stress-resultant T_2^0. The orders of the stress-resultants T_1^0, T_2^0 and S^0 are chosen in such a way that all of these stress-resultants affect the zeroth approximation for the boundary value problem when $t_1^0, t_2^0, t_3^0 \sim 1$.

We assume that the values d, g and t_i^0 do not depend on s. Then the boundary value problem in the zeroth approximation has the coefficients that are constant in s and we can write the following equation

$$\frac{k^2 g}{q^4}\frac{\partial^4 w_0}{\partial s^4} + d\,q^4 w_0 - \lambda_0 N\,w_0 = 0,$$
$$N\,w_0 = t_1^0\,\frac{\partial^2 w_0}{\partial s^2} + 2i\,q\,t_3^0\,\frac{\partial w_0}{\partial s} - q^2 t_2^0 w_0. \tag{9.1.2}$$

The general solution of equation (2) has the form

$$w_0 = \sum_{k=1}^{4} C_k e^{p_k s}, \qquad p_k = i\,q^2 x_k, \tag{9.1.3}$$

and for x_k we get the fourth order equation

$$x^4 + a_2 x^2 + a_3 x + a_4 = 0, \tag{9.1.4}$$

where

$$a_2 = \frac{\lambda_0 t_1^0}{g k^2}, \quad a_3 = \frac{2\lambda_0 t_3^0}{g k^2 q}, \quad a_4 = \frac{d}{g k^2} + \frac{\lambda_0 t_2^0}{g k^2 q^2}. \tag{9.1.5}$$

We represent the roots of equation (4) as

$$x_{1,2} = \alpha \pm i\gamma, \qquad x_{3,4} = -\alpha \pm \beta, \tag{9.1.6}$$

then

$$3\alpha^2 = \beta^2 - \frac{a_2}{2} + \left(3a_4 + \left(2\beta^2 + \frac{a_2}{2}\right)^2\right)^{1/2}, \quad \gamma^2 = 2\alpha^2 + \beta^2 + a_2, \tag{9.1.7}$$

$$a_3 = 2\alpha\left(\beta^2 + \gamma^2\right). \tag{9.1.8}$$

We only consider the boundary condition groups of clamped and simply supported types (see Section 8.4). If, on both edges $s = 0$ and $s = l$, the same boundary conditions apply then the equation which determines the parameter λ_0 has the form

$$\tan\beta_1 = r\left[\coth\gamma_1 - \frac{\cos 2\alpha_1}{\sinh\gamma_1\,\cos\beta_1}\right],$$
$$(\alpha_1, \beta_1, \gamma_1) = q_1\,(\alpha, \beta, \gamma), \qquad q_1 = q^2 l, \tag{9.1.9}$$

where

$$r = \frac{2\beta\,\gamma}{6\alpha^2 + a_2} \qquad \text{for} \quad w_0 = w_0' = 0, \qquad (9.1.10)$$

$$r = -\frac{8\alpha^2\beta\,\gamma}{(2\alpha^2 + 2\beta^2 + a_2)^2 + 4\alpha^2(2\alpha^2 + a_2)} \qquad \text{for} \quad w_0 = w_0'' = 0. \qquad (9.1.11)$$

If there are different boundary conditions on the edges ($w_0 = w_0' = 0$ for $s = 0$ and $w_0 = w_0'' = 0$ for $s = l$ or vice-verse), then instead of (9) we have the equation

$$\tan\beta_1 = \frac{\beta\,(2\beta^2 - 2\alpha^2 + a_2)}{\gamma\,(6\alpha^2 + 2\beta^2 + a_2)}\left[\tanh\gamma_1 + \frac{4\alpha\,\gamma\,\sin 2\alpha_1}{(2\beta^2 - 2\alpha^2 + a_2)\cos\beta_1\,\cosh\,\gamma_1}\right]. \qquad (9.1.12)$$

We will first consider the buckling of a circular cylindrical shell with constant parameters l, h, E and ν under pure torsion. In this case

$$t_1^0 = t_2^0 = a_2 = 0, \qquad d = g = k = t_3^0 = a_4 = 1. \qquad (9.1.13)$$

Then equation (9) coincides with the equation obtained in [159], and equation (12) coincides with the equation derived in [152]. Equations (9) and (12) may be easily solved by taking an arbitrary value of β, evaluating α and γ by formulae (7) and β_1 from equations (9) and (12), and then evaluating q and λ_0 by formulae (9), (5) and (8).

We note that in the neighbourhood of the critical load $\gamma_1 \sim 10$, and therefore we can simplify equations (9) and (12) replacing the square brackets in the right sides by unity. We represent the results obtained in the form proposed in [61]

$$\tau_0 = \frac{S^0}{h} = k_s\,E\left(\frac{h^2}{(1-\nu^2)\,R^2}\right)^{5/8}\left(\frac{R}{L}\right)^{1/2}, \qquad n_0 = \frac{q_0}{\varepsilon} = k_n\left(\frac{(1-\nu^2)\,R^6}{h^2L^4}\right)^{1/8}, \qquad (9.1.14)$$

where τ_0 and n_0 are the critical shear stress and the number of waves in the circumferential direction. The dimensionless coefficients k_s and k_n depend on the boundary conditions and for the clamped edge—clamped edge (CC), clamped edge—simply supported edge (CS) and simply supported edge—simply supported edge (SS) cases their values are

$$k_s = 0.770, \quad k_n = 4.41, \quad \beta = 0.320 \quad \text{(CC)},$$
$$k_s = 0.738, \quad k_n = 4.15, \quad \beta = 0.332 \quad \text{(CS)}, \qquad (9.1.15)$$
$$k_s = 0.703, \quad k_n = 3.87, \quad \beta = 0.345 \quad \text{(SS)},$$

and

$$\lambda_0 = 12^{5/8} k_s l^{-1/2}, \quad q_0 = 12^{-1/8} k_n l^{-1/2}, \quad k_n = 12^{1/8} q_1^{1/2}. \tag{9.1.16}$$

Comparing the critical stresses for shell pure torsion in the cases of simply supported and clamped boundary conditions we note that they differ by 10 %. It can be seen in Table 8.4, that the same difference for buckling under external normal pressure is 50 %.

The weak dependence on the boundary conditions noted above allows us to use the following approximate method [61, 175] in solving the torsion problem. In solution (3) we keep only two terms ($k = 3, 4$) and satisfy only one boundary condition, $w = 0$, on each shell edge. Then instead of (9) and (12) we get the equation

$$\tan \beta_1 = 0, \qquad \beta_1 = \pi. \tag{9.1.17}$$

As a result we obtain the values

$$k_s = 0.74, \qquad k_n = 4.2, \tag{9.1.18}$$

which usually are referred to as the classical values [61].

A reviews of the works on the problem under consideration are given in [61, 149]. We also note the works [3, 58], in the first, the influence of the boundary conditions on the critical load and in the second the influence of non-linear terms on the critical load were studied.

We can use formulae (14) only for shells of moderate length. The inequality $n_0 \gg 1$ (for example, $n_0 \geqslant 5$) may be the applicability criterion for formulae (14). We will not consider long shells, but it should be noted that for sufficiently long shells, $n_0 = 2$, and the critical load does not depend on the length but is determined by the relation [137]

$$\tau_0 = \frac{E}{4.24} \left(\frac{h^2}{(1 - \nu^2) R^2} \right)^{3/4}. \tag{9.1.19}$$

9.2 Cylindrical Shell under Combined Loading

We will now consider the buckling of a circular cylindrical shell of moderate length under combined torsion, internal pressure and axial force. We assume that the parameters E, ν and h are constant. The stress-resultants T_1^0, T_2^0 are assumed to be known and the stress-resultant S^0 to be unknown. Instead of (1.5) we denote

$$a_2 = t_1^0, \quad a_3 = \frac{2\lambda}{q} \quad (t_3^0 = 1), \quad a_4 = 1 + \frac{t_2^0}{q^2}. \tag{9.2.1}$$

Referring the initial stress-resultants to the corresponding critical values we introduce instead of (1)

$$a_2 = 2\,R_1, \quad a_3 = 7.0\,R_3 q_1^{-1/2}, \quad a_4 = 1 + 5.51\,\frac{R_2}{q_1}, \tag{9.2.2}$$

where

$$R_1 = \frac{\sigma_1}{\sigma_1^0}, \qquad R_2 = \frac{p}{p_0}, \qquad R_3 = \frac{\tau}{\tau_0},$$
$$\sigma_1^0 = 2E\,\varepsilon^4, \qquad p_0 = \frac{4\pi}{3^{3/4}}\,\frac{E\,h\,\varepsilon^6}{L}. \tag{9.2.3}$$

Here σ_1 is the axial tensile stress, σ_1^0 is the critical value under axial compression (see (3.4.3)), p is the internal pressure, p_0 is the critical value of external pressure for a simply supported shell (see (7.4.9) for $k_* = 1$), and τ_0 is the critical value for the tangential stress under torsion (see (1.14) for $k_s = 0.74$).

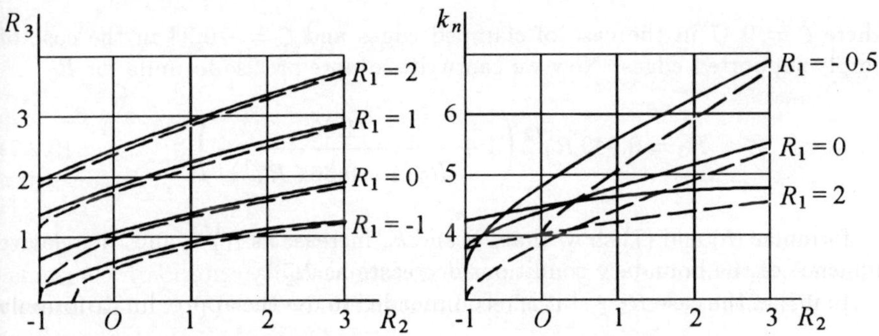

Figure 9.1: The functions $R_3(R_2)$ and $k_n(R_2)$.

In Figure 9.1 the dependencies of R_3 and k_n on R_2 for different values of R_1 are shown. The solid curves correspond to a shell with clamped edges, while the dashed curves correspond to a shell with simply supported edges. The results were obtained by the numerical solution of equation (1.9) followed by the minimization by the parameter q_1, connected with the number of waves, n, in the circumferential direction. The compressive stress-resultants T_1^0 and T_2^0 correspond to negative values of R_1 and R_2.

In Figure 9.1 one can see that the difference between the results for shells with clamped and simply supported edges decreases as R_1 and R_2 increase. For cases when this difference is not taken into account the numerical results are given in [61].

We can also write approximate formulae for R_3 and k_n assuming that R_1 or R_2 are large. The roots (1.6) of equation (1.4) we represent in the form of part of the negative power series in the large parameters R_1 or R_2. We find R_3 from the equation (1.9) in the same form.

First, let us consider the case of a large internal pressure ($R_2 \gg 1$, $R_1 = 0$). In this assumption $\beta \ll 1$. We assume that $\beta = 0$ in (1.7) and (1.8) and get

$$\alpha = \left(\frac{a_4}{3}\right)^{1/4}, \quad \gamma^2 = 2\alpha^2, \quad R_3 = \frac{4}{7}\, q_1^{1/2}\alpha^{3/4}. \tag{9.2.4}$$

Minimizing R_3 by q_1, we find

$$R_3 = 0.949\, R_2^{1/2}, \qquad k_n = 2.264\, R_2^{1/2}. \tag{9.2.5}$$

formulae (5) do not reveal the dependence of R_3 on the boundary conditions. To correct (5) we retain terms of order β^2. From equation (1.9) we find approximately that

$$\beta = \pi\,(q_1 - \zeta)^{-1}, \tag{9.2.6}$$

where $\zeta = 0.47$ in the case of clamped edges and $\zeta = -0.94$ in the case of simply supported edges. Now we can write a more precise formula for R_3

$$R_3 = 0.949\, R_2^{1/2}\left(1 + \frac{1.30}{R_2^2\,(1 - 0.36\,\zeta\, R_2^{-1})^2}\right). \tag{9.2.7}$$

formulae (5) and (7) show that R_3 and k_n increase as $R_2^{1/2}$, and the relative influence of the boundary conditions decreases as R_2^{-3}.

In [61] in the case $R_2 \lesssim 1$ it is recommended to use the approximate formula

$$R_3 = 1 + 0.207\, R_2 - 0.00175\, R_2^2, \tag{9.2.8}$$

obtained by the approximation of the dependence of R_3 on R_2, found numerically. We also note the similarity of formula (5) and formula $R_3^2 - R_2 = 1$ (see [61]) for large values of R_2.

Now we consider buckling under torsion for large axial expanding stresses ($R_1 \gg 1$, $R_2 \sim 1$). We represent the roots of equation (1.4) in the form of (1.6). Then for $a_2 \gg 1$ we get approximately

$$\alpha = \frac{a_3}{2a_2}, \quad \beta = \frac{\sqrt{a_3^2 - 4a_2a_4}}{2a_2}, \quad \gamma = \sqrt{a_2}. \tag{9.2.9}$$

From equation (1.9) we have approximately

$$\beta = \frac{\pi}{q_1\,(1 - \eta)}, \tag{9.2.10}$$

where the parameter η takes into account the influence of the boundary conditions and $\eta = (\pi R_1)^{-1}$ in the case when the shell edges are clamped and $\eta \simeq 0$ in the case when the shell edges are simply supported.

Equating the expressions obtained for β, we get

$$R_3 = \frac{2}{7}\left(\frac{a_2^2\pi^2}{(1-\eta)^2 q_1} + q_1 a_2 a_4\right)^{1/2}. \tag{9.2.11}$$

Minimizing R_3 by q_1, we find

$$R_3 = 1.21\left(\frac{R_1^{3/2}}{1-\eta} + 0.62\, R_1 R_2\right)^{1/2}, \quad k_n = 6.06\, R_1^{1/2}. \tag{9.2.12}$$

It follows from the last formula that the relative difference of R_3 for shells with clamped and simply supported edges decreases as R_1^{-1} as R_1 increases.

9.3 A Shell with Non-Constant Parameters Under Torsion

In this Section we will assume that $t_1^0 = t_2^0 = 0$, and the variables t_3^0, k, l, g and d may depend on φ. In other words, we will examine the torsion of a non-circular cylindrical shell of length, wall thickness and elastic properties all of which vary in φ. Since these variables do not depend on s we may use the results of Section 9.1.

By virtue of (1.5) and (1.8) we have

$$\lambda_0 = \frac{gk^2}{t_3^0}\, q\alpha\,(\beta^2 + \gamma^2), \tag{9.3.1}$$

where formulae (1.7) give the dependence of α and γ on β and φ_0 (through a_4). To make the study of the dependence of α and γ on φ_0 more convenient we make the substitution

$$(\alpha, \beta, \gamma) = a_4^{1/4}\,(\alpha_0, \beta_0, \gamma_0). \tag{9.3.2}$$

Then λ_0 may be represented as

$$\lambda_0 = f\,(q, \varphi_0) = F\,(\varphi_0)\, G\left[\,\beta_0(q, \varphi_0)\,\right], \tag{9.3.3}$$

where $\beta_0(q, \varphi_0)$ is the solution of the equation $\Psi\,(\beta_0) = \eta\,(\varphi_0)\, q^{-2}$, where we denote the right hand side of equation (1.9) or (1.12) by $r_1(\beta_0)$. Here

$$\begin{aligned}
&\Psi(\beta_0) = \beta_0\left(\pi + \arctan r_1(\beta_0)\right)^{-1},\\
&G(\beta_0) = \left[\psi(\beta_0)\right]^{-1/2}\alpha_0\left(\beta_0^2 + \gamma_0^2\right),\\
&3\alpha_0^2 = \beta_0^2 + \left(3 + 4\beta_0^4\right)^{1/2}, \qquad \gamma_0^2 = 2\alpha_0^2 + \beta_0^2,\\
&F(\varphi_0) = \frac{1}{t_3^0}\left(\frac{k^6 g^3 d^5}{l^4}\right)^{1/8}, \qquad \eta(\varphi_0) = \frac{1}{l}\left(\frac{gk^2}{d}\right)^{1/4}.
\end{aligned} \tag{9.3.4}$$

In transition to approximate equation (1.17) we should assume that $\Psi(\beta_0) = \beta_0\pi^{-1}$.

The weakest generatrix could be evaluated after the minimization of $F(\varphi_0)$ by φ_0. Keeping the designation φ_0 for this generatrix we normalize the determining parameters in such a way that relations (1.13) are satisfied. Then

$$F(\varphi_0) = l_0^{-1/2}, \quad \eta(\varphi_0) = l_0^{-1}, \quad l_0 = l(\varphi_0). \tag{9.3.5}$$

Minimizing now by q, we come to the same formulae (1.16), in which we should take $l = l_0$.

To evaluate the first correction

$$\varepsilon\lambda_1 = \frac{1}{2}\varepsilon\left(\lambda_{qq}\lambda_{\varphi\varphi} - \lambda_{q\varphi}^2\right)^{1/2} \tag{9.3.6}$$

for the load parameter λ_0^0 we calculate the second derivatives in (6) and obtain

$$\begin{aligned}
&\lambda_{qq} = b_{11}l_0^{1/2}, \qquad \lambda_{q\varphi} = -b_{12}l_0\frac{d\eta}{d\varphi_0},\\
&\lambda_{\varphi\varphi} = a_{22}\frac{d^2F}{d\varphi_0^2} + b_{22}l_0^{3/2}\left(\frac{d\eta}{d\varphi_0}\right)^2,
\end{aligned} \tag{9.3.7}$$

where the functions F and η are given by formulae (4). The constants a_{22} and b_{ij} depend only on the boundary conditions and are determined by the formulae

$$\begin{aligned}
&a_{22} = 12^{5/8}k_s, \qquad b_{11} = 4.12^{3/4}k_n^{-6}b, \qquad b_{12} = 2.12^{5/8}k_n^{-5}b,\\
&b_{22} = 12^{1/2}k_n^{-4}b, \quad b = \frac{d^2G}{d\beta_0^2}\left(\frac{d\psi}{d\beta_0}\right)^{-2},
\end{aligned} \tag{9.3.8}$$

where k_s and k_n have the values of (1.15). To finish the calculations we must find the value of b. For the cases described by (1.15) and (1.18) we obtain respectively

$$b = 385.6, \quad b = 265.5, \quad b = 172.7, \quad b = 283.0\,. \tag{9.3.9}$$

It follows from the form of the coefficients b_{ij} that the value of $\varepsilon\lambda_1$ does not depend on the terms containing $\dfrac{d\eta}{d\varphi_0}$.

The final result we write as

$$\tau = \tau_0 k_*, \quad k_* = 1 + c\left(\frac{h^2 L^6}{(1-\nu^2)\,R^8}\right)^{1/8}\left(\frac{d^2F}{d\varphi_0^2}\right)^{1/2} + O\left(\varepsilon^2\right), \qquad c = \frac{b^{1/2}12^{1/16}}{k_n^3 k_s^{1/2}}, \tag{9.3.10}$$

where τ_0 is determined by formula (1.14), and the multiplier k_* takes into account the variability of the determining parameters. Coefficient c depends on the boundary conditions and is equal to

$$c = 0.305, \quad c = 0.310, \quad c = 0.316, \quad c = 0.308, \tag{9.3.11}$$

Here the values of c are written in the same sequence as in (9). Since these values are close to each other, we can take an average value of $c = 0.310$ for all cases.

The correction term $O\left(\varepsilon^2\right)$ in (10) depends not only on the main but also on the additional boundary conditions. As we noted before, for a complete evaluation we should start with the system of equations which is more precise than (8.1.1).

By virtue of (7.2.3) the rate of decrease of the depth of the pits away from the weakest generatrix $\varphi = \varphi_0$ is determined by the multiplier $e^{\{-a_1(\varphi-\varphi_0)^2\}}$, where

$$a_1 = \frac{\Im a}{\varepsilon} = 1.75\left(\frac{R^4\left(1-\nu^2\right)}{L^2h^2}\right)^{1/8}\left(\frac{d^2F}{d\varphi_0^2}\right)^{1/2}. \tag{9.3.12}$$

For example, we consider the torsion of a non-circular cylindrical shell of the constant (moderate) length with constant parameters E, ν and h.

The torsion occurs due to axial moment M, applied to the body to which the shell is attached. Let the generatrix have an elliptic form with semi-axes a_0 and b_0 ($a_0 > b_0$). Two generatrices passing through the points of the ellipse B and B' (see Figure 9.2) are the weakest.

To satisfy relations (1.13) we take the radius of curvature of the ellipse at point B i.e. $R = a_0^2 b_0^{-1}$ as the unit length. Under torsion point N moves by an amount proportional to $r = ON$, and the stress-resultant S^0 is proportional to the projection of this displacement on the tangent to the ellipse at point N. In accordance with this fact we take

$$t_3^0 = \frac{r}{b_0}\cos\beta = \left(\frac{1+\delta^2\tan^2\psi}{1+\delta^4\tan^2\psi}\right)^{1/2}, \quad \delta = \frac{b_0}{a_0} < 1. \tag{9.3.13}$$

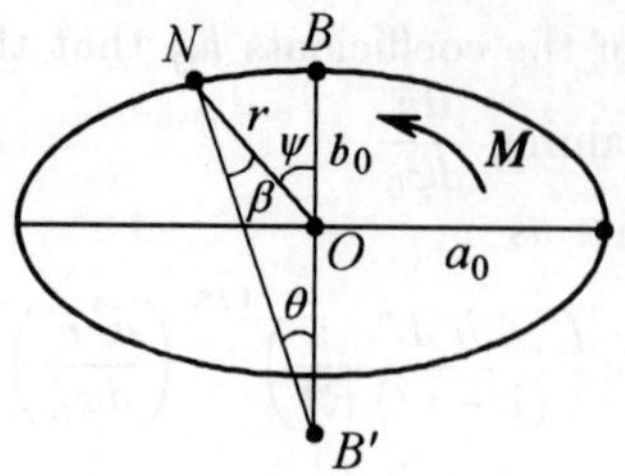

Figure 9.2: An elliptic cylinder in torsion.

For an ellipse we have

$$k = \delta^{-3}\left(\sin^2\theta + \delta^2\cos^2\theta\right)^{3/2} = \left(\frac{1+\delta^2\tan^2\psi}{1+\delta^4\tan^2\psi}\right)^{3/2} = \left(t_3^0\right)^3,$$

$$\tan\theta = \delta^2\tan\psi, \quad F(\varphi) = \left(t_3^0\right)^{5/4}, \quad \left.\frac{d\psi}{d\varphi}\right|_{\psi=0} = \frac{1}{\delta^2}, \tag{9.3.14}$$

$$\left.\frac{d^2F}{d\varphi^2}\right|_{\psi=0} = \frac{5\left(1-\delta^2\right)}{4\delta^2}, \quad M = \oint S^0 r\cos\beta\,\frac{R\,d\theta}{k}.$$

The critical value of tangential stresses, τ, on the weakest generatrix is evaluated by means of formula (10), and the critical value of the torsional moment is obtained after the evaluation of integral (14) and is equal to

$$\begin{aligned}
M &= 4K(\delta')\,a_0 b_0 h\,\tau, \qquad \delta' = \sqrt{1-\delta^2},\\
\tau &= k_s E\left(\frac{h^{10}b_0^6}{(1-\nu^2)\,a_0^{12}L^4}\right)^{1/8} k_*,\\
k_* &= 1 + 0.39\left(\frac{h^2L^6(1-\delta^2)^4}{(1-\nu^2)\,a_0^8}\right)^{1/8} + O\left(\varepsilon^2\right),\\
K(\alpha) &= \int_0^{\pi/2}\frac{dx}{(1-\alpha^2\sin^2 x)^{1/2}}.
\end{aligned} \tag{9.3.15}$$

Here depending on the boundary conditions, constant k_s takes one of the values in (1.15), and $K(\alpha)$ is the complete elliptic integral of the first kind (a table for $K(\alpha)$ can be found, for example, in [67]).

9.4 Bending of a Cylindrical Shell

Let us consider the bending by a transverse force, P_1, (see Fig. 1.3) of a console circular cylindrical shell of moderate length. The initial stress state is considered to be membrane and

$$T_1^0 = \frac{P_1 s}{\pi RL} \cos\varphi, \qquad T_2^0 = 0, \qquad S^0 = \frac{P_1}{\pi R} \sin\varphi, \\ 0 \leqslant s \leqslant l = \frac{L}{R}, \qquad 0 \leqslant \varphi \leqslant 2\pi. \tag{9.4.1}$$

Parameters E, ν and h are assumed to be constant. In accordance with (1.1) take

$$\lambda E h \varepsilon^5 = \frac{P_1}{\pi R}, \qquad t_3^0 = \sin\varphi. \tag{9.4.2}$$

For a moderately long shell, buckling leads to the formation of pits in the neighbourhood of the two weakest generatrices $\varphi_0 = \pm\frac{\pi}{2}$ where the shear stresses are maximum. To evaluate the critical value of force P_1 we use formulae (1), (1.14) and (3.10). Then

$$P_1^0 = \pi R h \tau_0 k_*, \quad k_* = 1 + 0.31 \left(\frac{h^2 L^4}{(1-\nu^2) R^6}\right)^{1/8} + O(\varepsilon^2), \tag{9.4.3}$$

where τ_0 is the critical value of the tangential stresses under torsion (see (1.14)), and k_* takes into account the variability of the shear stresses. Formula (3) does not include the influence of the stress-resultant T_1^0 which is of order ε^2.

We note that the critical value of P_1^0 is asymptotically tetra-multiple, i.e. four linearly independent buckling modes are possible and the values of P_1 which are very close to P_1^0 correspond to them. As a rule, the eigenvalues are asymptotically double (see Remark 4.2). Here the tetra-multiplicity is connected with the existence of the two weakest generatrices $\varphi_0 = \pm\frac{\pi}{2}$.

It is interesting to compare formula (3) with the results (see [61]), obtained numerically by the Bubnov–Galerkin method which approximates bending by a piece of a double trigonometric series. For $1 \leqslant \frac{L}{R} \leqslant 5$, $100 \leqslant \frac{R}{h} \leqslant 500$ the difference between the results is 1–3 %, and formula (3) gives the smaller values of stress than [61]. For $\frac{R}{h} = 1000$ the difference is 4 % which probably means that the difficulties of approximating bending by a piece of a trigonometric series increases as the shell thickness decreases. The buckling mode given in [61] qualitatively agrees with the form determined by expression (7.2.25).

For long shells, due to (1), the compressive stress-resultants T_1^0 increase and shell buckling occurs in the neighbourhood of point $s = l$, $\varphi = \pi$. Comparing the critical load under axial compression and torsion in the zeroth approximations we get the applicability condition for formula (3)

$$L < 0.6\,(1-\nu^2)^{1/4}\, h^{-1/2} R^{3/2}, \tag{9.4.4}$$

where, in evaluating τ_0 by (1.14), we take $k_s = 7.4$.

We will now consider the buckling of a cylindrical shell under bending by transverse force, P_1, and torsion by axial moment, M, applied to the shell edge, $s = 0$. Instead of (1) and (2), in this case we get

$$\begin{aligned} S^0 &= \frac{P_1}{\pi R}\sin\varphi + \frac{M}{2\pi R^2}, \qquad & t_3^0 &= \frac{\sin\varphi + m}{1+m}, \\ m &= \frac{M}{2QR}, & P_1^0 &= \frac{\pi R\, h\, \tau_0}{1+m}\, k_*, \end{aligned} \tag{9.4.5}$$

where τ_0 has the same value as in (1.14), and

$$k_* = 1 + 0.31\left(\frac{h^2 L^4}{(1-\nu^2)\,(1+m)^4\, R^6}\right)^{1/8} + O\left(\varepsilon^2\right). \tag{9.4.6}$$

Here, unlike the case when $M = 0$ for $M > 0$ only one generatrix $\varphi_0 = \frac{\pi}{2}$ (see Figure 9.3) is the weakest.

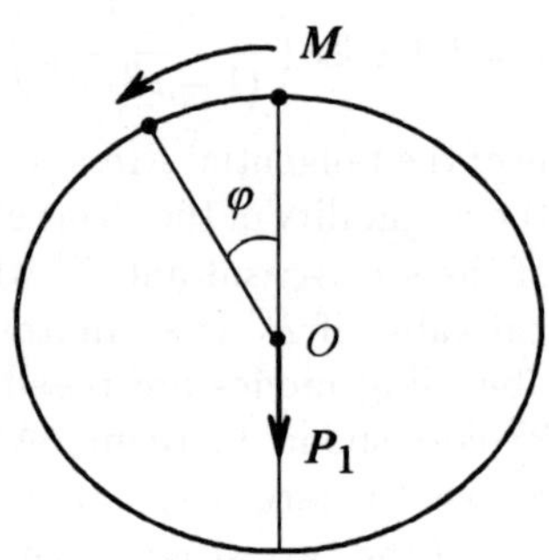

Figure 9.3: A force and a moment acting on a shell.

The numerical results obtained in solving this problem by the Bubnov–Galerkin method can be found in [61]. They agree with formula (5) to the same degree as in the case $M = 0$.

9.5 The Torsion and Bending of a Conic Shell

In this Section we will consider the buckling of a straight circular conic shell with constant values of E, ν and h under bending by transverse force, P_1, and torsion by axial moment, M, (see Figure 9.3) applied to the largest base of the shell. We denote the base radius and the shell generatrix length by R and L respectively. Then according to (1.4.6) we have membrane initial stress-resultants

$$T_1^0 = \frac{P_1}{\pi R \sin\alpha} \cdot \frac{\varkappa y \cos\varphi}{(1-\varkappa y)^2}, \quad S^0 = \frac{P_1 (\sin\varphi + m)}{\pi R (1-\varkappa y)^2}, \quad T_2^0 = 0, \tag{9.5.1}$$

where designations (8.7.3) and (4.5) are used

$$s = s_2(1-\varkappa y), \quad l = s_2 - s_1, \quad \varkappa = l\, s_2^{-1}, \quad m = \frac{M}{2RP_1}, \quad l = \frac{L}{R}. \tag{9.5.2}$$

We also take

$$S^0 = E\,h\,\varepsilon^5 \lambda\, t_3, \qquad t_3 = \frac{\sin\varphi + m}{(1+m)\,(1-\varkappa y)^2}. \tag{9.5.3}$$

We assume that $P_1 \geqslant 0$, $M \geqslant 0$, then the critical values P_1 and M are related to λ by the expression

$$2RP_1^0 + M^0 = 2\pi R^2 E\,h\,\varepsilon^5 \lambda. \tag{9.5.4}$$

The system of equations of the zeroth approximation (8.1.4) we can write in the form

$$\begin{gathered} \frac{\partial^2 w_0}{\partial y^2} + \frac{a^4 \Phi_0}{(1-\varkappa y)^3} = 0, \\ \frac{\partial^2 \Phi_0}{\partial y^2} - \frac{a^4 w_0}{(1-\varkappa y)^3} - \frac{2i\,a\,b}{(1-\varkappa y)^3}\left((1-\varkappa y)\frac{\partial w_0}{\partial y} + \varkappa\, w_0\right) = 0, \end{gathered} \tag{9.5.5}$$

where

$$q = \left(\frac{\cos\alpha}{\varkappa^2 \sin^2\alpha}\right)^{1/4} a, \quad \lambda = \frac{(m+1)\,b}{m+\sin\varphi}\left(\frac{\cos^3\alpha \sin^2\alpha}{\varkappa^2}\right)^{1/4}. \tag{9.5.6}$$

For a fixed value of $\varkappa$, let the smallest positive eigenvalue be $b = b\,(a)$. We then obtain

$$b_0 = \min_a b\,(a) = b\,(a_0). \tag{9.5.7}$$

formulae (8.1.2) and (8.1.11) give

$$\lambda = \left(\frac{\cos^3\alpha \sin^2\alpha}{\varkappa^2}\right)^{1/4} b_0 \Bigg(1 + \\ + \frac{\varepsilon}{2}\left(\frac{\varkappa^2}{\sin^2\alpha \cos\alpha}\right)^{1/4}\left(\frac{b_{aa}}{(m+1)\,b_0}\right)^{1/2} + O\,(\varepsilon^2)\Bigg). \tag{9.5.8}$$

In [98] the numerical integration of system (5) for two boundary condition variants, i.e. clamped—clamped (CC) and simply supported—simply supported (SS) has been done

$$\begin{aligned} w_0 &= \frac{\partial w_0}{\partial y} = 0 \quad \text{for} \quad y = 0, \quad y = 1, \\ w_0 &= \Phi_0 = 0 \quad \text{for} \quad y = 0, \quad y = 1 \end{aligned} \tag{9.5.9}$$

and graphs of the functions $a_0(\varkappa)$, $b_0(\varkappa)$ and $b_{aa}(\varkappa) = \dfrac{d^2 b}{da_0^2}$ have been plotted in Figure 9.4.

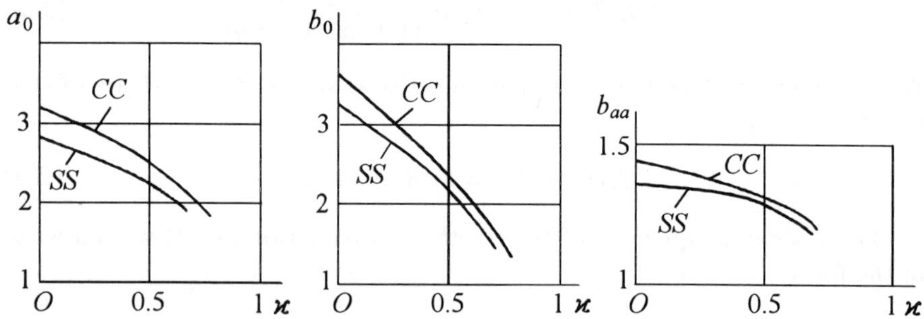

Figure 9.4: The functions $a_0(\varkappa)$, $b_0(\varkappa)$, and $b_{aa}(\varkappa)$.

For $m = 0$ and $m = \infty$ formulae (4) and (8) give the solutions of the bending and pure torsion of a conic shell. As in the case of a cylindrical shell (see Section 9.4), for $m = 0$, the eigenvalue λ is asymptotically tetra-multiple and for $0 < m < \infty$ it is asymptotically double. The number of waves in the circumferential direction, n, under pure torsion ($m = \infty$) is equal to

$$n = \left(\frac{12\,(1-\nu^2)\,R^6 \cos^2\alpha}{L^4 h^2} \right)^{1/8} a_0(\varkappa). \tag{9.5.10}$$

The area of applicability of the formulae obtained above is determined by two inequalities, the first being $n \gg 1$, and the second being similar to inequality (4.4). The latter means that the initial stress-resultant T_1^0 does not exceed its critical value under axial compression

$$L < 7.4\,(1-\varkappa)^2\,(1+m)^2\,b_0^{-2}\,(1-\nu^2)^{1/4} \left(\frac{R^3 \cos\alpha}{h} \right)^{1/2}. \tag{9.5.11}$$

For $\alpha = \varkappa = 0$ due to $\varkappa R = L \sin\alpha$, formulae (4) and (8) evolve to formulae (4.5) and (4.6) for a cylindrical shell.

9.6 Problems and Exercises

9.1. Derive approximate relation (2.7) describing the effect of the initial pressure on the critical load under torsion. Integrate equation (9.2) numerically for shells with (i) clamped and (ii) simply supported edges and compare the numerical and asymptotic results.

9.2. Derive approximate relation (2.12) describing the effect of the initial tensile stress-resultant on the critical load under torsion. Integrate equation (9.2) numerically for shells with (i) clamped and (ii) simply supported edges and compare the numerical and asymptotic results.

9.3. Consider the buckling under torsion of a circular cylindrical shell with a sloped edge with constant parameters E, ν and h.

Answer
We take

$$l = l_0 + (\cos\varphi - 1)\tan\alpha, \quad t_3^0 = 1, \quad F(\varphi_0) = l^{-1/2}, \quad l_0 = \frac{L}{R}.$$

Formula (10) for the critical value of τ gives

$$\tau = \tau_0 k_*, \quad k_* = 1 + 0.22\left(\frac{h^2\tan^4\alpha}{(1-\nu^2)\,R^2}\right)^{1/8} + O(\varepsilon^2),$$

where α is the slope angle and the stress τ_0 is determined by formula (1.14) for a shell of constant length L. The critical value of the torsional moment is equal to $M = 2\pi R^2 h\tau$. The buckling pits are localized in the neighbourhood of the weakest generatrix which is also the longest.

9.4. Derive relation (4.6) for the critical load under combined loading by torsion and bending by force.

9.5. Study buckling of circular cylindrical shell under torsion moment M and bending moment M_1 which are applied on the edge $s = l$ (see Fig. 1.3).
Hint Use relations (1.4.6) and (3.10).

Chapter 10

Nearly Cylindrical and Conic Shells

In Chapters 7 through 9 we considered the buckling modes of moderately long cylindrical and conic shells. Those modes are characterized by pits which are stretched in the direction of the generatrix. It is clear that a buckling mode cannot vary greatly if the bending of the generatrix is small. In this Chapter the results of Chapters 7-9 are extended to shells which are nearly cylindrical and conic.

We will consider a shell such that the deviation, i.e. difference between its geometry and that of a shell of zero curvature, affects the critical load in the zeroth approximation. It proves to be, that under this condition the amplitude of deviation must be of order $(Rh)^{1/2}$. The function describing the deviation should not increase significantly under differentiation because otherwise the proposed method of construction of the solution is not valid. Similar problems for free vibrations are solved in [105, 106].

10.1 Basic Relations

Let us first describe the geometry of the shell to be considered. The neutral surface of the shell is a nearly conic surface and is referred to as the basic surface. On the basic conic surface we introduce a system of orthogonal curvilinear coordinates s, φ similar to those used in Section 7.1.

Let $\boldsymbol{e}_1^0$, $\boldsymbol{e}_2^0$, $\boldsymbol{n}^0$ be the unit vectors of this coordinate system ($\boldsymbol{n}^0 = \boldsymbol{e}_1^0 \times \boldsymbol{e}_2^0$). We denote the deviation from the basic surface by $F_0(s,\varphi)$ which is equal to the distance along the normal $\boldsymbol{n}^0$ from a point $\boldsymbol{r}_0$ on the basic surface to a point $\boldsymbol{r}$ on the neutral surface of the shell (see Figure 10.1).

Then we can write

$$\boldsymbol{r} = \boldsymbol{r}_0 + F_0 \boldsymbol{n}^0 = R\left(s\,\boldsymbol{e}_1^0 + \mu_0 F\,\boldsymbol{n}^0\right), \qquad F_0 = \mu_0 R\,F, \tag{10.1.1}$$

where R is the characteristic size of the neutral surface and μ_0 is a small parameter.

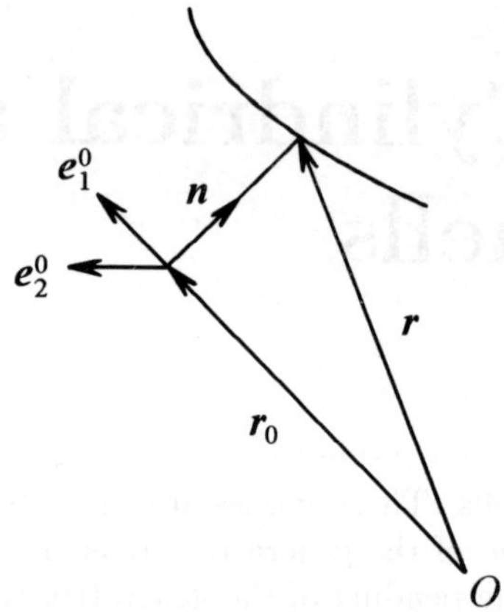

Figure 10.1: A nearly conic surface.

Coordinates s and φ are the curvilinear coordinates in the neutral surface (in the general case they are non-orthogonal). By means of formulae (1.1.4) we get:

$$\begin{aligned}
\frac{\partial \boldsymbol{r}}{\partial s} &= R\left(\boldsymbol{e}_1^0 + \mu_0 F_s \boldsymbol{n}^0\right), \\
\frac{\partial \boldsymbol{r}}{\partial \varphi} &= R\Big[\left(s - \mu_0 \boldsymbol{k}\,F\right)\boldsymbol{e}_2^0 + \mu_0 F_\varphi \boldsymbol{n}^0\Big], \\
\frac{\partial^2 \boldsymbol{r}}{\partial s^2} &= R\,\mu_0 F_{ss} \boldsymbol{n}^0, \\
\frac{\partial^2 \boldsymbol{r}}{\partial s\,\partial \varphi} &= R\Big[\left(1 - \mu_0 \boldsymbol{k}\,F_s\right)\boldsymbol{e}_2^0 + \mu_0 F_{s\varphi} \boldsymbol{n}^0\Big], \\
\frac{\partial^2 \boldsymbol{r}}{\partial \varphi^2} &= R\Big[\left(\mu_0 \boldsymbol{k}\,F - s\right)\boldsymbol{e}_1^0 - \mu_0\left(2\boldsymbol{k}\,F_\varphi + \boldsymbol{k}_\varphi F\right)\boldsymbol{e}_2^0 + \\
&\qquad + \Big(\mu_0 F_{\varphi\varphi} + \boldsymbol{k}\left(s - \mu_0 \boldsymbol{k}\,F\right)\Big)\boldsymbol{n}^0\Big],
\end{aligned} \tag{10.1.2}$$

where function $\boldsymbol{k} = \boldsymbol{k}(\varphi)$ is the same as in Section 7.1.

The first quadratic form of the surface is equal to

$$\begin{aligned} d\sigma^2 &= R^2\left(A^2 ds^2 + 2AB\cos\chi\, ds\, d\varphi + B^2 d\varphi^2\right), \\ A^2 &= 1 + \mu_0^2 F_s^2, \qquad B^2 = (s - \mu_0 k\, F)^2 + \mu_0^2 F_\varphi^2, \\ \cos\chi &= \frac{\mu_0^2 F_s F_\varphi}{A\, B}. \end{aligned} \tag{10.1.3}$$

In the general case we obtain very complicated expressions for the radii of curvature which are not represented here. In order to simplify these expressions we will take advantage of the close proximity of the shell neutral surface to the basic surface and assume that $\mu_0 \ll 1$, and function F is of order 1 and does not increase significantly under differentiation. In other words, we also assume that the derivatives F_s, F_φ, F_{ss}, $F_{s\varphi}$ and $F_{\varphi\varphi}$ have the orders not larger than unity.

From (3) we find that $\cos\chi = O(\mu_0^2)$, i.e. the coordinate system s, φ in the neutral surface is almost orthogonal with an error of order μ_0^2.

We obtain for the radii of curvature

$$\begin{aligned} \frac{R}{R_1} &= \mu_0 k_1 + O(\mu_0^2), \qquad \frac{R}{R_{12}} = \mu_0 k_{12} + O(\mu_0^2), \\ \frac{R}{R_2} &= \frac{k}{s} + \mu_0 k_2 + O(\mu_0^2), \end{aligned} \tag{10.1.4}$$

where functions k_1, k_{12}, and k_2 depend on s, φ and are equal to

$$\begin{aligned} k_1 &= F_{ss}, \qquad k_{12} = s^{-2}\left(F_\varphi - s\, F_{s\varphi}\right), \\ k_2 &= s^{-2}\left(k^2 F + s\, F_s + F_{\varphi\varphi}\right). \end{aligned} \tag{10.1.5}$$

As the basic system we choose the same system as (7.1.2)

$$\begin{aligned} \varepsilon^4 \Delta\left(d\Delta\, w\right) + \lambda\,\varepsilon^2 \Delta_t w - \Delta_k \Phi &= 0, \\ \varepsilon^4 \Delta\left(g^{-1}\Delta\Phi\right) + \Delta_k w &= 0, \end{aligned} \tag{10.1.6}$$

in which the operators Δ, Δ_t and Δ_k unlike (7.1.3) have the form

$$\begin{aligned}
\Delta w &= \frac{1}{B_0}\frac{\partial}{\partial\varphi}\left(\frac{1}{B_0}\frac{\partial w}{\partial\varphi}\right)+\frac{1}{B_0}\frac{\partial}{\partial s}\left(B_0\frac{\partial w}{\partial s}\right), \quad B_0 = s-\mu_0 kF,\\
\Delta_t w &= \frac{1}{B_0}\frac{\partial}{\partial\varphi}\left(\frac{t_2}{B_0}\frac{\partial w}{\partial\varphi}\right)+\frac{1}{B_0}\frac{\partial}{\partial s}\left(t_3\frac{\partial w}{\partial\varphi}\right)+\\
&+\frac{1}{B_0}\frac{\partial}{\partial\varphi}\left(t_3\frac{\partial w}{\partial s}\right)+\frac{1}{B_0}\frac{\partial}{\partial s}\left(B_0 t_1\frac{\partial w}{\partial s}\right),\\
\Delta_k w &= \frac{\mu_0}{B_0}\frac{\partial}{\partial\varphi}\left(\frac{k_1}{B_0}\frac{\partial w}{\partial\varphi}\right)+\frac{\mu_0}{B_0}\frac{\partial}{\partial s}\left(k_{12}\frac{\partial w}{\partial\varphi}\right)+\\
&+\frac{\mu_0}{B_0}\frac{\partial}{\partial\varphi}\left(k_{12}\frac{\partial w}{\partial s}\right)+\frac{1}{B_0}\frac{\partial}{\partial s}\left(B_0\left(\frac{k}{s}+\mu_0 k_2\right)\frac{\partial w}{\partial s}\right),
\end{aligned} \tag{10.1.7}$$

and the other notation is the same as in (7.1.3).

We choose a characteristic deviation, F_0, so as to get the effect of it in the zeroth approximation of the iterative process, which has been developed in Sections 7.2 and 8.1. We assume for that

$$\mu_0 = \varepsilon^2 \quad \left(\varepsilon^8 = \frac{h_0^2}{12\,(1-\nu^2)\,R^2}\right), \tag{10.1.8}$$

i.e. we assume that the characteristic value of F_0 is of order $(Rh_0)^{1/2}$.

10.2 Boundary Problem in the Zeroth Approximation

We seek a solution of system (1.6) in the same form as (8.1.2). First let us consider the case when $t_2 > 0$ (at least in part of the neutral surface), and t_1 and t_3 are not larger than t_2 respectively. Then in the zeroth approximation, we obtain the system of equations

$$\begin{aligned}
k\frac{\partial^2\Phi_0}{\partial s^2}-k_1\frac{q^2}{s}\Phi_0-\frac{q^4 d}{s^3}w_0+\lambda_0 N w_0 &= 0,\\
k\frac{\partial^2 w_0}{\partial s^2}-k_1\frac{q^2}{s}w_0+\frac{q^4}{gs^3}\Phi_0 &= 0, \quad N w_0 = \frac{q^2 t_2 w_0}{s},
\end{aligned} \tag{10.2.1}$$

which differs from system (8.1.4) only by the second terms.

Remark 10.1. If the shear stresses are dominant for buckling then it should be assumed in (1) that

$$N w_0 = -\frac{q^2 t_2^0 w_0}{s} + i\,q\left(t_3^0\,\frac{\partial w_0}{\partial s} + \frac{\partial}{\partial s}\left(t_3^0 w_0\right)\right) + \frac{\partial}{\partial s}\left(s t_1^0\,\frac{\partial w_0}{\partial s}\right), \qquad (10.2.2)$$

where t_i^0 are introduced by formulae (9.1.1).

In solving system (1) it is necessary to define two main boundary conditions at each shell edge $s = s_k(\varphi)$ (see Section 8.4). As in Section 7.2 the smallest eigenvalue λ_0 is a function of two parameters q and φ_0

$$\lambda_0 = f\,(q, \varphi_0). \qquad (10.2.3)$$

We can find the zeroth and the first approximations for the load parameter by function f under the same assumptions as in Section 7.2.

$$\begin{aligned} \lambda &= \lambda_0^0 + \varepsilon\,\lambda_1 + O\,(\varepsilon^2), \\ \lambda_0^0 &= \min_{q,\varphi_0}\{f\,(q,\varphi_0)\} = f\left(q_0, \varphi_0^0\right), \\ \lambda_1 &= \frac{1}{2}\left(f_{qq} f_{\varphi\varphi} - f_{q\varphi}^2\right)^{1/2}. \end{aligned} \qquad (10.2.4)$$

Thus, we can obtain most of the information on the buckling type by analyzing the function $f\,(q, \varphi_0)$. If this function has a distinct minimum at point (q_0,φ_0^0) (see (7.2.12)), then the buckling concavities are localized in the neighbourhood of line $\varphi = \varphi_0^0$, and the load parameter, λ, is determined by formula (4). The minimum is termed distinct if function $f\,(q, \varphi_0)$ has the only minimum in the interval $|q - q_0| \sim \mu^{1/2}$. At the same time the concavities are stretched along lines $\varphi = \text{const}$.

If the coefficients k, k_1, d, g and t_2, and the value of s_k do not depend on φ, then function f does not also depend on φ_0, i.e.

$$\lambda_0 = f\,(q). \qquad (10.2.5)$$

Here the buckling concavities spread over the entire shell surface, and we should minimize f by the values $q = n\varepsilon$, where n is the integer wave number in the circumferential direction. This case corresponds to the buckling under axisymmetric loading of a shell of revolution which is nearly conic.

Let us now consider the special cases.

10.3 Buckling of a Nearly Cylindrical Shell

Let the shell neutral surface be nearly a cylinder, the parameters of which are defined by relations (7.4.1). System of equations (2.1) has the form

$$\begin{aligned} k\,\frac{\partial^2 \Phi_0}{\partial s^2} - k_1 q^2 \Phi_0 - q^4 d\, w_0 + \lambda_0 q^2 t_2 w_0 &= 0, \\ k\,\frac{\partial^2 w_0}{\partial s^2} - k_1 q^2 w_0 + q^4 g^{-1} \Phi_0 &= 0. \end{aligned} \tag{10.3.1}$$

If the functions k_1, d, g and t_2 do not depend on s, then system (1) has an explicit solution.

The function k_1 does not depend on s, if lines $\varphi = \text{const}$ in the neutral surface have constant curvature $R_1^{-1} = R^{-1}\varepsilon^2 k_1$. In this case $F_0 = F_{00} + s^2 F_{02}$, where F_{00} and F_{02} may depend on φ (see (1.1) and (1.5)). We note that for $F_0 \sim (Rh)^{1/2}$ the surfaces the cross sections of which $\varphi = \text{const}$ are circles or parabolas could not be destinguished in the zeroth and first approximation in ε.

Functions d and g do not depend on s, in particular, if E, ν and h do not depend on s or if they are constant (in the last case we may assume that $d = g = 1$).

Let k_1, d, g and t_2 not depend on s. In the case of simply supported edges we can write

$$w_0 = \Phi_0 = 0 \quad \text{at} \quad s = s_k(\varphi), \tag{10.3.2}$$

and the solution of system (1) is represented in the simplest form. We now have

$$\begin{aligned} &w_0 = \sin\frac{\alpha\,(s - s_1)}{l}, \qquad l\,(\varphi) = s_2 - s_1, \qquad \alpha = \pi, \\ &\lambda_0 = f\,(q, \varphi_0) = \frac{dq^2}{t_2} + \frac{g}{t_2 q^6}\left(k_1 q^2 + \frac{k\alpha^2}{l^2}\right)^2. \end{aligned} \tag{10.3.3}$$

We write the function f in the form

$$\begin{aligned} &f = b\,F\,(x, a), \qquad F = x^2 + \frac{(\alpha^2 + a x^2)^2}{x^6}, \qquad x = cq, \\ &a = k_1 l\left(\frac{g}{dk^2}\right)^{1/4}, \qquad b = \frac{d}{t_2 l}\left(\frac{gk^2}{d}\right)^{1/4}, \qquad c = \left(\frac{dl^4}{gk^2}\right)^{1/8}. \end{aligned} \tag{10.3.4}$$

For an axisymmetrically loaded shell of revolution the functions a, b and c do not depend on φ_0. The critical load is evaluated by minimizing F by $q = q_n = n\varepsilon$ (or by x_n). In Figure 10.2 the minimum value of F and the corresponding value of x are represented by curve SS which depends on parameter a characterizing the curvature of the generatrix.

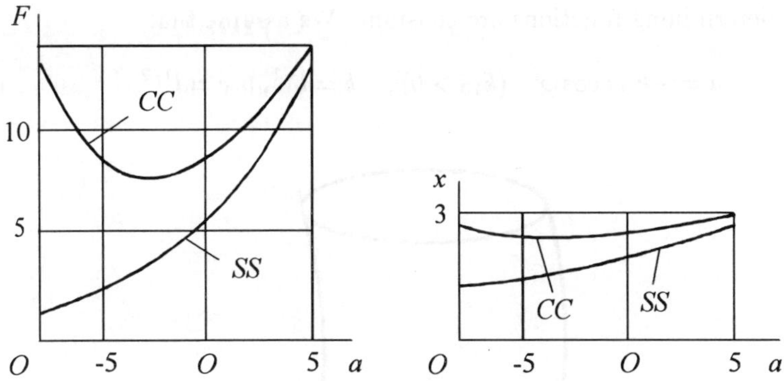

Figure 10.2: Functions $F(a)$ and $x(a)$.

For convex shells ($a > 0$) both the critical load bF and the wave number n in the circumferential direction ($n = q\varepsilon^{-1}$) increase as the curvature k_1 increases. For shells of negative Gaussian curvature ($a < 0$) the load and the wave number decrease as $|k_1|$ increases.

Now, let functions a, b and c depend on φ_0. Then the parameters q_0 and φ_0^0 are determined from the minimum condition for function f by x, φ_0, i.e. by the equations

$$F_x = 0, \qquad b_\varphi F + b\, F_a a_\varphi = 0. \tag{10.3.5}$$

To evaluate the correction λ_1 by formula (2.4) we calculate the second derivatives

$$\begin{aligned} f_{qq} &= b\, c^2 F_{xx}, \\ f_{q\varphi} &= b\, c\, (F_{xx} q\, c_\varphi + F_{xa} a_\varphi)\,, \\ f_{\varphi\varphi} &= b_{\varphi\varphi} F + (2 a_\varphi b_\varphi + a_{\varphi\varphi} b)\, F_a - b\, q^2 c_\varphi^2 F_{xx} + b\, a_\varphi^2 F_{aa}. \end{aligned} \tag{10.3.6}$$

Remark 10.2. For the applicability of formula (2.4) it is necessary to verify the fulfilment of two conditions: $\varepsilon^{-1} q_0 \gg 1$ (for example, $\varepsilon^{-1} q_0 \geqslant 4-6$) and $\varepsilon^{-4} \lambda_1 f_{qq}^{-1} \geqslant 1$. The first condition means that there are many concavities, and the second means that they damp sufficiently fast in the φ direction.

Example 10.1. We will now consider a shell obtained as a result of slightly bending the axis of symmetry of a circular cylindrical shell (see Figure 10.3) under homogeneous external pressure. We assume that the shell edges are simply supported. In the case under consideration $k_1 = -k_{10} \cos\varphi$ and the other determining functions are constant. We assume that

$$a = -k_{10} l \cos\varphi \quad (k_{10} > 0)\,, \quad b = l^{-1}, \quad c = l^{1/2}. \tag{10.3.7}$$

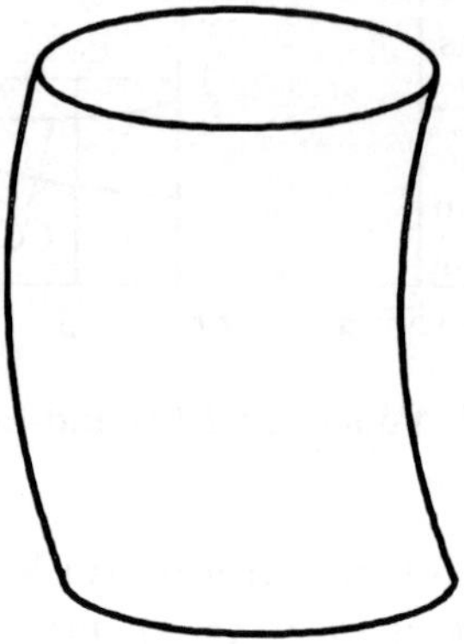

Figure 10.3: A Bent Cylindrical Shell.

It is clear (see curve SS for function $F(a)$ in Figure 10.2) that line $\varphi_0 = 0$ for which the parameter a is a minimum is the weakest. Due to (4) and (6) we get from formula (2.4) for the critical load that

$$\begin{aligned} \lambda &= f_0 + \frac{\varepsilon}{2}\left(f_{qq} f_{\varphi\varphi}\right)_0^{1/2} + O\left(\varepsilon^2\right), \quad f_0 = l^{-1} F, \\ f_{qq} &= 48\,\alpha^2\left(\alpha^2 + a\,x^2\right) x^{-8} + 8\,a^2 x^{-4}, \\ f_{\varphi\varphi} &= 2\,k_{10}\left(\alpha^2 + a\,x^2\right) x^{-4}. \end{aligned} \tag{10.3.8}$$

We can calculate by means of formulae (8) at $\alpha = \pi$ and $a = -k_{10} l$. For the values of a we take the values of F and x from Figure 10.2.

To illustrate, we will take the following numerical values of the shell parameters: $\nu = 0.3$, $R/h = 500$, $l = L/R = 2$ and $k_{10} = 1$. The maximum value of deflection of the generatrix $F_{\max}$ is related to parameter a by the following equation

$$F_{\max} = \frac{a\,L}{8}\left(\frac{h^2}{12\left(1-\nu^2\right) R^2}\right)^{1/4} \tag{10.3.9}$$

and for the shell being considered, $F_{\max} = 6.15\,h$. For $a = -2$ we find $F = 3.6$ and $x = 1.76$ from Figure 10.2. Now we can evaluate $f_{qq} = 22.2$, $f_{\varphi\varphi} = 0.766$, $f_0 = 1.8$ and $\lambda = 2.12$ by formulae (8).

Let us verify the fulfilment of the inequalities given in Remark 10.2. We have $\varepsilon^{-1}q_0 = 8$, $\varepsilon^{-1}\lambda_1 f_{qq}^{-1} = 0.6$, from that we find that the second inequality in Remark 10.2 is not valid, i.e. the depth of the pits decreases too slowly in the circumferential direction.

Compare the load parameter $\lambda = 2.12$ with the corresponding values of λ for the other three problems. For a simply supported circular $(k_1 = 0)$ cylindrical shell $\lambda = 2.755$. For a concave shell of revolution $(k_1 = -1)$ $\lambda = 1.8$ and lastly for a convex shell of revolution $(k_1 = 1)$ $\lambda = 4.1$. It follows from these data that, at first, the small amount of bending of the shell generatrix causes a significant change in the critical load and secondly the critical load of the shell with a curved axis is close to the critical load of the corresponding shell with negative Gaussian curvature but slightly larger.

Now we proceed to discussion of the case of a shell with clamped edges. Instead of (2) we have

$$w_0 = \frac{\partial w_0}{\partial s} = 0 \quad \text{at} \quad s = s_k\,(\varphi). \tag{10.3.10}$$

Calculate the function $f\,(q, \varphi_0)$ by the same formula (3) or (4), in which $\alpha \neq \pi$, but α depends on the parameters of the problem and may be determined from equation

$$\alpha \tan\frac{\alpha}{2} + \beta \tanh\frac{\beta}{2} = 0, \qquad \beta^2 = \alpha^2 + 2a\,x^2, \tag{10.3.11}$$

in which we use the same notation as in (4). To evaluate x from equation (5), $F_x = 0$ we do the following. First, from equation (11) we numerically find the function $\alpha = \alpha\,(\zeta)$, where $\zeta = a\,x^2$ and then

$$F = x^2 + \left(\alpha^2 + \zeta\right)^2 x^{-6} \tag{10.3.12}$$

and the equation $F_x = 0$ gives

$$x^8 = 3\left(\alpha^2 + \zeta\right)^2 - 2\zeta\left(\alpha^2 + \zeta\right)\left(2\alpha\alpha_\zeta + 1\right). \tag{10.3.13}$$

Therefore, three values, a, x and F are functions of ζ. Changing ζ, we can find the functions $F\,(a)$ and $x\,(a)$. These functions are marked by letters CC in Figure 10.2.

From Figure 10.2 we find that for $a > 0$, i.e. for convex generatrices, function F increases and becomes closer to function F for a simply supported shell as a grows. For $a < 0$, function F first decreases until the value $F = 7.55$ is reached at $a = -2.3$, and then it increases as $|a|$ increases, in contrast to the case of simply supported shell. The wave numbers in the circumferential direction are a little larger for a clamped shell than for a simply supported shell.

For a shell of revolution with the same parameters as in the previous example we obtain the following results. For a convex shell with $k_1 = 1$ we get $\lambda = 5$. For a cylindrical shell $\lambda = 4.15$ and for the concave shell with $k_1 = -1$ we obtain $\lambda = 3.8$. Hence, for a clamped shell, the deformation of its generatrix affects the critical load relatively less than for a simply supported shell.

10.4 Torsion of a Nearly Cylindrical Shell

Now we will assume that the shear stress-resultant S^0 dominates buckling and so we introduce the load parameter λ by formula (9.1.1). Then in the zeroth approximation we have the system of equations

$$\begin{aligned} k\,\frac{\partial^2\Phi_0}{\partial s^2} - k_1 q^2\Phi_0 - d\,q^4 w_0 + \lambda_0 N w_0 &= 0, \\ k\,\frac{\partial^2 w_0}{\partial s^2} - k_1 q^2 w_0 + g^{-1}q^4\Phi_0 &= 0, \\ N w_0 = t_1^0\,\frac{\partial^2 w_0}{\partial s^2} + 2i\,q\,t_3^0\,\frac{\partial w_0}{\partial s} - q^2 t_2^0 w_0. & \end{aligned} \tag{10.4.1}$$

We assume that the functions k_1, d, g and t_k^0 do not depend on s. Then system (1) may be solved by the same method as in Section 9.1. Taking its solution in the form of (9.1.3) and (9.1.4), we obtain the expressions (instead of (9.1.5)) for coefficients a_k

$$a_2 = \frac{\lambda_0 t_1^0}{gk^2} + \frac{2k_1}{kq^2}, \quad a_3 = \frac{2\lambda_0 t_3^0}{gk^2 q}, \quad a_4 = \frac{d}{gk^2} + \frac{\lambda_0 t_2^0}{gk^2q^2} + \frac{k_1^2}{k^2q^4}. \tag{10.4.2}$$

To determine the critical load we may use equations (9.1.9) or (9.1.12) according to the type of shell edge support.

We will present the results of the solution of equation (9.1.9) only for the case of pure torsion ($t_1^0 = t_2^0 = 0$) of a shell of revolution with constant parameters E, ν, h and k_1. In this case formulae (2) have the form

$$a_2 = \frac{2a}{q_1}, \quad a_3 = 7.0\,R_3\,q_1^{-1/2}, \quad a_4 = 1 + \frac{a^2}{q_1^2}, \tag{10.4.3}$$

where

$$q_1 = q^2 l, \qquad R_3 = \frac{\tau}{\tau_0}, \qquad a = k_1 l, \tag{10.4.4}$$

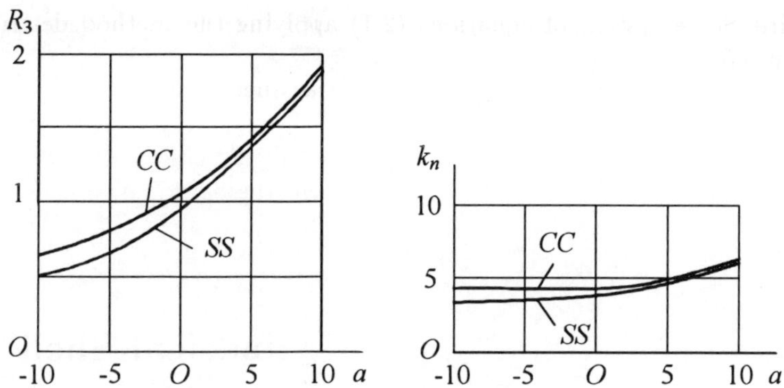

Figure 10.4: Functions $R_3(a)$ and $k_n(a)$.

and τ_0 is defined by formula (9.1.14) at $k_s = 0.74$.

Figure 10.4 shows plots of functions $R_3(a)$ and $k_n(a)$, where $k_n = 12^{1/8}q_1^{1/2}$ is the coefficient in formula (9.1.14). In this Figure the curves corresponding to shells with simply supported and clamped edges are marked respectively by SS and CC.

10.5 Problems and Exercises

10.1. Consider nearly cylindrical shell of revolution under external normal pressure. One of the shell edges is simply supported, the other is clamped. Plot graphics of the function $F(a)$ for the critical loading and $x(a)$ for the number of waves in the circumferential direction (see 10.3.4). Compare the graphs with those given in Figure 10.2.

Hint Assume $\alpha = 3.92$ in (3.3) in accordance with formula (8.6.3).

10.2. Consider a nearly conic shell of revolution with the generatrix of constant small curvature k_1 under external normal pressure. Study numerically the effect of the problem parameters on the critical load and the number of waves in the circumferential direction for the buckling mode.

Hint Solve system of equations (2.1) applying the method described in Section 8.7.

10.3. Consider a nearly conic shell of revolution with the generatrix of constant small curvature k_1 under torsion. Study numerically the effect of the problem parameters on the critical load and the number of waves in the circumferential direction for the buckling mode. Compare the results with those for nearly cylindrical shell under torsion given in Figure 10.4.

Hint Solve system of equations (2.1) applying the method described in Section 9.5.

Chapter 11

Shells of Revolution of Negative Gaussian Curvature

The stability of the membrane axisymmetric stress state of shells of revolution with negative Gaussian curvature will be considered in this Chapter. Assuming that the Gaussian curvature is not small, the buckling modes of this class of shell differs from those of shells which have positive or zero Gaussian curvature.

The localization of the deformation in the neighbourhood of lines (Chapter 4) or points (Chapter 6) is an important characteristic for shells of positive curvature while the deformation pits stretched along the generatrices (Chapters 7-10) are characteristics of shells of zero curvature, for example, under external normal pressure. This property is caused by the fact that the deformations have a tendency to spread along the asymptotic lines of the neutral surface.

In contrast, shells of revolution with negative Gaussian curvature have two systems of asymptotic lines. This fact leads to buckling modes, under axisymmetric loading, which spread along the entire neutral surface and results in a system of buckling pits which is reminiscent of a chess board.

All of the problems considered in this Chapter (in contrast to those examined in Chapters 6 through 10) allow separation of the variables and may be reduced to a system of eighth order ordinary differential equations. It is not difficult to solve this system by means of a numerical method [73, 115, 170]. In this Chapter we wish to clarify the qualitative side of the buckling of shells with negative Gaussian curvature by means of an asymptotic solution.

The buckling of an axisymmetrically loaded shell of alternating Gaussian curvature, the generatrix of which has a point of inflection is also considered.

In preparing this Chapter the results of [90, 91, 92, 130, 165] have been used.

It should be noted that at the present time an asymptotic solution of the problem of the buckling of a non-axisymmetrically loaded shell of revolution with negative Gaussian curvature has not yet been developed.

11.1 Initial Equations and Their Solutions

The critical load and buckling mode of a shell of revolution with negative Gaussian curvature depend on whether or not the tangential boundary conditions guarantee the absence of infinitesimally small neutral surface bending. First, we assume that there is no bending of the neutral surface. Then, it follows from (3.6.15) that the index of variation of the additional stress state under buckling $t = 1/3$ and we may use the system of equations of shallow shells (4.3.1) which we can rewrite in the form

$$\begin{aligned} \varepsilon_0^3 \Delta\,(d\,\Delta\,w) + \varepsilon_0 \lambda\,\Delta_t w - \Delta_k \Phi &= 0, \\ \varepsilon_0^3 \Delta\,(g^{-1} \Delta\,\Phi) + \Delta_k w &= 0, \end{aligned} \tag{11.1.1}$$

where

$$\varepsilon_0^6 = \frac{h_0^2}{12\,(1-\nu_0^2)\,R^2}, \qquad (T_1^0, T_2^0, S^0) = -\lambda\, E_0 h_0 \varepsilon_0^4\,(t_1, t_2, t_3)\,, \tag{11.1.2}$$

and the other notation is the same as in (4.3.1). Introducing the parameter λ by formulae (2) we take into account the expected order of the initial stress-resultants (see (3.6.15)).

The coefficients of system (1) do not depend on φ and that is why it is possible to separate the variables. We find the solution of system (1) with m waves in the circumferential direction as a formal power series in ε_0

$$w\,(s,\varphi) = \sum_{n=0}^{\infty} \varepsilon_0^n w_n(s) \exp\left\{\frac{i}{\varepsilon_0}\left(\rho\varphi + \int q\, ds\right)\right\}, \tag{11.1.3}$$

where $m = \varepsilon_0^{-1}\rho$. By virtue of (3.6.15) $t = 1/3$, that is why $\rho \sim 1$.

The substitution of (3) into (1) leads to a sequence of systems of equations. We find from the system of the zeroth approximation $\Delta_k \Phi_0 = \Delta_k w_0 = 0$

$$q = q_{1,2} = \pm\left(-\frac{k_1 \rho^2}{k_2 b^2}\right)^{1/2}, \qquad q_1 > 0, \quad q_2 = -q_1. \tag{11.1.4}$$

In the first approximation we obtain the system of equations for the evaluation of w_0 and Φ_0

$$\begin{aligned} 2i\sqrt{\frac{k_2 q}{b}}\,\frac{d}{ds}\left(\sqrt{b\,q\,k_2}\,w_0\right)+\frac{\rho^4}{gb^4}\left(\frac{k_1}{k_2}-1\right)^2\Phi_0 &= 0,\\ 2i\sqrt{\frac{k_2 q}{b}}\,\frac{d}{ds}\left(\sqrt{b\,q\,k_2}\,\Phi_0\right)-\frac{\rho^4}{gb^4}\left(\frac{k_1}{k_2}-1\right)^2 w_0+ &\\ +\lambda_0\left(q^2 t_1+2\,\frac{q\rho}{b}\,t_3+\frac{\rho^2}{b^2}\,t_2\right)w_0 &= 0. \end{aligned} \tag{11.1.5}$$

For the evaluation of w_n, Φ_n for $n \geqslant 1$ we obtain the non-homogeneous system of equations, the left sides of which coincide with the left sides of equations (5), and the right sides are expressed through previously known functions w_j and Φ_j where $j < n$.

Noting that q has two values (4), and system (5) is of the second order, we may construct four linearly independent solutions of system (1) of the type of (3). We call these solutions the *main* solutions. Four other (*additional*) solutions are edge effect solutions

$$w\,(s,\varphi)=\sum_{n=0}^{\infty} w_n^{\mathrm{e}}\,(s)\,\varepsilon_0^{n/2}\exp\Big\{\varepsilon_0^{-3/2}\int_{s_0}^{s}\left(r+\varepsilon_0 r^1\right)ds+\frac{i\rho\varphi}{\varepsilon_0}\Big\}, \tag{11.1.6}$$

where

$$r^4=-\frac{g\,k_2^2}{d}. \tag{11.1.7}$$

We may find the eigenvalues of the load parameter λ by substituting the linear combination of solutions (3) and (6) into the boundary conditions which are introduced on shell edges $s = s_1$ and $s = s_2$. However, it is more convenient, similar to Section 8.4, to separate the boundary conditions at each of the shell edges and to determine two *main boundary conditions* which must be satisfied under construction of *the main integrals.*

11.2 Separation of the Boundary Conditions

For all of the functions which determine the stress-strain state (in particular the functions contained in the formulation of the boundary conditions) we can represent the main solutions and the solutions of the edge effect respectively

Table 11.1: Indices of intensity

I	II	III	IV	V	VI
x	β_0	x_0^0	β_e	x_0^{e}	Δ_β
u	1	$-ik_1q^{-1}w_0^0$	$\frac{3}{2}$	$(k_1+\nu k_2)\,r^{-1}w_0^{\mathrm{e}}$	$-\frac{1}{2}$
v	1	$-ibk_2\rho^{-1}w_0^0$	2	$((2+\nu)\,k_2-k_1)\frac{i\rho}{br^2}w_0^{\mathrm{e}}$	-1
w	0	w_0^0	0	w_0^{e}	0
γ_1	-1	iqw_0^0	$-\frac{3}{2}$	qw_0^{e}	$\frac{1}{2}$
T_1	1	$-\frac{\rho^2}{b^2}\,\Phi_0^0$	1	$\frac{gk_2\rho^2}{r^2b^2}\,w_0^{\mathrm{e}}$	0
S	1	$\frac{q\rho}{b}\,\Phi_0^0$	$\frac{1}{2}$	$\frac{ig\rho k_2}{rb}\,w_0^{\mathrm{e}}$	$\frac{1}{2}$
Q_{1*}	3	$\frac{idq\rho^2}{b^2}\left(2-\nu-\frac{k_1}{k_2}\right)w_0^0$	$\frac{9}{2}$	$-dr^3w_0^{\mathrm{e}}$	$\frac{3}{2}$
M_1	4	$\frac{d\rho^2}{b^2}\left(\frac{k_1}{k_2}-\nu\right)w_0^0$	3	$dr^2w_0^{\mathrm{e}}$	1

in the form

$$
\begin{aligned}
x^0\left(s,\varphi,\varepsilon_0\right) &= x^0\left(s,\varepsilon_0\right)\exp\Big\{\frac{i}{\varepsilon_0}\Big(\rho\varphi+\int q\,ds\Big)\Big\},\\
x^0\left(s,\varepsilon_0\right) &= \varepsilon_0^{\beta_0(x)}\sum_{n=0}^{\infty}\varepsilon_0^n x_n^0\left(s\right),
\end{aligned}
\tag{11.2.1}
$$

$$
\begin{aligned}
y^{\mathrm{e}}\left(s,\varphi,\varepsilon_0\right) &= y^{\mathrm{e}}\left(s,\varepsilon_0\right)\exp\Big\{\frac{i\rho\varphi}{\varepsilon_0}+\varepsilon_0^{-3/2}\int_{s_0}^{s}\left(r+\varepsilon_0 r^1\right)ds\Big\},\\
y^{\mathrm{e}}\left(s,\varepsilon_0\right) &= \varepsilon_0^{\beta_e(x)}\sum_{n=0}^{\infty}\varepsilon_0^{n/2}y_n^{\mathrm{e}}\left(s\right).
\end{aligned}
\tag{11.2.2}
$$

Functions x_0^0 and y_0^{e} and the corresponding indices of intensity $\beta_0(x)$ and $\beta_e(x)$ are presented in Table 11.1. Table 11.1 gives the dimensionless functions introduced by formulae (8.2.1) and contained in the formulation of the boundary conditions.

Table 11.2: The boundary conditions

I	*II*	*III*	*IV*
1	$u = v = 0$	1111	3/2
		1110	3/2
		1101	2
		1100	5/2
		1011	1/2
		1010	1/2
2	$T_1 = v = 0$	0111	1/2
		0110	1
		0101	3/2
		0100	3/2
		0011	1/2
		0010	1/2
3	$u = S = 0$	1001	1
		1000	3/2
4	$T_1 = S = 0$	0001	1
		0000	1

We note that $u^0, v^0 \sim \varepsilon_0 w^0$ and all of the tangential additional stress-resultants for the main solutions (i.e. T_1^0, T_2^0 and S^0) have the same order.

Remark 11.1. Table 11.1 differs only in notation from Table 5.3 in [57], which contains the solutions of non-axisymmetric vibrations of shells of revolution with negative Gaussian curvature which have $m \sim \varepsilon_0^{-1}$ waves in the circumferential direction. This is caused by the fact that both the load term $\varepsilon_0 \lambda \Delta_t w$ in system (1.1) and the inertial terms in the vibration problem [57], are not the most important and are included only in the first approximation.

While the low-frequency vibration problem and the buckling problem for shells of revolution with negative Gaussian curvature are very similar, differential equations (1.5) for w_0 and Φ_0 and (13.10) from [57] are different. The problems also differ since the main interest in the buckling problem is in the smallest value of parameter λ.

The boundary conditions can be separated using the same method as in Section 8.4. We consider 16 boundary condition variants obtained as a result of equating to zero one of the values in each column

$$\begin{array}{ccccc} u & v & w & \gamma_1 & (1) \\ T_1 & S & Q_{1*} & M_1 & (0) \end{array} \tag{11.2.3}$$

We mark these variants by four-digit series of ones and zeros as was done in Section 1.2.

The results of the separation of the boundary conditions are remarkably similar to those obtained in Section 8.4. In the zeroth approximation the boundary conditions separate into four groups: the clamped support group, the simple support group, the weak support group and the free edge group. Each group contains the same conditions as in Section 8.4.

The boundary conditions which form a group are presented in column *III* of Table 11.2. The numbers α such that the error incurred in transforming from the complete boundary value problem to the simplified problem does not exceed ε_0^α are contained in column *IV*. The simplified problem is solved by substituting the linear combination of the main solutions (1) into the main boundary conditions which are shown in column *II* of Table 11.2.

Since solutions (1) are expanded in an integer power series in ε_0, and solutions (2) are expanded in a power series in $\varepsilon_0^{1/2}$ the parameter λ is also expanded in a power series in $\varepsilon_0^{1/2}$

$$\lambda = \lambda_0 + \varepsilon_0^{1/2}\lambda_1 + \varepsilon_0\lambda_2 + \varepsilon_0^{3/2}\lambda_3 + \cdots \tag{11.2.4}$$

If $\alpha \geqslant 1$ then $\lambda_1 = 0$, since the contribution of the solutions of the edge effect begins from terms of order ε_0^α. The parameter λ_1 differs from zero for five boundary condition variants for which $\alpha = 1/2$ (see Table 11.2). For these variants the transformation to the simplified problem has a maximum error of order $\varepsilon_0^{1/2}$.

As in Section 8.4, by eliminating the edge effect solutions we may reduce the incurred error. We do not consider all of the cases in detail but limit the discussion to one example. We consider at edge $s = s_2$, the boundary conditions 0010 or

$$T_1 = S = w = M_1 = 0. \tag{11.2.5}$$

As can be seen in Table 11.2, the replacement of conditions (5) by the conditions $T_1 = v = 0$ for the main solutions has an error of order $\varepsilon_0^{1/2}$. We can calculate the corrected conditions by means of formulae (8.4.9) and (8.4.10). We then obtain

$$T_1^0 + \frac{\rho^2}{\varepsilon_0^2 b k_2} M_1^0 = 0, \quad w_0 + \frac{i\sqrt{2}\, b\, |r|}{\sqrt{\varepsilon_0}\, \rho\, g\, k_2} S^0 = 0. \tag{11.2.6}$$

Taking into account the relation between the orders of w^0 and S^0 (see Table 11.1), we note that in the second of conditions (6), the second term has order $\varepsilon_0^{1/2}$ with respect to the first one. In the first condition of (6) the correction term has the relative order of $\varepsilon_0^{3/2}$ (and so it may be neglected). To prove this we must consider that M_1^0 is proportional to w^0, and by virtue of the second of conditions (6), w^0 has the order $\varepsilon_0^{1/2}$. In the example considered, $\lambda_1 \neq 0$ in expansion (4).

11.3 Boundary Problem in the Zeroth Approximation

We can reduce system of equations (1.5) to a second order equation

$$\frac{d}{ds}\left(f(s)\,\frac{dy_k}{ds}\right) + \left(\lambda_0\rho^4 f_k(s) - \frac{\rho^6}{f(s)}\right) y_k = 0, \quad k = 1, 2, \tag{11.3.1}$$

in which

$$\begin{aligned} f &= \frac{2gq_1b^4k_2^3}{\rho\,(k_1-k_2)^2}, \qquad f_k = \frac{1}{2k_2}\left(\frac{t_1q_1}{\rho} + \frac{2q_kt_3}{bq_1} + \frac{\rho t_2}{b^2q_1}\right), \\ y_k &= \sqrt{bk_2q_1}\; w_{0k}, \qquad \Phi_{0k} = -\frac{2igq_kb^4k_2^2}{\rho^4\,(k_2-k_1)^2}\left(\frac{k_2}{bq_1}\right)^{1/2}\frac{dy_k}{ds}. \end{aligned} \tag{11.3.2}$$

In equation (1) only function $f_k(s)$ depends on the sign of q_k at $t_3 \neq 0$, i.e. under torsion. If $t_3 = 0$ then equation (1) coincides with the equation obtained in [91], if, in contrast $t_1 = t_2 = 0$, then it coincides with that obtained in [92] which considered torsional buckling. We note that in [91, 92], in contrast to system of equations (1.1), more precise initial system in displacements have been taken.

We can now write the expressions for the displacements u, v and w and stress-resultants T_1 and S in the zeroth approximation which are contained in the formulation of the boundary conditions (see Table 11.2). Due to (2.1) and

Table 11.1 we find

$$u = u^* \left(y_1 e^{i\psi} - y_2 e^{-i\psi}\right) e^{im\varphi}, \qquad u^* = \frac{-ik_1}{\left(bk_2 q_1^3\right)^{1/2}};$$

$$v = v^* \left(y_1 e^{i\psi} + y_2 e^{-i\psi}\right) e^{im\varphi}, \qquad v^* = \frac{-i}{\rho} \left(\frac{bk_2}{q_1}\right)^{1/2};$$

$$w = w^* \left(y_1 e^{i\psi} + y_2 e^{-i\psi}\right) e^{im\varphi}, \qquad w^* = \varepsilon_0^{-1} \left(bk_2 q_1\right)^{-1/2};$$

$$T_1 = T_1^* \left(y_1' e^{i\psi} - y_2' e^{-i\psi}\right) e^{im\varphi}, \qquad T_1^* = \frac{2igb^2 k_2^2}{\rho^2 (k_1 - k_2)^2} \left(\frac{k_2 q_1}{b}\right)^{1/2},$$

$$S = S^* \left(y_1' e^{i\psi} + y_2' e^{-i\psi}\right) e^{im\varphi}, \qquad S^* = -\frac{2igb^3 k_2^2}{\rho^3 (k_1 - k_2)^2} \left(\frac{k_2 q_1^3}{b}\right)^{1/2};$$

$$\psi = \psi(s) = \int_{s_1}^{s} q_1 ds, \qquad y_k' = \frac{dy_k}{ds}. \tag{11.3.3}$$

Here y_1 and y_2 are the solutions of equations (1) for $k = 1, 2$. The multipliers u^*,v^*, T_1^* and S^* are cancelled after substitution into the boundary conditions and further we do not need to know their explicit form. The boundary conditions which the solutions y_1 and y_2 of equation (1) must satisfy depend on the main boundary condition variant being considered.

Remark 11.2. We indicated two tangential conditions which characterize the group and set in column *II* of Table 11.2. At the same time, it follows from Table 11.2 that condition $v = 0$ means that at least one of conditions $v = 0$ or $w = 0$ is introduced.

If at the shell edge (for example, at $s = s_1$) $u = v = 0$, then by virtue of formulae (3) we obtain $y_1 = y_2 = 0$.

If $T_1 = S = 0$, then $y_1' = y_2' = 0$.

However, if we consider one of the cases where $T_1 = v = 0$ or $u = S = 0$, then the functions y_1 or y_2 are not separated in formulating the boundary conditions. In the case where $T_1 = v = 0$ at $s = s_1$, the boundary conditions have the form $y_1 + y_2 = 0$, $y_1' - y_2' = 0$, and in the case where $u = S = 0$ they have the form $y_1 - y_2 = 0$, $y_1' + y_2' = 0$. In both cases we obtain the non-linear condition

$$(y_1 y_2)' = 0 \quad \text{for} \quad s = s_1, \tag{11.3.4}$$

connected the functions y_1 or y_2.

Let us now formulate and discuss the boundary value problems in which parameter λ_0 is the smallest eigenvalue. At the same time we will consider here the general case when the initial shear stresses differ from zero, i.e. $t_3 \not\equiv 0$ (2). The case where $t_3 \equiv 0$ is considered in Section 11.4.

If the conditions of shell edges $s = s_1$ or $s = s_2$ are clamped $u = v = 0$, then we come to two Sturm–Liouville problems for the functions y_1 and y_2, which satisfy equations (1) and the conditions

$$y_k = 0 \qquad \text{for} \qquad s = s_1, \quad s = s_2. \tag{11.3.5}$$

We may also evaluate the eigenvalues for this problem, $\lambda^{(k)}$, by minimization of the functional

$$\lambda^{(k)} = \min_y \left\{ \frac{I_1 + \rho^6 I_2}{\rho^4 I_3^{(k)}} \right\}, \qquad k = 1, 2, \tag{11.3.6}$$

where

$$I_1 = \int_{s_1}^{s_2} f\,(y')^2 ds, \quad I_2 = \int_{s_1}^{s_2} f^{-1} y^2 ds, \quad I_3^{(k)} = \int_{s_1}^{s_2} f_k y^2 ds. \tag{11.3.7}$$

The unknown value λ_0 is obtained as a result of minimization by ρ and k, i.e.

$$\lambda_0 = \min_{\rho, k} \left\{ \lambda^{(k)} \right\}. \tag{11.3.8}$$

Let $t_3 > 0$ in f_k for definiteness. Then $I_3^{(1)} > I_3^{(2)}$, and $\lambda^{(1)} < \lambda^{(2)}$. That is why

$$\lambda_0 = \min_{y, \rho} \left\{ \frac{I_1 + \rho^6 I_2}{\rho^4 I_3^{(1)}} \right\}. \tag{11.3.9}$$

By virtue of (2) and (1.4), functions f and f_k do not depend on ρ. That is why in (6) and (9), I_1, I_2 and $I_3^{(k)}$ do not also explicitly depend on ρ. Calculating in (9) the minimum by ρ we obtain

$$\lambda_0 = 3 \cdot 2^{-2/3} \min_y \left\{ \frac{\left(I_1 I_2^2\right)^{1/3}}{I_3^{(1)}} \right\}, \quad \rho_0 = m_0 \varepsilon_0 = \left(\frac{2 I_1}{I_2} \right)^{1/6}. \tag{11.3.10}$$

Consequently, the minimum is attained for finite values of ρ_0, different from zero, and buckling occurs for the integer values of m, which is the closest to m_0.

If clamped conditions are introduced at edge $s = s_1$, and edge $s = s_2$ is free (i.e. $T_1 = S = 0$) then all of the formulae written above are valid with the only difference being that functions y_k satisfy the conditions

$$y_k(s_1) = 0, \qquad y_k'(s_2) = 0. \tag{11.3.11}$$

In both cases considered here the buckling deformation in the zeroth approximation has the form

$$w = w^* y_1 \cos\beta(s,\varphi), \quad \beta = m(\varphi - \varphi_0) + \frac{1}{\varepsilon_0}\int_{s_1}^{s} q_1 ds, \tag{11.3.12}$$

where y_1 satisfies equation (1) and boundary conditions (5) or (11) and φ_0 is an arbitrary constant.

Assuming in (12) that $\beta =$ const, we note that the function $w(s,\varphi)$ describes the set of m pits which are inclined at a variable angle

$$\gamma(s) = -\arctan\left(-\frac{k_1}{k_2}\right)^{1/2} \tag{11.3.13}$$

to the generatrix.

Let edge $s = s_1$ be clamped, $u = v = 0$, and further let there be only one tangential support ($u = S = 0$ or $v = T_1 = 0$) introduced at edge $s = s_2$.

First, consider the case when $v = T_1 = 0$. We now arrive at the boundary value problem in which we must determine the smallest value of λ_0 such that the non-trivial solutions y_1 and y_2 of equations (1) exist and satisfy the boundary conditions

$$y_1 = y_2 = 0 \quad \text{at} \quad s = s_1, \tag{11.3.14}$$

$$y_1 e^{i\psi_2} + y_2 e^{-i\psi_2} = 0, \quad y_1' e^{i\psi_2} - y_2' e^{-i\psi_2} = 0 \quad \text{at} \quad s = s_2, \tag{11.3.15}$$

where $\psi_2 = \psi(s_2)$ (see formula (3)). It is clear that the functions y_1 and y_2 are complex-valued. We assume

$$y_1 = y_1^0 e^{-i\psi_2}, \qquad y_2 = y_2^0 e^{i\psi_2}. \tag{11.3.16}$$

Then for real functions y_1^0 and y_2^0, conditions (14) at $s = s_1$ and condition (4) at $s = s_2$ must be fulfilled. The deformation mode has the form

$$\begin{aligned} w(s,\varphi) &= w^*(y_1 \cos\beta_1 + y_2 \cos\beta_2), \\ \beta_k &= m(\varphi - \varphi_0) + \frac{1}{\varepsilon_0}\int_{s_1}^{s} q_k ds \end{aligned} \tag{11.3.17}$$

and represents the superposition of two systems of pits which overlap and are inclined at angles γ and $-\gamma$ to the generatrix. At the same time $y_1^0 + y_2^0 = 0$ at $s = s_2$.

In the case of conditions $u = S = 0$ at $s = s_2$ for the determination of λ_0 we have the same boundary value problem, however, in contrast to the case where $v = T_1 = 0$ one should take $y_1^0 - y_2^0 = 0$ at $s = s_2$.

As in the case when both shell edges are clamped the value λ_0 may be determined by minimizing the functional

$$\lambda_0 = \min_{\rho, y_k^0} \left\{ \frac{I_1^{(1)} + I_1^{(2)} + \rho^6 \left(I_2^{(1)} + I_2^{(2)} \right)}{\rho^4 \left(I_3^{(1)} + I_3^{(2)} \right)} \right\}, \tag{11.3.18}$$

where

$$I_1^{(k)} = \int_{s_1}^{s_2} f \left(y_k^{0\prime} \right)^2 ds, \quad I_2^{(k)} = \int_{s_1}^{s_2} f^{-1} \left(y_k^0 \right)^2 ds, \quad I_3^{(k)} = \int_{s_1}^{s_2} f_k \left(y_k^0 \right)^2 ds, \tag{11.3.19}$$

and

$$\begin{aligned} y_1^0 = y_2^0 = 0 \quad &\text{at} \quad s = s_1, \\ y_1^0 - y_2^0 = 0, \quad y_1^{0\prime} + y_2^{0\prime} = 0 \quad &\text{at} \quad s = s_2. \end{aligned} \tag{11.3.20}$$

We note that the last of boundary conditions (20) is natural and may be omitted when we choose the functions y_1^0 and y_2^0.

As in the case when both shell edges are clamped, it follows from (18) that the minimum is attained for a finite value of ρ which is not equal to zero.

Boundary conditions which allow non-trivial bending of the neutral surface are considered in Section 11.5.

11.4 Buckling Modes Without Torsion

In the case where $t_3 \equiv 0$ by virtue of (3.2) coefficient f_k in (3.1) does not depend on the choice of q_1 or q_2 and that is why the functions y_1 and y_2 satisfy the same equation (3.1). This fact causes the buckling mode to change as will be discussed below.

Let us consider the case when conditions $u = v = 0$ are introduced on both shell edges. To determine λ_0 we arrive at boundary value problem (3.1) and (3.5), however the buckling mode is not determined uniquely and is represented by formula

$$w(s,\varphi) = w^* y_1 \Big[C_1 \cos\Big(m(\varphi - \varphi_1) + \psi(s)\Big) + \\ + C_2 \cos\Big(m(\varphi - \varphi_2) - \psi(s)\Big)\Big], \qquad (11.4.1)$$

where C_1, C_2, φ_1 and φ_2 are arbitrary.

Deformation (1) represents the result of the overlapping of two systems of pits which are stretched along the asymptotic lines of the neutral surface.

We may turn from deformation form (1) to the form

$$w(s,\varphi) = w^* y \cos m(\varphi - \varphi_0) \cos\big(\psi(s) - \psi_0\big), \qquad (11.4.2)$$

where in the zeroth approximation φ_0 and ψ_0 are arbitrary. Deformation (2) describes a chess board system of pits. The constant φ_0 also remains arbitrary under the development of a higher approximation, and the constant ψ_0 may be found.

Let one of the four boundary condition variants for which $\alpha > 1$ (see Table 11.2) is introduced at each of the shell edges $s = s_1$, $s = s_2$. Then when developing the next approximation we may neglect the non-tangential boundary conditions. In expansion (2.4) $\lambda_1 = 0$, coefficient λ_2 and constant ψ_0 have two values $\lambda_2^{(1)}$, $\psi_0^{(1)}$ and $\lambda_2^{(2)}$, $\psi_0^{(2)}$, and $\lambda_2^{(2)} = -\lambda_2^{(1)}$. This construction for the vibration problem of a shell of negative curvature has been done in [89].

Therefore, the critical value λ is given by the approximate formula $\lambda = \lambda_0 + \varepsilon_0 \lambda_2^{(1)}$, $\lambda_2^{(1)} < 0$ and by virtue of the arbitrariness of φ_0, two independent buckling modes correspond to it. The value $\lambda = \lambda_0 + \varepsilon_0 \lambda_2^{(2)}$ differs from the critical one by a small amount of order ε_0 and two buckling modes correspond to it.

Now let conditions $u = v = 0$ be introduced at edge $s = s_1$, and conditions $v = T_1 = 0$ be introduced at $s = s_2$. As in Section 11.3, in the zeroth approximation we come to boundary value problem (3.1) with boundary conditions (3.14) and (3.15). However, since y_1 and y_2 satisfy the same equation, it follows from (3.14) that $y_2 = C y_1$. Then we obtain from conditions (3.15)

$$y_1 \left(e^{i\psi_2} + C\, e^{-i\psi_2}\right) = 0, \quad y_1' \left(e^{i\psi_2} - C\, e^{-i\psi_2}\right) = 0. \qquad (11.4.3)$$

As a result, the boundary value problem splits into two problems. They consist of equation (3.1) and the boundary conditions

$$y_1(s_1) = y_1(s_2) = 0, \qquad C = e^{2i\psi_2} \qquad (11.4.4)$$

or conditions

$$y_1(s_1) = y_1'(s_2) = 0, \qquad C = -e^{2i\psi_2}. \qquad (11.4.5)$$

It is clear that problem (3.1), (5) gives the smallest value of λ_0.

The deformation mode in the zeroth approximation is determined uniquely

$$w(s,\varphi) = w^* y_1 \cos m(\varphi - \varphi_0) \sin\left(\frac{1}{\varepsilon_0}\int_{s_2}^{s} q_1 ds\right) \tag{11.4.6}$$

but φ_0 is still arbitrary. As in the case of condition $u = v = 0$ on both edges, deformation (6) describes a system of pits resembling a chessboard. However, in contrast to the case of condition $u = v = 0$ on both edges the boundary value problem for fixed m has no eigenvalues close to the critical one.

11.5 The Case of the Neutral Surface Bending

Let the boundary conditions belong to the second or third group (see Table 11.2 and Remark 11.2) and let them be introduced at both shell edges. Under the construction of the main solutions in the zeroth approximation we satisfy the boundary conditions

$$v = T_1 = 0 \qquad \text{or} \qquad u = S = 0. \tag{11.5.1}$$

At some characteristic shell sizes (*eigensizes*) (see [20, 49, 172]) the bending of the neutral surface is permitted, and it leads to the decrease in the order of the critical load. That is why the equation for the determination of λ_0 essentially differs from the ones obtained in Sections 11.3 and 11.4.

To formulate this equation at $t_3 \neq 0$ (i.e. under torsion) we introduce the fundamental system of solutions y_{k1} and y_{k2} of equation (3.1), which at $s = s_1$ satisfies the initial conditions

$$y_{k1} = 0, \quad y'_{k1} = 1; \quad y_{k2} = 1, \quad y'_{k2} = 0, \quad k = 1, 2. \tag{11.5.2}$$

Then, in order to determine λ_0 we come to the equation

$$\left(y_{11}y_{22} + y_{12}y_{21}\right)'\Big|_{s=s_2} = \pm\frac{f(s_1)}{f(s_2)}\, 2\cos 2\psi_2, \tag{11.5.3}$$

in which we take the "plus" sign if the same boundary conditions (1) are introduced at both shell edges, and the "minus" sign if the conditions are different.

Later, we will arrive at the general case where $t_3 \neq 0$, but now let us assume that $t_3 = 0$. Then $y_{1j} = y_{2j}$ and equation (3) in the case of the same boundary conditions as (1) has the form

$$y_{11}y'_{12} = -\frac{f(s_1)}{f(s_2)}\sin^2\psi_2, \tag{11.5.4}$$

and in the case of different boundary conditions has the form

$$y_{11}y'_{12} = -\frac{f(s_1)}{f(s_2)}\cos^2\psi_2, \tag{11.5.5}$$

where

$$\psi_2 = \psi(s_2) = m\int_{s_1}^{s_2}\left|\frac{k_1}{k_2 b^2}\right|^{1/2} ds. \tag{11.5.6}$$

We will now consider in more detail the case when both shell edges are simply supported ($v = T_1 = 0$). The other cases may be considered in a similar fashion.

Let the parameters of the neutral surface and the number m be such that $\sin\psi_2 = 0$. We then have again the Sturm–Liouville problem, which consists of equation (3.1) and the boundary conditions

$$y'_1(s_1) = y'_1(s_2) = 0. \tag{11.5.7}$$

For the given m let the minimum eigenvalue for this problem be $\lambda_0(m)$. For its approximate evaluation we assume in (3.6) that $y \equiv 1$. Then

$$\lambda_0(m) = \frac{\rho^2 L_1}{L_2}, \quad L_1 = \int_{s_1}^{s_2}\frac{ds}{f}, \quad L_2 = \int_{s_1}^{s_2} f_1 ds, \quad \rho = \varepsilon_0 m. \tag{11.5.8}$$

To determine if the value $\lambda_0(m)$ is critical or not we must consider the other values of m.

Let now $\sin\psi_2 \neq 0$. The functions y_{k1} and y_{k2} satisfy the Volterra integral equations

$$\begin{aligned} y_{k1} &= \int_{s_1}^{s}\frac{f(s_1)\,dt}{f(t)} - \int_{s_1}^{s}\left(\int_{s_1}^{t} F_k(t_1)\, y_{k1}(t_1)\, dt_1\right)\frac{dt}{f(t)}, \\ y_{k2} &= 1 - \int_{s_1}^{s}\left(\int_{s_1}^{t} F_k(t_1)\, y_{k2}(t_1)\, dt_1\right)\frac{dt}{f(t)}, \end{aligned} \tag{11.5.9}$$

where

$$F_k(s) = \lambda_0\rho^4 f_k(s) - \rho^6 f^{-1}(s), \qquad k = 1, 2. \tag{11.5.10}$$

We can solve equation (9) by means of an iterative method starting with $y^0_{k1} = y^0_{k2} = 0$. This method converges faster with smaller values of the functions $|F_k|$. For $t_3 = 0$ we take only the functions y_{11} and y_{12} (since $y_{2j} = y_{1j}$). Assuming that $|F_k| \ll 1$ we limit the discussion to approximate solutions (9) in the form

$$y_{11}(s) = \int_{s_1}^{s} \frac{f(s_1)}{f(t)}\,dt, \quad y_{12} = 1 - \int_{s_1}^{s} \left(\int_{s_1}^{t} F_1(t_1)\,dt_1 \right) \frac{dt}{f(t)}. \tag{11.5.11}$$

After substituting into (4) we obtain the approximate formula

$$\lambda_0(m) \simeq \frac{\sin^2 \psi_2}{\rho^4 L_1 L_2} + \frac{\rho^2 L_1}{L_2}, \tag{11.5.12}$$

which generalizes formula (8).

The critical value of parameter λ_0 is $\lambda_0 = \min \lambda_0(m)$. The cases when $\sin \psi_2$ is close to zero for a small value of m are the least favourable for shell stability. We note that $\sin \psi_2 = 0$ approximately corresponds to the case when the neutral surface has bending satisfying the conditions $v = 0$ at $s = s_1$ and $s = s_2$.

Now we obtain a formula of the same type as (12) for $t_3 \neq 0$. First let us take the approximate solutions of equations (9) in the form

$$\begin{aligned} y_{k1} &= \int_{s_1}^{s} \frac{f(s_1)}{f(t)}\,dt - \int_{s_1}^{s} \left(\int_{s_1}^{t} F_k(t_1) \int_{s_1}^{t_1} \frac{dt_2}{f(t_2)}\,dt_1 \right) \frac{f(s_1)}{f(t)}\,dt, \\ y_{k2} &= 1 - \int_{s_1}^{s} \left(\int_{s_1}^{t} F_k(t_1)\,dt_1 \right) \frac{dt}{f(t)}. \end{aligned} \tag{11.5.13}$$

The substitution into equation (3) after neglecting terms of order $|F_k|^2$ leads to the same equation (12) in which L_1 has the previous value and

$$L_2 = \frac{1}{2} \int_{s_1}^{s_2} (f_1 + f_2)\,ds. \tag{11.5.14}$$

Recalling (3.2) we note that the value of L_2 does not depend on t_3, consequently the effect of torsion is not detected in the approximation being considered.

If the shell is loaded only under torsion (i.e. $t_1 = t_2 = 0$, $t_3 \neq 0$) then $L_2 = 0$ and formula (12) is not valid.

To obtain a more accurate relation in the case when all $t_i \neq 0$, we retain terms of order $|F_k|^2$ in solutions of the same type as (13). Then equation (3) may be written in the form

$$\lambda_0 \rho^4 L_1 L_2 + \lambda_0^2 \rho^8 L_1 L_3 = \sin^2 \psi_2 + \rho^6 L_1^2, \tag{11.5.15}$$

where L_1 and L_2 are given by formulae (8) and (14) and

$$L_3 = L_1 \int_{s_1}^{s_2} \left(\int_{s_1}^{s} \frac{f_1 - f_2}{2}\, dt \right)^2 \frac{ds}{f} - \left(\int_{s_1}^{s_2} \left(\int_{s_1}^{s} \frac{f_1 - f_2}{2}\, dt \right) \frac{ds}{f} \right)^2. \qquad (11.5.16)$$

For $L_2 \neq 0$, $L_3 \neq 0$ equation (15) allows us to account for the effect of torsion on the critical load.

However, for $L_2 = 0$ (i.e. under pure torsion) equation (15) leads to an unexpected result: if the neutral surface size is close to the eigensize for small m then it does not cause (in contrast to the case where $t_3 = 0$, $t_1, t_2 \neq 0$) a decrease in the order of the critical load. Indeed, let $\sin \psi_2 = 0$ for small m ($\rho \ll 1$). Then $\lambda_0(m) \sim \rho^{-1}$, i.e. decreasing m (starting from the same value) leads to an increase of $\lambda_0(m)$. We will discuss this question again in Sections 11.6 and 12.2.

We may make some remarks about formulae (12) and (15). These formulae were developed under the assumption that $|F_k| \ll 1$, $m \gg 1$, which may be not fulfilled simultaneously. For small m (for example, $m = 2$ or $m = 3$) constructed solution (1.3) does not guarantee the required accuracy since both system (1.1) and the solution method are not sufficiently accurate. Cases when the neutral surface has bending for small m will be considered in Chapter 12.

It follows from (12) and (15) that the critical value of the parameter λ_0 for a shell of revolution of negative curvature with support ($v = 0$) at each shell edge may change in range from values of order unity to values of order ε_0^2 depending on the size of the shell.

Let σ be the characteristic value of the membrane initial shell stresses under buckling. We consider the dimensionless value $\sigma_* = \sigma/(Eh_*)$, where $h_* = h_0/R$. Then for strongly supported shells with negative curvature, $\sigma_* \sim h_*^{1/3}$ (see formula (1.2) and Sections 11.3 and 11.4).

For shells which have one support $v = 0$ or $u = 0$ at each edge, the order of σ_* varies from $h_*^{1/3}$ to h_* depending on the size of the shell. It follows from Table 11.2 that the same result is obtained if we replace condition $v = 0$ by $w = 0$ or if $v = w = 0$ simultaneously, however, the composition of conditions $u = w = 0$ gives $\sigma_* \sim h_*^{1/3}$.

In order to compare the results we recall that for the strongly supported shells of positive curvature $\sigma_* \sim 1$ (see (3.1.3)), and for the shells of zero curvature under external pressure $\sigma_* \sim h_*^{1/2}$ (see (3.5.5)).

11.6 The Buckling of a Torus Sector

To illustrate the results of Sections 11.3-11.5 we will now consider the buckling of a torus sector which has negative Gaussian curvature, under the axial force $\boldsymbol{P}$ and the torsional moment $\boldsymbol{M}$, applied to shell edges s_1 and s_2 (see Figure 11.1). The torus is obtained by rotation of an arc of a circle of radius R around axis OO'. Let R_0 be a distance between the centre C of this circle and the axis. We take R as the unit length. Then

$$k_1 = -1, \quad b = a - \cos s, \quad k_2 = b^{-1} \cos s, \quad a = \frac{R_0}{R} > 1. \tag{11.6.1}$$

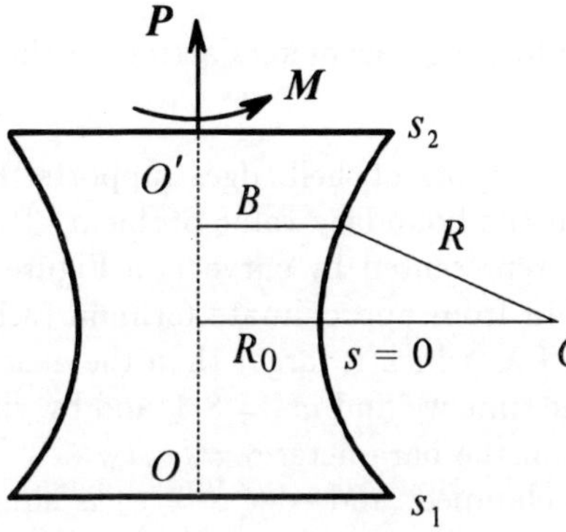

Figure 11.1: A torus sector.

By virtue of (1.4.6) the initial stress-resultants are the following

$$T_1^0 = \frac{P}{2\pi R b^2 k_2}, \quad T_2^0 = \frac{T_1^0}{k_2}, \quad S^0 = \frac{M}{2\pi R^2 b^2}. \tag{11.6.2}$$

We take in (1.2)

$$t_1 = \frac{l_P}{b^2 k_2}, \quad t_2 = \frac{l_P}{b^2 k_2^2}, \quad t_3 = \frac{l_M}{b^2}, \tag{11.6.3}$$

then

$$\begin{aligned} P &= 2\pi R E h \varepsilon_0^4 R_P, & R_P &= -\lambda l_P, \\ M &= -2\pi R^2 E h \varepsilon^4 R_M, & R_M &= \lambda l_M. \end{aligned} \tag{11.6.4}$$

We introduce l_P and l_M and seek the critical values of the parameter $\lambda > 0$. We will consider both compressive ($l_P > 0$) and tensile ($l_P < 0$) stress-resultants T_1^0, T_2^0, however due to symmetry we can only consider the case when $l_M \geqslant 0$.

The numerical parameters in the example are: $a = 1.5$, $\nu = 0.3$, $s_2 = -s_1 = \pi/4$ and $h/R = 0.003$, then $\varepsilon_0 = 0.0968$.

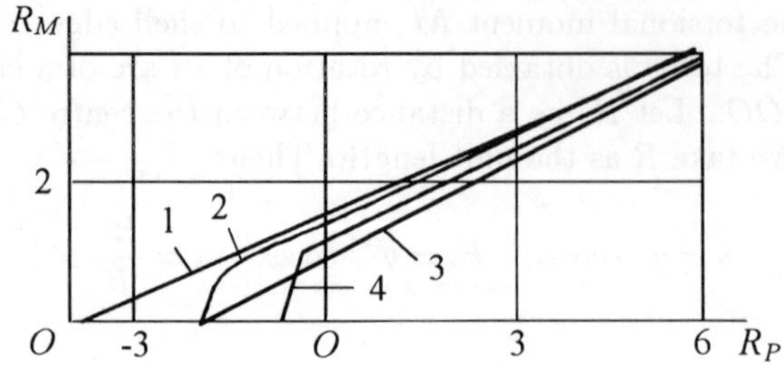

Figure 11.2: The loading parameters acting on the torus at buckling.

Consider the various types of shell edges supports. First, let both edges be clamped. This is then the boundary value problem (3.1), (3.5) for $k = 1$. The numerical solution is represented by curve 1 in Figure 11.2. Buckling occurs for $m = 8$. We obtain from approximate formula (3.10) at $R_P \leqslant 3$ that at $y = \cos 2s$ the value of λ_0 which is larger than the exact solution by not more than 1%. At the same time we find $m_0 = 8.3$, and by virtue of (3.10) the value m_0 does not depend on the parameter $\gamma = l_P/l_M$.

If edge $s = s_1$ is clamped, and edge $s = s_2$ is simply supported then on edge $s = s_2$ we arrive at condition (3.4). The results of numerical integration are shown as curve 2 in Figure 11.2. If $R_P < 0.7$, buckling occurs for $m = 7$, and if $R_P > 0.7$ buckling occurs for $m = 8$.

If edge $s = s_1$ is clamped, and edge $s = s_2$ is free, we come to boundary value problem (3.1), (3.11) for $k = 1$. The corresponding results are shown as curve 3 in Figure 11.2. If $R_P \leqslant 5$, buckling occurs for $m = 7$.

As the tensile axial stress ($R_P > 0$) increases, curves 1, 2 and 3 in Figure 11.2 approach each other. To explain this effect, note that by virtue of (3.2) and (3)

$$f_k(s) = \frac{l_P}{\cos^2 s}\Big(\cos s\,(a - \cos s)\Big)^{-1/2} - \frac{(-1)^k l_M}{\cos s\,(a - \cos s)^2}. \tag{11.6.5}$$

For $l_P < 0$ we have $f_2(s) < 0$ for all s, and $f_1(s) > 0$ for all s, if $\gamma > -0.84$. For $-0.84 > \gamma > -2.8$ the function $f_1(s)$ changes its sign in the interval (s_1, s_2) while at the same time near the endpoints of this interval $f_1(s) < 0$. It follows that the solutions of the corresponding boundary value problems are damped as we approach edges s_1 and s_2 and the solution is less dependent on the type of support. Localization of the buckling mode near parallel $s = 0$ occurs in a manner similar to that described in Section 4.2.

If we change the wave number m in the circumferential direction then the eigenvalue $\lambda_0(m)$ of the corresponding boundary value problem has a minimum for $m = 7$ or $m = 8$ in all of the cases considered (see, for example, curve 4 in

Figure 11.3, which corresponds to the case when $l_P = l_M = 1$ with clamped edge $s = s_1$ and free edge $s = s_2$).

We have another picture in the case of simply supported edges $s = s_1$ and $s = s_2$. It follows from approximate formulae (5.12) and (5.15) that the value of $\lambda_0(m)$ depends on $\sin\psi_2(m)$, and the critical load depends on whether or not $\sin\psi_2$ is close to zero for a small value of m. We recall that if $\sin\psi_2$ is close to zero it means that the shell size is close to the eigensize.

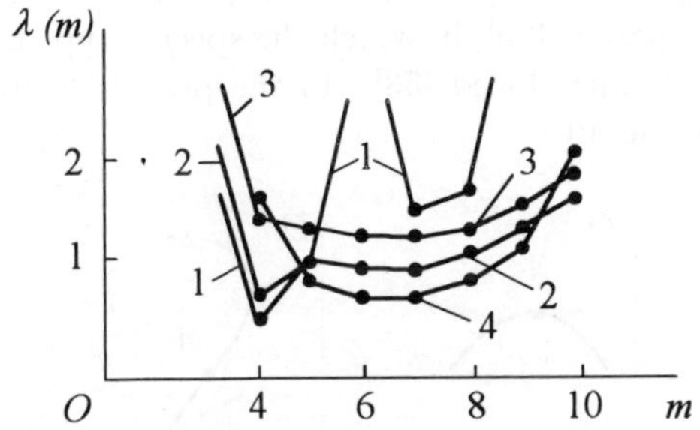

Figure 11.3: The values of $\lambda(m)$.

For simply supported edges we come to boundary value problem (3.1), (5.2), (5.3). The solution, after minimization by m are shown as curve 4 in Figure 11.2. We have $\sin\psi_2 = 0.18$ at $m = 3$ and that is why without torsion ($l_M = 0$) while under small torsion ($\gamma = \dfrac{l_P}{l_M} > 0.4$) buckling occurs for $m = 3$. Under pure torsion (without any axial force) shell buckling takes place for $m = 7$. As was noted in Section 11.5, if the shell size is close to eigensize then the critical load and buckling mode are not affected. The calculation shows that the same case also takes place under tensile stresses P and small compressive stresses ($\gamma < 0.4$).

It follows from above that curve 4 in Figure 11.2 has two distinct parts. The left hand portion of the curve corresponds to $m = 3$ and to the buckling mode close to surface bending, and the right portion of the curve corresponds to $m = 6 - 9$.

We will now consider another shell with the same parameters and $s_2 = -s_1 = 0.86$. For this shell $\sin\psi_2 = 0$ if $m = 4$. Let the shell edges be simply supported. The graphs of $\lambda_0(m)$ for $l_P = 1$, $l_M = 0$ (curve 1), for $l_P = 0.5$, $l_M = 1$ (curve 2) and for $l_P = 0$, $l_M = 1$ (curve 3) are plotted in Figure 11.3. For comparison, curve 4 gives $\lambda_0(m)$ for a shell, one edge of which is clamped and the other is free for $l_P = l_M = 1$.

Curves 1 and 2 have two minimums at $m = 4$ and $m = 7$, the first is caused by bending. One can see that the effect of bending decreases as the torsion becomes relatively larger.

11.7 Shell with Gaussian Curvature of Variable Sign

There are two types of shells of revolution with alternating Gaussian curvature. Toroidal shells, the curvature k_2 of which changes sign at some specific line (point A in Figure 11.4a) is the first type. The development of asymptotic solutions of the equations of statics, dynamics and buckling of this type of shell has been the subject of many studies (see, for example, [12, 24, 82, 119]). We note especially the papers [20], in which the specific type of quasi-bending of the neutral surface is introduced [53]. In the present Section we will not discuss this problem in detail.

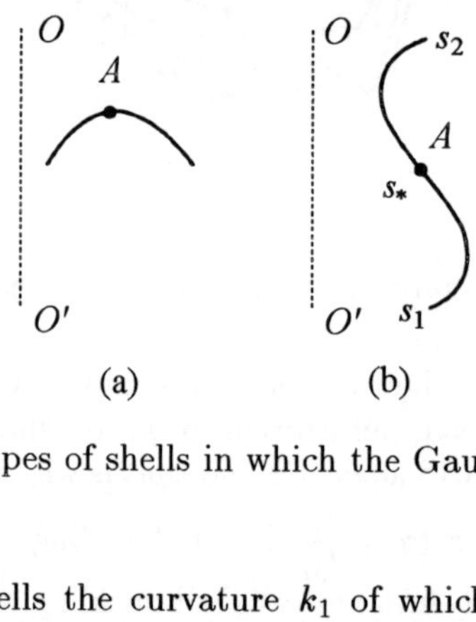

Figure 11.4: The two types of shells in which the Gaussian curvature changes sign.

We will consider shells the curvature k_1 of which changes sign at some parallel $s = s_*$ (point A in Figure 11.4b). The generatrix has a point of inflection at point A. We assume that $k_1'(s_*) < 0$, i.e. the part $s_* < s \leqslant s_2$ of the shell has negative curvature.

We begin with system of equations (1.1) and seek its solution in the form of (1.3) or (3.3). It follows from (1.4), that $q_1 = q_2 = 0$ at $s = s_*$, i.e. point s_* is the turning point for system of equations (1.1). In the neighbourhood of $s = s_*$ by virtue of (1.4), $q_1 \sim (s - s_*)^{1/2}$ and equation (3.1) has a singular point $s = s_*$. We consider only the case without torsion ($t_3 \equiv 0$). Then, in the neighbourhood of point $s = s_*$ equation (3.1) has two solutions of the form

$$\begin{aligned} y^{(1)} &= 1 + a_1(s - s_*) + \cdots , \\ y^{(2)} &= (s - s_*)^{1/2}\,[1 + b_1(s - s_*) + \cdots]. \end{aligned} \tag{11.7.1}$$

At the same time by virtue of (3.3), all solutions of system of equations (1.1) are non-regular.

We seek the regular solutions of system (1.1) in the form

$$w(s,\varphi) = \left[w^{(1)}(s,\varepsilon_0)\, U(\eta) + w^{(2)}(s,\varepsilon_0)\, \frac{dU}{d\eta} \right] e^{im\varphi}, \tag{11.7.2}$$

where $U(\eta)$ is the solution of the Airy equation

$$\frac{d^2U}{d\eta^2} + \eta\, U = 0, \qquad \eta = \varepsilon_0^{-2/3} \zeta(s), \tag{11.7.3}$$

and the functions $w^{(k)}$ are the formal power series in ε_0^2

$$w^{(k)} = \varepsilon_0^{\gamma_k(w)} \sum_{l=0}^{\infty} \varepsilon_0^{2l} w^{(k,l)}(s). \tag{11.7.4}$$

In the same form (2) we seek the rest of the functions, determining the stress-strain state of the shell.

At this point we will discuss further constructions very briefly. The details may be found in [130, 165], where the initial system of equations more accurate than (1.1) has been considered. However, this circumstance does not effect the zeroth approximation.

Unknown functions $\zeta(s)$, $w^{(k,l)}(s)$ are determined as a result of substituting solutions (2) into system of equations (1.1). They are regular at $s = s_*$, and

$$\eta(s) = \left(\frac{3}{2\varepsilon_0} \int_{s_*}^{s} q_1\, ds \right)^{2/3}. \tag{11.7.5}$$

We assume that the shell edges are sufficiently far from point $s = s_*$, i.e.

$$\varepsilon_0^{-1} \int_{s_*}^{s_2} q_1\, ds \gg 1, \qquad \varepsilon_0^{-1} \int_{s_1}^{s_*} |q_1|\, ds \gg 1. \tag{11.7.6}$$

Then $\eta(s_2) \gg 1$, $-\eta(s_1) \gg 1$ and substituting solutions (2) into the boundary conditions we may use the asymptotic formulae for the Airy functions as $\eta \to \pm\infty$ (see [15, 63]),

$$\begin{gathered} U_1 \sim \pi^{-1/2} \eta^{-1/4} \cos\gamma, \qquad U_2 \sim \pi^{-1/2} \eta^{-1/4} \sin\gamma, \qquad \eta \to \infty; \\ U_1 \sim (1/2)\, \pi^{-1/2} (-\eta)^{-1/4} e^{-\delta}, \quad U_2 \sim \pi^{-1/2} (-\eta)^{-1/4} e^{\delta}, \quad \eta \to -\infty, \\ \gamma = \frac{2}{3} \eta^{3/2} - \frac{\pi}{4}, \qquad \delta = \frac{2}{3} (-\eta)^{3/2}. \end{gathered} \tag{11.7.7}$$

In searching for $w^{(k,l)}$ we arrive at second order differential equations. Therefore, noting that equation (3) also has two linearly independent solutions, U_1 and U_2, in total four solutions of system (1.1) of type (2) have been developed.

It follows from (7) that for $s > s_*$, all solutions (2) oscillate. For $s < s_*$ two solutions corresponding to $U = U_1$, decrease exponentially away from the turning point, and the other two solutions corresponding to $U = U_2$, increase exponentially away from the turning point.

If at least one support $u = 0$, $v = 0$ or $w = 0$ is introduced at edge $s = s_1$, which borders the part of the shell which has positive curvature, this edge cannot cause buckling. To satisfy the boundary conditions at edge $s = s_2$ we must take two solutions (2) for $U = U_1$, assuming that, since they decrease exponentially for $s < s_*$, they approximately satisfy any boundary conditions at $s = s_1$.

We can now also present the explicit form of the main (real) terms of solutions (2) for $U = U_1$ using asymptotic formulae (7) as $\eta \to \infty$ (i.e. for $s > s_*$)

$$\begin{aligned}
u &= \left(-C_1 y^{(1)} \sin\psi_* + C_2 y^{(2)} \cos\psi_*\right) |u^*| \cos m(\varphi - \varphi_0), \\
v &= \left(C_1 y^{(1)} \cos\psi_* + C_2 y^{(2)} \sin\psi_*\right) |v^*| \sin m(\varphi - \varphi_0), \\
w &= \left(C_1 y^{(1)} \cos\psi_* + C_2 y^{(2)} \sin\psi_*\right) |w^*| \cos m(\varphi - \varphi_0), \\
T_1 &= \left(-C_1 y^{(1)\prime} \sin\psi_* + C_2 y^{(2)\prime} \cos\psi_*\right) |T_1^*| \cos m(\varphi - \varphi_0), \\
S &= \left(C_1 y^{(1)\prime} \cos\psi_* + C_2 y^{(2)\prime} \sin\psi_*\right) |S^*| \sin m(\varphi - \varphi_0), \\
\psi_*(s) &= \frac{1}{\varepsilon_0}\int_{s_*}^{s} q_1 ds - \frac{\pi}{4},
\end{aligned} \tag{11.7.8}$$

where C_1, C_2 and φ_0 are arbitrary constants, $y^{(1)}$ and $y^{(2)}$ are the solutions of equation (3.1) which have expansions (1) at $s = s_*$ and u^*, $v^* \ldots$ are determined by formulae (3.3).

Under the above assumptions for determining λ_0 we should substitute solution (8) into the main boundary conditions at $s = s_2$.

Let clamped conditions $u = v = 0$ be introduced at $s = s_2$. Then the above substitution leads to the equation

$$y^{(1)}(s_2, \lambda_0)\, y^{(2)}(s_2, \lambda_0) = 0, \tag{11.7.9}$$

from which it follows that the critical value λ_0 is equal to the smallest roots of equations $y^{(1)} = 0$, $y^{(2)} = 0$, into which equation (9) may be split. It can be shown that it is the equation $y^{(1)}(s_2, \lambda_0) = 0$ which gives the critical value of λ_0.

The buckling mode in the zeroth approximation has the form

$$w(s,\varphi) = w_0 \cos m(\varphi - \varphi_0), \quad w_0 = \varepsilon_0^{-1/6} w_1(s)\, U_1[\eta(s,\varepsilon_0)], \tag{11.7.10}$$

where $w_1(s)$ does not depend on ε_0. For the case where $s > s_*$, by virtue of (7)

$$w_0 = y^{(1)} |w^*| \cos\psi_*. \tag{11.7.11}$$

For the case when $s < s_*$, the function w_0 exponentially decreases as $|s - s_*|$ increases, and for $s > s_*$, it oscillates as shown in the graph of $w_0(s)$ in Figure 11.5. The portion of the shell with negative Gaussian curvature is covered by a system of pits that resembles a chess board. In the neighbourhood where $s = s_*$, the pits are deepest.

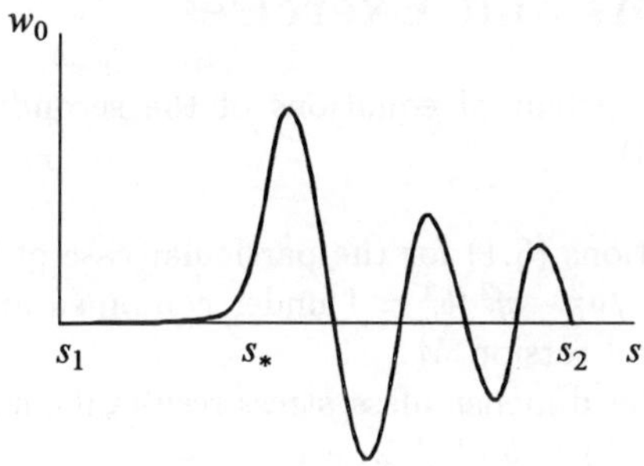

Figure 11.5: Schematic of function $w_0(s)$.

Now, let the conditions $v = T_1 = 0$ be introduced at $s = s_2$. Then the equation for the determination of λ_0 has the form

$$y^{(1)} y^{(2)\prime}\big|_{s_2} \cos^2\psi_{2*} + y^{(2)} y^{(1)\prime}\big|_{s_2} \sin^2\psi_{2*} = 0, \quad \psi_{2*} = \psi_2(s_*). \tag{11.7.12}$$

By virtue of the identity $f\left(y^{(1)}y^{(2)\prime} - y^{(2)}y^{(1)\prime}\right) = \text{const}$ equation (12) is written in the form

$$y^{(1)\prime} y^{(2)} = -\frac{c_f}{f(s_2)} \cos^2\psi_{2*}, \quad c_f = g b^3 \left(-k_1' k_2\right)^{1/2}, \tag{11.7.13}$$

where c_f is calculated at $s = s_*$.

In this case as in Section 11.5, we may obtain the approximate formula for λ_0 if $\cos\psi_{2*}$ is close to zero for small m. Using the inequality $|F_1| \ll 1$, we take the approximate solutions of equation (3.1) in a form similar to (5.11)

$$y^{(1)} = 1 - \int_{s_*}^{s} \left(\int_{s_*}^{t} F_1(t_1)\, dt_1 \right) \frac{dt}{f(t)}, \quad y^{(2)} = c_f \int_{s_*}^{s} \frac{dt}{f(t)}. \tag{11.7.14}$$

Now equation (13) gives

$$\lambda_0(m) = \frac{\cos^2\psi_{2*}}{\rho^4 L_{1*} L_{2*}} + \frac{\rho^2 L_{1*}}{L_{2*}},$$
$$L_{1*} = \int_{s_*}^{s_2} \frac{ds}{f}, \qquad L_{2*} = \int_{s_*}^{s_2} f_1\, ds. \tag{11.7.15}$$

In the case of the boundary conditions $u = S = 0$ at $s = s_2$ we obtain similar results with the only difference being that $\cos^2\psi_{2*}$ is replaced by $\sin^2\psi_{2*}$ in (13) and (15).

11.8 Problems and Exercises

11.1. Derive the system of equations of the second approximation for functions $w_1(s)$ in (1.3)

11.2. Derive equations (3.1) for the particular case of hyperboloid of revolution of one sheet $x^2/a_0^2 - y^2/b_0^2 = 1$ under combined axial compression P, external pressure q^0 and torsion M.

Hint To obtain the dimensionless stress-resultants use relations (1.4.6). Geometric functions in (3.1) and (3.2) are

$$k_2 = a_0^{-2}\sqrt{a_0^2 \sin^2\theta - b_0^2 \cos^2\theta}, \qquad b = k_2^{-1}\sin\theta,$$

$$k_1 = -a_0^4 b_0 - 2k_2^3, \qquad d(\ )/ds = -k_1\, d(\ )/d\theta,$$

where θ is the angle between the axis of revolution and the normal.

11.3. Find the critical load and the wave number in the circumferential direction for hyperboloid of revolution of one sheet $x^2/a_0^2 - y^2/b_0^2 = 1$ under axial compression P. Shell edges $\theta_1 = 2\pi/3$ and $\theta_2 = 4\pi/3$ are clamped. Assume $a_0/h = 0.005$, $b_0 = a_0$, $\nu = 0.3$, and $d = g = 1$.

Hint For arbitrary m solve boundary value problem (3.1) with the boundary conditions $y(\theta_1) = y(\theta_2) = 0$. Then find the minumum of function $\lambda(m)$.

11.4. Use approximate relations (3.10) to determine λ_0 and m_0 for Problem 11.3 assuming $y = (\theta_2 - \theta)(\theta - \theta_1)$. Compare the result with that of Problem 11.3.

11.5. Consider Problem 11.3 assuming shell edges $\theta_{1,2} = \pi/2 \pm \theta_0$, $\theta_0 < \pi/4$ to be simply supported. Find the eigensizes of the shell.

Hint For arbitrary m and θ solve boundary value problem (3.1) with the boundary conditions $y'(\theta_1) = y'(\theta_2) = 0$. Obtain the function $\lambda(m, \theta)$ and then find its minumum.

11.6. Use approximate relation (5.12) to solve Problem 11.5 and compare the results.

Chapter 12

Surface Bending and Shell Buckling

In this Chapter the buckling of a membrane axisymmetric state of a weakly supported shell of revolution will be considered. The case when non-trivial bending of the neutral surface exists which, as a rule, leads to a decrease in the order of the critical load, are also considered.

We will also consider cases without pure bending that satisfy the boundary conditions but with deformations of the neutral surface that are close to bending. Such deformations lead to a decrease in the order of the critical load compared to that for a well supported shell.

12.1 The Transformation of Potential Energy

It is noted above that the consideration of deformations which are close to neutral surface bending imposes a special requirement on the accuracy of the initial system of equations. That is why the expressions which specify (2.3.10) and (2.3.11) are introduced here.

As in Section 2.2, we represent the unknown functions in the form of sums $u_i^0 + u_i$, $w_i^0 + w_i$, $T_i^0 + T_i$, $M_i^0 + M_i$, ... and expand potential energy (1.10.1) as the sum of homogeneous terms with respect to the additional unknowns u_i, w, ... (henceforth, for brevity we will refer to the initial displacements u_i^0 and to the additional displacements u_i and in the same manner we refer to other unknowns).

$$\Pi = \Pi_\varepsilon + \Pi_\varkappa = \Pi_\varepsilon^0 + \Pi_\varkappa^0 + \Pi_\varepsilon^1 + \Pi_\varkappa^1 + \Pi_\varepsilon^2 + \Pi_\varkappa^2 + \cdots , \qquad (12.1.1)$$

where Π_ε^0 and $\Pi_\varkappa^0$ are calculated by formulae (1.10.1), (1.1.10) and (1.1.13) for displacements u_i^0 while at the same time, Π_ε^0 is non-linear in u_i^0. Further we can write

$$\Pi_\varepsilon^1 = \iint_\Omega (T_1^0 \varepsilon_{11}^1 + S^0 \varepsilon_{12}^1 + T_2^0 \varepsilon_{22}^1)\, d\Omega,$$
$$\Pi_\varkappa^1 = \iint_\Omega (M_1^0 \varkappa_1 + 2\, H^0 \tau + M_2^0 \varkappa_2)\, d\Omega, \tag{12.1.2}$$

where T_i^0 and M_i^0 are the initial stress-resultants and stress-couples and $\varkappa_i$ and τ are calculated by formulae (1.1.13) for displacements u_i, and

$$\varepsilon_{ii}^1 = \varepsilon_i + \varepsilon_i^0 \varepsilon_i + \omega_i^0 \omega_i + \gamma_i^0 \gamma_i,$$
$$\varepsilon_{12}^1 = \omega + \varepsilon_1^0 \omega_2 + \omega_2^0 \varepsilon_1 + \varepsilon_2^0 \omega_1 + \omega_1^0 \varepsilon_2 + \gamma_1^0 \gamma_2 + \gamma_2^0 \gamma_1. \tag{12.1.3}$$

Here ε_i^0, ω_i^0 and γ_i^0 are calculated by formulae (1.1.7) for displacements u_i^0, and ε_i, ω_i and γ_i are calculated by formulae (1.1.7) for displacements u_i.

The second order terms in u_i are equal to

$$\Pi_\varepsilon^2 = \Pi_\varepsilon^{21} + \Pi_\varepsilon^{2T}, \tag{12.1.4}$$

$$\Pi_\varepsilon^{2T} = \frac{1}{2} \iint_\Omega \Big[T_1^0 \left(\varepsilon_1^2 + \omega_1^2 + \gamma_1^2\right) + T_2^0 \left(\varepsilon_2^2 + \omega_2^2 + \gamma_2^2\right) + S^0 \left(\varepsilon_1 \omega_2 + \varepsilon_2 \omega_1 + \gamma_1 \gamma_2\right) \Big]\, d\Omega, \tag{12.1.5}$$

where Π_ε^{21} is calculated by formula (1.10.1), in which deformations ε_{ij} are taken from (3). In (2) $\Pi_\varkappa^2$ is calculated for displacements u_i.

Let δA be the elementary work of external forces (1.10.4) on displacements δu_i. We represent δA as a sum

$$\delta A = \delta A_1 + \delta A_2, \tag{12.1.6}$$

where δA_1 and δA_2 depend on u_i^0 and δu_i while at the same time δA_1 does not depend on u_i, and δA_2 linearly depends on u_i.

Then, in order to determine the initial state, u_i^0, one can use equation (1.10.5)

$$\delta\, \Pi_\varepsilon^1 + \delta\, \Pi_\varkappa^1 = \delta A_1 \tag{12.1.7}$$

and the additional state u_i is determined from the equation

$$\delta\, \Pi_\varepsilon^{21} + \delta\, \Pi_\varepsilon^{2T} + \delta\, \Pi_\varkappa^2 = \delta A_2. \tag{12.1.8}$$

In the case of a conservative external load we may introduce the work of the external forces, $A(u_i^0 + u_i)$ which depend on the displacements. Let us represent it as in (1), in the form

$$A = A_0 + A_1 + A_2 + \cdots, \tag{12.1.9}$$

where A_0 does not depend on u_i, A_1 linearly depends on u_i, and A_2 has quadratic terms in u_i (i.e. $u_i^2, u_i u_j$).

If the external load is "dead", i.e. the surface and boundary loads maintain the same value and direction when the shell deforms, then the work, A_0, linearly depends on u_i^0, A_1 linearly depends on u_i, and the other terms in (9) are absent $(A_2 = A_3 = \cdots = 0)$.

For a conservative load we may substitute differential relation (8) in the finite relation

$$\Pi_\varepsilon^{21} + \Pi_\varepsilon^{2T} + \Pi_\varkappa^2 = A_2. \tag{12.1.10}$$

The discussions of this Section have had a general character until now. At this point we will consider the buckling of a shell of revolution under a membrane axisymmetric stress state, with some simplifying assumptions. Namely, we will ignore any pre-buckling shell displacements. We assume that under buckling, the bifurcation into a non-axisymmetric form takes place. We also assume that the external load q and the initial stress-resultants T_i^0 and S^0 linearly depend on the loading parameter λ

$$q = \lambda R^{-1} T q^0, \quad (T_i^0, S^0) = -\lambda T(t_i, t_3), \quad T = E_0 h_0 \mu^2, \tag{12.1.11}$$

where we assume that λ, $q^0 = q^0(s)$ and $t_i = t_i(s)$ are dimensionless.

Under these assumptions, in the case of a conservative external load we may write the buckling problem in a variational form similar to (3.6.1)

$$\lambda = \min_{u_i}{}^{(+)} \frac{\Pi'_\varepsilon + \mu^4 R^2 \Pi'_\varkappa}{\mu^2 \Pi'_T}, \quad \mu^4 = \frac{h_0^2}{12(1-\nu^2) R_0^2}, \tag{12.1.12}$$

where we calculate the minimum by all u_i, which satisfy the geometric boundary conditions and for which the value of Π'_T is positive. Here Π'_ε and $\Pi'_\varkappa$ are the same as in (3.6.1), and

$$\Pi'_T = \left(A_2 - \Pi_\varepsilon^{2T}\right)(\lambda T)^{-1}. \tag{12.1.13}$$

Analyzing the buckling modes close to bending, we may neglect in (5) terms containing ε_i and ω, and according to (11) we can write

$$\Pi'_T = \frac{1}{2} \iint\limits_\Omega \Big[t_1(\gamma_1^2 + \omega_1^2) + t_2(\gamma_2^2 + \omega_2^2) + 2 t_3 \gamma_1 \gamma_2 + \\ + \frac{\delta_c q^0}{2R}(\gamma_1 u + \gamma_2 v)\Big] d\Omega, \tag{12.1.14}$$

where $\delta_c = 0$ for "dead" external loads and $\delta_c = 1$ for a follower normal pressure which is also a conservative load. Furthermore, we assume that $\delta_c = 1$.

Let the additional stress-strain state under buckling have high variability (the index of variation $t > 0$, see (1.3.4)). Then with an error of order h_*^{2t} we may neglect the terms with ω_i^2 and δ_c in (14) (see [55], and also (2.3.11), (3.6.2)). In some of the cases considered below, $t = 0$, and if we do not keep these terms in determining the critical load we incur an error which, while small, does not converge to zero with h_*.

12.2 Pure Bending Buckling Mode of Shells of Revolution

Let us consider the system of equations $\varepsilon_1 = \varepsilon_2 = \omega = 0$ and seek its solution in the form

$$u = u(s)\cos m\varphi, \quad v = v(s)\sin m\varphi, \quad w = w(s)\cos m\varphi. \tag{12.2.1}$$

Then, this system has the form

$$\frac{du}{ds} - \frac{w}{R_1} = 0, \quad \frac{B'}{B}u + \frac{m}{B}v - \frac{w}{R_2} = 0, \quad B\frac{d}{ds}\left(\frac{v}{B}\right) - \frac{m}{B}u = 0 \tag{12.2.2}$$

and leads to the equation

$$\frac{d}{ds}\left(BR_2\frac{d}{ds}\left(\frac{v}{B}\right)\right) - m^2\,\frac{R_2^2}{B^2R_1}v = 0. \tag{12.2.3}$$

Let non-trivial bending (i.e. the solutions of system (2) which satisfy all of the geometrical boundary conditions) exist. Then in (1.12) $\Pi'_\varepsilon = 0$ and

$$\lambda = \mu^2 \min_{u_i(m)}{}^{(+)} J'(m, u_i(m)), \quad J' = \frac{R^2\Pi'_\varkappa}{\Pi'_T}, \tag{12.2.4}$$

where the minimum is calculated for all possible cases of bending.

As was noted in Section 3.6, the existence of bending decreases the critical load. In cases where there is no bending or the deformations closely resemble bending $\lambda \sim 1$ (as follows from (1.12)). At the same time with bending $\lambda \sim h_*$.

The method used here is similar to the method of studying low frequency shell vibrations (see [57]). The difference is that in the case of vibration there is a quantity which is proportional to the shell kinetic energy in the denominator of functional (4). Also, the solution of buckling problems for weakly supported shells is less important, from a practical point of view, than the solution of vibration problems.

Listed below are the main cases of support under which a shell of revolution has non-trivial bending.

A shell in the shape of a cupola has bending if its edge is free.

A shell with two edges has bending if only one support is applied at one of its edges.

A shell with two edges has bending if one of the edges ($s = s_1$) is supported by a diaphragm, which is absolutely rigid in-plane and absolutely flexible out-of-plane. In this case the geometrical boundary conditions have the form

$$w \sin\theta - u\cos\theta = 0, \qquad v = 0 \quad \text{for} \quad s = s_1, \tag{12.2.5}$$

where θ is the angle between the axis of revolution and the shell normal. Here we should actually satisfy only the condition $v = 0$. At the same time the first of conditions (5) is automatically fulfilled by virtue of $\varepsilon_2 = 0$.

A shell with two free edges has for each m two-parametric set of bending and we should minimize functional (4) in this set.

A shell which has positive Gaussian curvature with two edges, has no bending for $m \neq 0$, if at least one support is introduced at each of its edges.

The question of the existence of bending for a shell of revolution of zero curvature may be easily solved, since one can write explicitly the set of solutions of system (2). For a cylindrical shell

$$u = C_2 \frac{1}{m^2}, \quad v = C_1 \frac{1}{m} + C_2 \frac{s}{mR}, \quad w = C_1 + C_2 \frac{s}{R} \tag{12.2.6}$$

and for a conical shell

$$\begin{aligned} u &= C_2 \frac{\sin\alpha}{m^2}, \\ v &= C_1 \frac{s}{m\, s_2} - C_2 \frac{1}{m}, \\ w &= C_1 \frac{s}{s_2 \cos\alpha} + C_2 \frac{\sin^2\alpha - m^2}{m^2\cos\alpha}, \end{aligned} \tag{12.2.7}$$

where 2α is the angle at the top of the cone, C_1 and C_2 are arbitrary constants, and s is the distance from the top of the cone.

A shell of negative Gaussian curvature may have bending if at each of its edges only one support is introduced (or the edge is supported by a diaphragm). Bending exists for some shell eigensizes for which boundary value problem (3) has eigenvalues m^2 where m is an integer for corresponding boundary conditions which depend on the support type.

A shell with torsional moments applied to its edges will not buckle by pure bending modes. It follows from this fact, that by virtue of (1.14), $\Pi'_T = 0$ in (4). In other words, the initial stress-resultant, S^0, does not do work on the additional displacements (1) which agrees with the finding of Section 11.5.

Let us now consider some examples.

Let a shell be under an external follower pressure $q^*(s)$. We denote $q^{**} = \max\limits_s\{q^*(s)\}$. Then for the critical pressure, we find by formulae (4) and (1.11) that

$$q^{**} = \frac{E\,h_0^3}{12\,(1-\nu^2)\,R^3}\,\min J', \tag{12.2.8}$$

where the functional J' is the same as in (4).

Example 12.1. Consider a long cylindrical shell under constant external pressure q^*. Let the boundary supports not be introduced or at least let their effect be negligible. Then, by virtue of (5) the following are bending deformations

$$u = 0, \qquad v = \frac{1}{m}, \qquad w = 1 \tag{12.2.9}$$

and formula (4) gives $\lambda_0 = \left(m^2-1\right)\mu^2$. For $m = 2$ we come to the well-known Grashof–Bress formula (3.5.7). In this problem, $\omega_2 = 0$, and the correction introduced in Π'_T is not revealed.

Example 12.2. Next, consider a moderately long cylindrical shell of length L and radius R under external pressure $q^*(s)$. Let edge $s = 0$ be simply supported ($T_1 = v = w = M_1 = 0$), and edge $s = L$ be free. The bending deformations are then

$$u = \frac{1}{m^2}, \qquad v = \frac{s}{Rm}, \qquad w = \frac{s}{R}. \tag{12.2.10}$$

Then by (4) we obtain

$$\lambda_0 = \frac{\mu^2\,(m^2-1)}{J_q}\left(1 + \frac{6\,(1-\nu)}{m^2 l^2}\right), \quad l = \frac{L}{R}, \tag{12.2.11}$$

$$J_q = \frac{3}{L^3}\int_0^L s^2 q^0\,(s)\,ds. \tag{12.2.12}$$

If, in (1.14) we omit terms with ω_i^2, then we obtain instead of (12)

$$J_q = \frac{3}{L^3}\int_0^L s^2 q^0\,(s)\,ds - \frac{3\,R^2}{L^3 m^2\,(m^2-1)}\int_0^L q^0(s)\,ds. \tag{12.2.13}$$

In particular, for homogeneous external pressure $q^0 \equiv 1$ and for $m = 2$ we find by (11) and (12) that

$$J' = 3\left(1 + \frac{3\,(1-\nu)}{2\,l^2}\right)J_q^{-1}, \quad J_q = 1. \tag{12.2.14}$$

We note that as the shell length ($l \to \infty$) increases, formula (14) evolves into the Grashof–Bress formula.

We obtain from formula (13) for $m = 2$

$$J_q = 1 - \frac{1}{4\,l^2}. \tag{12.2.15}$$

For hydrostatic pressure which increases linearly with depth

$$q^0\,(s) = 1 - \frac{s}{L}, \quad J_q = \frac{1}{4}, \quad J_q = \frac{1}{4} - \frac{1}{8\,l^2}, \tag{12.2.16}$$

where the values of J_q are obtained respectively by (12) and (13). This last problem models the buckling of an empty cylindrical pail immersed in a container of liquid.

12.3 The Buckling of a Weakly Supported Shell of Revolution

In this Section we will consider cases when the shell edge supports prevent pure bending of the neutral surface, but there are shell deformations that closely resemble bending. In [57] which considers low-frequency shell vibrations, the possible tangential bending of the neutral surface, i.e. shell deformations which satisfy (2.2) and the tangential supports, are introduced. It was shown in [57] that in the case of vibration, the existence of possible tangential bending leads to a decrease in the order of the lowest frequency. It will become clear in the Section below that this result is not always valid in buckling problems.

Let us begin with (1.12) and represent the additional stress-strain state as a sum of three terms: the bending, membrane and edge effect

$$w = w^{\mathrm{b}} + w^{\mathrm{m}} + w^{\mathrm{e}}. \tag{12.3.1}$$

The bending deformations satisfy equation (2.2) and the membrane deformations satisfy the system of membrane equations

$$\begin{aligned}
&\frac{dT_1}{ds} + \frac{B'}{B}\,(T_1 - T_2) + \frac{m}{B}\,S = 0, \\
&\frac{d}{ds}\,(B^2 S) - B^2 m\,T_2 = 0, \qquad \frac{T_1}{R_1} + \frac{T_2}{R_2} = 0, \\
&Eh\varepsilon_1 = T_1 - \nu\,T_2, \quad Eh\varepsilon_2 = T_2 - \nu\,T_1, \quad Eh\omega = 2\,(1+\nu)\,S
\end{aligned} \tag{12.3.2}$$

and in this system $\Pi'_\varepsilon > 0$.

We assume that $w^{\mathrm{b}} \sim 1$, $w^{\mathrm{m}} \sim \mu^{\alpha_{\mathrm{m}}}$, $w^{\mathrm{e}} \sim \mu^{\alpha_{\mathrm{e}}}$. Then $\alpha_{\mathrm{m}} > 0$, since otherwise the bending deformation is not dominant.

Due to (1) we can represent all the variables in (1.12) as the sum of three terms, for example,

$$\varepsilon_1 = \varepsilon_1^{\mathrm{b}} + \varepsilon_1^{\mathrm{m}} + \varepsilon_1^{\mathrm{e}} \quad \left(\varepsilon_1^{\mathrm{b}} = 0\right), \quad \varkappa_1 = \varkappa_1^{\mathrm{b}} + \varkappa_1^{\mathrm{m}} + \varkappa_1^{\mathrm{e}}. \tag{12.3.3}$$

Then

$$\begin{aligned} \Pi'_\varepsilon &= \Pi_\varepsilon^{\mathrm{m}} + \Pi_\varepsilon^{\mathrm{e}} + \Pi_\varepsilon^{\mathrm{me}}, \\ \Pi'_\varkappa &= \Pi_\varkappa^{\mathrm{b}} + \Pi_\varkappa^{\mathrm{e}} + \Pi_\varkappa^{\mathrm{be}}, \\ \Pi'_T &= \Pi_T^{\mathrm{b}} + \Pi_T^{\mathrm{e}} + \Pi_T^{\mathrm{be}}. \end{aligned} \tag{12.3.4}$$

Here $\Pi_\varepsilon^{\mathrm{m}}$, $\Pi_\varkappa^{\mathrm{b}}$, Π_T^{b}, $\Pi_\varepsilon^{\mathrm{e}}$, $\Pi_\varkappa^{\mathrm{e}}$ and Π_T^{e} are constructed by substituting into (3.6.1), (1.14) and (1.15) the bending and membrane deformations and deformations due to the edge effect, $\Pi_\varepsilon^{\mathrm{me}}$ ($\Pi_\varkappa^{\mathrm{be}}$, Π_T^{be}) are bi-linear integrals of a form which consists of membrane (or bending) and edge effect deformations.

It can be shown that the order of the last terms in (4) is always less than the sum of the first two term. Below, we find only the main term in the expansion λ by the powers of μ and that is why we omitted the last terms in (4). For the same reason, by virtue of $\alpha_{\mathrm{m}} > 0$ we omit the membrane compared to the bending deformations.

As a result (1.12) has the form

$$\lambda = \min_{u_i}{}^{(+)} \frac{\Pi_\varepsilon^{\mathrm{m}} + \mu^4 R^2 \Pi_\varkappa^{\mathrm{b}} + Z^{\mathrm{e}}}{\mu^2 \left(\Pi_T^{\mathrm{b}} + \Pi_T^{\mathrm{e}}\right)}, \qquad Z^{\mathrm{e}} = \Pi_\varepsilon^{\mathrm{e}} + \mu^4 R^2 \Pi_\varkappa^{\mathrm{e}}, \tag{12.3.5}$$

while at the same time, the orders of the variables in (5) for $m \sim 1$ are the following

$$\Pi_\varepsilon^{\mathrm{m}} \sim \mu^{2\alpha_{\mathrm{m}}}; \quad \Pi_\varkappa^{\mathrm{b}}, \Pi_T^{\mathrm{b}} \sim 1; \quad Z^{\mathrm{e}} \sim \mu^{2\alpha_{\mathrm{e}}+1}. \tag{12.3.6}$$

Estimating the orders of $\Pi_\varepsilon^{\mathrm{e}}$ and $\Pi_\varkappa^{\mathrm{e}}$ we take into account that the edge effect solutions are essential only in a band whose width is of order μ. The order of Π_T^{e} depends on the edge conditions in the neighbourhood of the edge $s = s_0$ being considered, namely

$$\begin{aligned} &\Pi_T^{\mathrm{e}} \sim \mu^{2\alpha_{\mathrm{e}-1}} && \text{for} \quad t_1(s_0) \neq 0, \\ &\Pi_T^{\mathrm{e}} \sim \mu^{2\alpha_{\mathrm{e}}} && \text{for} \quad t_1(s_0) = 0, \quad \left|t'_1(s_0)\right| + \left|t_3(s_0)\right| \neq 0, \\ &\Pi_T^{\mathrm{e}} \sim \mu^{2\alpha_{\mathrm{e}+1}} && \text{in the other cases}. \end{aligned} \tag{12.3.7}$$

At the same time we assume that all $|t_i| \lesssim 1$.

We can now present some more useful formulae for the edge effect solutions

$$y_j^e(s,\mu) = \tilde{y}_j^e(s,\mu)\, e^{\mu^{-1} r_j (s-s_0)}, \qquad r_j^4 + \frac{1}{R^2 R_2^2(s_0)} = 0, \tag{12.3.8}$$

where y is any unknown function, $\tilde{y}_j^e$ are power series in μ, the higher order terms of which are constant, while the others are polynomials in powers of s.

The following estimates and relations for the higher terms of series (8) are then valid

$$\tilde{w}^e, \tilde{\gamma}_2^e \sim \mu^{\alpha_e}; \quad \tilde{u}^e, \tilde{\omega}_1^e, \tilde{\omega}_2^e, \tilde{\omega}^e \sim \mu^{\alpha_e+1}; \quad \tilde{\varkappa}_2^e, \tilde{\tau}^e \sim \mu^{\alpha_e-1};$$
$$\tilde{\gamma}_1^e = -\frac{r}{\mu}\tilde{w}^e, \quad \tilde{\varepsilon}_2^e = -\frac{\tilde{w}^e}{R_2}, \quad \tilde{\varepsilon}_1^e = \frac{\nu \tilde{w}^e}{R_2}, \quad \tilde{\varkappa}_1^e = \frac{r^2}{\mu^2}\tilde{w}^e. \tag{12.3.9}$$

Here, the subscript j is omitted.

We can write the solution of the edge effect in the zeroth approximation

$$w^e = C_1 \exp\frac{r_1(s-s_0)}{\mu} + C_2 \exp\frac{r_2(s-s_0)}{\mu}, \tag{12.3.10}$$

where C_1 and C_2 are constants to be determined later, s_0 is the edge parallel in a neighbourhood of which the edge effect ($\Re\, r_j < 0$ for $s_0 = s_1 < s_2$) is constructed. Then

$$Z^e = \mu B C_1 C_2 \left(\frac{2R}{R_2^3}\right)^{1/2} \left(1 + O(\mu)\right),$$
$$\mu^2 \Pi_T^e = \frac{\mu B}{4}\left(r_1 C_1^2 + 4 r_1 r_2 C_1 C_2 (r_1 + r_2)^{-1} + \right.$$
$$\left. + r_2 C_2^2\right)\left(t_1(s_0) + O(\mu)\right). \tag{12.3.11}$$

The multiplier π, which appears in the integration by φ in the numerator and denominator (5), is omitted.

We write Z^e (with an error of order μ) for different types of support conditions

$$Z^e = \mu^3 B \left(\gamma_1^b\right)^2 \left(\frac{R^3}{8R_2}\right)^{1/2}$$
$$\text{for} \quad Q_1^e = 0, \quad \gamma_1^b + \gamma_1^e = 0; \tag{12.3.12}$$

$$Z^e = \mu B \left(w^b\right)^2 \left(\frac{R}{8R_2^3}\right)^{1/2}$$
$$\text{for} \quad w^b + w^e = 0, \quad M_1^e = 0; \tag{12.3.13}$$

$$Z^e = \mu B \left(w^b\right)^2 \left(\frac{R}{2R_2^3}\right)^{1/2} + \mu^2 B w^b \gamma_1^b \frac{R}{R_2} + \mu^3 B \left(\gamma_1^b\right)^2 \left(\frac{R^3}{2R_2}\right)^{1/2}$$
$$\text{for} \quad w^b + w^e = 0, \quad \gamma_1^b + \gamma_1^e = 0, \tag{12.3.14}$$

where all of the values are calculated at $s = s_0$. The value of Π_T^{e} is comparatively small in all of the cases considered later.

Beginning with a shell of positive curvature with one edge, $s = s_0$, i.e. the shell is a cupola. For fixed m system (2.2) has a unique solution u^{b}, v^{b}, w^{b}, which is regular in the top of the cupola.

First we introduce the support $\gamma_1(s_0) = 0$. Then, by virtue of (6), (9), (12) we find

$$\alpha_{\mathrm{e}} = 1, \qquad \alpha_{\mathrm{m}} = 2, \qquad \lambda = \frac{Z^{\mathrm{e}}}{\mu^2 \Pi_T^{\mathrm{b}}}\Big(1 + O(\mu)\Big) \sim \mu;$$
$$\Pi_T^{\mathrm{b}} = \frac{1}{2}\int_0^{s_0}\Big[t_1\left(\omega_1^2 + \gamma_1^2\right) + t_2\left(\omega_2^2 + \gamma_2^2\right) + \qquad (12.3.15)$$
$$+ \frac{\delta_c q^0}{R}(u\,\gamma_1 + v\,\gamma_2)\Big] B\,ds,$$

where Z^{e} is determined by (12), and in order to calculate Π_T^{b} we take the bending deformations.

Remark 12.1. For a strongly supported convex shell $\lambda \sim 1$, for a shell which has non-trivial bending $\lambda \sim \mu^2$. It follows from (15) that support introduced $\gamma_1(s_0) = 0$ leads to an increase in the order of the parameter λ ($\lambda \sim \mu$).

Now we can show that the minimum value λ is obtained for $m = 2$. For this we assume that $m \gg 1$, and find a solution of system (2.2) in the form

$$x^{\mathrm{b}}(s,m) = m^{\beta(x)}\sum_{k=0}^{\infty} m^{-k} x_k(s)\, e^{\pm m\int p\,ds}, \quad p = \frac{1}{B}\left(\frac{R_2}{R_1}\right)^{1/2}, \qquad (12.3.16)$$

where x is any unknown function.

Let $w^{\mathrm{b}} \sim 1$, then

$$u^{\mathrm{b}}, v^{\mathrm{b}} \sim m^{-1}; \quad \gamma_1^{\mathrm{b}}, \gamma_2^{\mathrm{b}} \sim m; \quad \varkappa_1^{\mathrm{b}}, \varkappa_2^{\mathrm{b}}, \tau^{\mathrm{b}} \sim m^2. \qquad (12.3.17)$$

Let us study the case when the sign is positive in formula (16). Then functions (16) increase fast with the growth of s. That is why in calculating the integrals of the functions of type (16), the order of the integral is m times less than the order of the integrand. Indeed, for $\psi'(s) > 0$ we have [15, 37]

$$\int_{s_1}^{s_2} f(s)\,\exp(m\,\psi(s))\,ds = \left[\frac{f(s_2)}{m\,\psi'(s_2)} + O\left(\frac{1}{m^2}\right)\right] e^{m\psi(s_2)}. \qquad (12.3.18)$$

It is not difficult to find the order of the variables in (14)

$$\Pi_T^{\mathrm{b}} = O\,(m), \qquad \lambda \sim \mu\, m, \tag{12.3.19}$$

from which it follows that under buckling, $m = 2$.

Now let the support $w\,(s_0) = 0$ be introduced (or simultaneously two supports $w\,(s_0) = \gamma_1(s_0) = 0$). The same developments show that in this case buckling occurs for $m \sim \mu^{-1}$ while at the same time $\lambda \sim 1$, i.e. the existence of the possible tangential bending (under its construction the non-tangential supports $w = 0$ and $\gamma_1 = 0$ are neglected) does not lead to a decreasing in the order of the critical load. This means that the buckling problem differs from the vibration problem, for which in this case a decrease in the order of the natural frequency does take place.

The difference mentioned is related to the denominator in (1.12). In the vibration problem, the main term in Π_T' is w^2 (see [57]), while in the case of buckling, the dominant term is the quantity $t_1\gamma_1^2 + 2t_3\gamma_1\gamma_2 + t_2\gamma_2^2$, which is of order m^2w^2.

Consider a convex shell with two edges and limit the discussion to the case when edges $s = s_1$ and $s = s_2$ are simply supported. As was noted above, if we introduce one or two supports described by (2.5), the neutral surface has non-trivial bending. We assume that the supports do not allow non-trivial bending. The free vibrations of such a shell are considered in [57] and we will use a similar method here for the buckling problem.

Let us take the bending solution (16)

$$\begin{aligned} w_1^{\mathrm{b}} &= \left(w_0(s) + O\left(m^{-1}\right)\right) \exp\left\{m \int_{s_2}^{s} p\, ds\right\}, \\ w_0 &= (R_1\, R_2)^{1/4}\,, \end{aligned} \tag{12.3.20}$$

which damps far away from the free edge $s = s_2$.

To satisfy the boundary conditions at $s = s_1$ we add to the solution the functions w_2^{b}, w^{m} and w^{e} which are respectively the bending solution, membrane solution and the edge effect solution which damps far away from edge $s = s_1$.

By virtue of (20), $w_1^{\mathrm{b}} \sim 1$. We can calculate the values of the variables in (5) for $m \gg 1$. The values of $\Pi_\varkappa^{\mathrm{b}}$ and Π_T^{b} are determined by the contribution of w_1^{b} in the neighbourhood of edge $s = s_2$, where the displacements and bending deformations are most essential.

We now have

$$\Pi^{\mathrm{b}}_{\varkappa} = \frac{m^3 R_1}{4\,B^2}\left[\left(1+\frac{R_2}{R_1}\right)^2 - 4\,\nu\,\frac{R_2}{R_1}\right]\left(1+O\left(m^{-1}\right)\right),$$
$$\Pi^{"}_{T} = \frac{m\,R_1}{4}\left(t_0+O\left(m^{-1}\right)\right), \quad t_0 = t_1\,\frac{R_2}{R_1}+t_2, \tag{12.3.21}$$

where R_i, B and t_0 are calculated at $s = s_2$. The other variables in (5) depend on the form of the boundary conditions at $s = s_1$.

We will consider only the case when $u = v = 0$ at $s = s_1$. Using relations developed in [57] we obtain

$$\Pi^{\mathrm{m}}_{\varepsilon} = \frac{4\,B^2 e^{-mc}}{m\,R_1}\left[\left(1+\frac{R_2}{R_1}\right)^2 + 4\,\nu\,\frac{R_2}{R_1}\right]^{-1}\left(1+O\left(m^{-1}\right)\right),$$
$$c = 2\int\limits_{s_1}^{s_2} p\,ds, \tag{12.3.22}$$

where R_i and B are calculated at $s = s_1$, and the value of $\Pi^{\mathrm{m}}_{\varepsilon}$ is determined by the contribution of the tangential deformations in the neighbourhood of edge $s = s_1$. In the case under consideration the non-tangential boundary conditions in the zeroth approximation are not essential and the values $\Pi^{\mathrm{e}}_{\varepsilon}$, $\Pi^{\mathrm{e}}_{\varkappa}$ and Π^{e}_{T} in (5) are secondary.

We find in the zeroth approximation

$$\lambda\,(m) = A_1\mu^{-2}m^{-2}e^{-mc} + A_2\mu^2 m^2, \tag{12.3.23}$$

where the constants A_1 and A_2 which do not depend on μ and m are obtained by substitution of (21) and (22) into (5).

By minimization of $\lambda\,(m)$ by m we get

$$\lambda_0 \sim \mu^2 m_0^2, \qquad m_0 \sim c^{-1}\ln\left(\mu^{-4}\right). \tag{12.3.24}$$

For shells of positive curvature with free edges the decrease in the order of the critical load is typical and is caused by the difference between the mathematical rigidity of the surfaces and the physical rigidity of the shells. This problem has been analyzed in [54, 55].

Estimates (24) are obtained under particular assumptions on the boundary conditions at edge $s = s_1$ and the type of load. For other boundary conditions which do not allow pure bending, estimates (24) are valid and only the constants in these estimates are changed (for example, constant M in the estimate $m_0 \sim M\ln\mu^{-1}$).

The effect of the initial stress-resultants t_i is more essential. Formula (21) for Π_T^b, and also estimate (24) are valid only for $t_0(s_2) \neq 0$. We note that the value t_0 is proportional to the external normal pressure.

Let $t_0(s_2) = 0$ and $t_0'(s_2) \neq 0$, which is valid for an empty shell entirely immersed in a liquid (see Figure 12.1a). In this case, instead of (21) and (24) we obtain the estimates

$$\Pi_T^b \sim 1, \quad m_0 \sim \ln \mu^{-1}, \quad \lambda_0 \sim \mu^2 m_0^3, \tag{12.3.25}$$

i.e. of order λ_0 increases compared to (24).

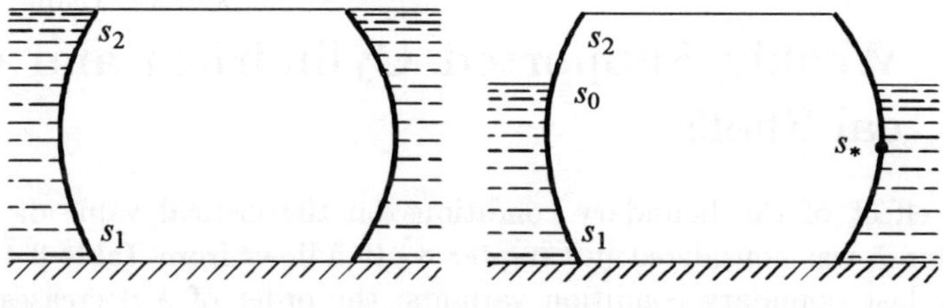

Figure 12.1: An empty convex shell immersed in a liquid (water).

Now, let the top portion of the shell be not loaded, i.e.

$$\begin{aligned} t_1 = t_2 = t_0 = 0 \qquad &\text{for} \quad s_0 < s \leqslant s_2, \\ t_0 > 0 \qquad &\text{for} \quad s_1 \leqslant s < s_0 \end{aligned} \tag{12.3.26}$$

(see Figure 12.1b). In this case the unloaded part of the shell acts as a support. The order of λ_0 increases as $s_2 - s_0$ gets larger and under condition (29) buckling modes close to bending do not occur.

Indeed, instead of (21) and (23) we now have

$$\begin{aligned} &\Pi_T^b \sim m^\beta e^{-m\,c_0}, \qquad c_0 = 2\int_{s_0}^{s_2} p\,ds, \\ &\lambda(m) \sim A_2 \mu^{\alpha_2} m^{\beta_2} e^{mc_0} + A_1 \mu^{-\alpha_1} m^{-\beta_1} e^{-m(c-c_0)}, \end{aligned} \tag{12.3.27}$$

where α_i, β, β_i and A_i depend on the boundary conditions and characteristics of the loading. For conditions $u = v = 0$ at $s = s_1$ and $t_0(s_0) \neq 0$ we have $\beta = 1$, $\alpha_i = \beta_i = 2$. The minimization by m gives

$$\begin{aligned} &\lambda_0 \sim \mu^{\alpha_0} m_0^{\beta_0}, \qquad m_0 \sim \ln \mu^{-1}, \\ &\alpha_0 = \frac{\alpha_2\,(c - c_0) - \alpha_1 c_0}{c}, \quad \beta_0 = \frac{\beta_2\,(c - c_0) - \beta_1 c_0}{c}, \end{aligned} \tag{12.3.28}$$

at the same time estimates (28) are valid only for $\alpha_0 > 0$. But, if

$$\alpha_0 \leqslant 0 \tag{12.3.29}$$

we then obtain $\lambda_0 \sim 1$, $m_0 \sim \mu^{-1}$, i.e. the loading order does not decrease.

We note that condition $\alpha_0 > 0$ imposes restrictions only on the location of parallel s_0 which separates the loaded and unloaded portions of the shell. In particular, for $\alpha_1 = \alpha_2$ for a shell which is symmetrical with respect to plane $s = s_* = 1/2\,(s_2 - s_1)$, condition $\alpha_0 > 0$ has the form $s_0 > s_*$.

12.4 Weakly Supported Cylindrical and Conical Shells

The effect of the boundary conditions on the critical value of the load parameter λ was considered in Chapter 8. It follows from Table 8.4 that for the four last boundary condition variants, the order of λ decreases. These variants are considered in more detail below.

We begin with the case of a cylindrical shell. Let the parameter λ be introduced by (1.11). Then by virtue of (7.1.3) for a "well" supported cylindrical shell

$$\lambda \sim \mu; \quad \lambda = \frac{\mu R}{L}\,\eta + O\,(\mu^2), \quad m_0 \sim \mu^{-1/2}, \tag{12.4.1}$$

where the parameter η is given in the last column of Table 8.4. In this case the shell edge condition is considered to be "well" supported if $\eta \neq 0$.

The other variants of (weak) support we divide into two groups: supports which do and do not, allow pure bending. For the first group we get

$$\lambda \sim \mu^2, \quad m_0 = 2, \quad \lambda = \mu^2 J' + O\,(\mu^3), \tag{12.4.2}$$

where J' is introduced by (2.4). If the shell is under an external normal pressure (see examples 12.1 and 12.2) then $J' = 3\,J_q^{-1}$ or J' is represented by (2.11) depending on whether bending case (2.9) or (2.10) satisfies the given supports. It occurs that all of the variants in lines 8 – 10 of Table 8.4, and the variants in line 7, without restriction on the angle of rotation γ_1 neither for $s = 0$ nor for $s = L$ allow the bending. In the first case we consider bending (2.9), in the second case bending (2.10). Reading Table 8.4 one should use formulae (8.4.13-16) in which we list the corresponding complete boundary conditions.

Next let us consider the type of support for which $\eta = 0$ and which, on the other hand, do not allow pure bending. Let one of the boundary condition variants of simple support be introduced on edge $s = 0$, and one of the variants of boundary conditions from the free edge group be introduced on edge $s = L$ (see line 7 of Table 8.4 and formulae (8.4.14) and (8.4.16)).

We take bending (2.10) which satisfies all of the supports except support by the angle of rotation ($\gamma_1 = 0$), if it is introduced. That is why we add to bending (2.10) (if necessary) the edge effect solutions in the neighbourhood of edges $s = 0$ and $s = L$

$$w = w^{\mathrm{b}} + w^{\mathrm{e}0} + w^{\mathrm{e}L}, \quad w^{\mathrm{b}} \sim 1, \quad w^{\mathrm{e}} \sim \mu m. \tag{12.4.3}$$

We are considering a moderately long shell and that is why we can neglect the interaction of the edge effect solutions. Unlike (3.1) in (3) the membrane deformation w^{m} is not introduced since it does not contribute the zeroth approximation, the only one which we are considering here.

Substituting (3) into (1.12) or into (3.5) and keeping the main terms we obtain

$$\lambda = \frac{Z^{\mathrm{e}0} + Z^{\mathrm{e}L} + \mu^4 R^2 \Pi_{\varkappa}^{\mathrm{b}}}{\mu^2 \Pi_T^{\mathrm{b}}}. \tag{12.4.4}$$

That by virtue of (3.12) and (3.14) gives us

$$\lambda(m) = \left[\sqrt{2}\,(\alpha_0 + \alpha_L)\,\mu^3 + \right.$$
$$\left. + 1/6\,(m^2 - 1)^2\, l^3 \mu^4 \left(1 + \frac{6\,(1-\nu)}{m^2 l^2}\right)\right] \left(\mu^2 \Pi_T^{\mathrm{b}}\right)^{-1}, \tag{12.4.5}$$

where $l = L/R$,

$$\alpha_0 = \begin{cases} 0 & \text{for } 0110,\ 0100,\ 0010, \\ 1/4 & \text{for } 0101,\ 0011, \\ 1/2 & \text{for } 0111, \end{cases} \tag{12.4.6}$$
$$\alpha_L = \begin{cases} 0 & \text{for } 0000, \\ 1/4 & \text{for } 0001 \end{cases}$$

(see the notation for the boundary conditions introduced in Section 8.4 and Table 1.1).

In (5) the first term is proportional to the energy of deformation caused by the edge effect solutions in the neighbourhood of edges $s = 0$ and $s = L$, and the second term is proportional to the bending deformation energy.

Let $\alpha_0 + \alpha_L > 0$. By virtue of (3.15) for $t_2 > 0$ we have

$$\Pi_T^{\mathrm{b}} = m^2 \left(\Pi_{T0}^{\mathrm{b}} + O\left(m^{-2}\right)\right), \quad \Pi_{T0}^{\mathrm{b}} = \frac{1}{2R^3} \int_0^L s^2 t_2 \, ds. \tag{12.4.7}$$

Now, minimization (5) by m yields

$$\lambda_0 = \frac{\sqrt{2}\,l^3\,(\alpha_0+\alpha_L)\mu^{3/2}}{3\,\Pi^{\mathrm{b}}_{T0}}\left(1+O\left(\mu^{1/2}\right)\right), \qquad m_0 = \left(\frac{6\sqrt{2}\,(\alpha_0+\alpha_L)}{\mu\, l^3}\right)^{1/4}. \tag{12.4.8}$$

Thus, we obtain the estimates

$$\lambda_0 \sim \mu^{3/2}, \qquad m_0 \sim \mu^{-1/4}, \tag{12.4.9}$$

which falls between estimates (1) and (2).

The accuracy of (8) for λ_0 is not high. Formula (5) for integer m which is the closest to m_0, is more accurate and has an error of order μ.

We can now write the expression for Π^{b}_T in the case when the shell is loaded by external pressure $q^*(s)$ and axial force P

$$\Pi^{\mathrm{b}}_T = \frac{(m^2-1)}{2\,R^3}\int_0^L s^2 q^0(s)\,ds + \frac{(m^2+1)\,t_1 l}{2}, \tag{12.4.10}$$

where t_1 is introduced by (1.11) for $P = 2\pi R T^0_1$. In the case of compression, $t_1 > 0$. If we additionally load the shell by a torsional moment then it does not effect the parameter λ within the accepted limits of accuracy since the integral of the term $2t_3\gamma_1\gamma_2$ (1.14) is cancelled ($\gamma_1\gamma_2 \sim \sin m\varphi \cos m\varphi$ and the integral by φ is equal to zero).

Now let us consider conical shells. As in the case of cylindrical shell estimates (1) are also valid for "well" supported conical shell. If the supports allow pure bending then estimates (2) are valid. We will consider the intermediate cases. The total displacement we will represent in form (3) again.

For different boundary conditions for $s = s_1$ and $s = s_2$ we can take different linear combinations of bending (2.7)

$$u = \frac{\sin\alpha}{m^2}, \quad v = \frac{s-s_k}{m s_k}, \quad w = \frac{s-s_k}{s_k\cos\alpha} + \frac{\sin^2\alpha}{m^2\cos\alpha}$$
$$(0111,\ 0110,\ 0101,\ 0100 \quad \text{for} \quad s = s_k); \tag{12.4.11}$$

$$u = \frac{\sin\alpha}{m^2-\sin^2\alpha}, \quad v = \frac{s}{m s_k} - \frac{m}{m^2-\sin^2\alpha}, \quad w = \frac{s-s_k}{s_k\cos\alpha}$$
$$(0011,\ 0010 \quad \text{for} \quad s = s_k); \tag{12.4.12}$$

$$u = 0, \quad v = \frac{s}{m s_2}, \quad w = \frac{s}{s_2\cos\alpha}$$
$$(1001,\ 1000 \quad \text{for} \quad s = s_1, \quad s = s_2); \tag{12.4.13}$$

We will use formulae (11) and (12) in the case when the conditions of the simple support group are introduced on edge $s = s_k$ and the conditions of the free edge group apply on the other edge, i.e. conditions 0000 or 0001. We use bending (11) when support $v(s_k) = 0$ is introduced, and bending (12) when it is not, but support $w(s_k) = 0$ does exist. We use bending (13) in the case when weak support conditions are introduced on both edges $s = s_1$ and $s = s_2$.

Let us now write the expressions for the potential energy due to bending (11)-(13).

$$R^2\Pi^{\mathrm{b}}_{\varkappa} = \frac{(m^2-1)^2}{2\sin\alpha\cos^2\alpha}\left[\frac{1-\varkappa^2}{2\varkappa^2}\zeta^2 - \frac{2(1-\varkappa)}{\varkappa^{3-k}}\zeta + \varkappa^{-2(2-k)}\ln\varkappa^{-1} + \frac{(1-\nu)\sin^2\alpha(1-\varkappa^2)\zeta^2}{2m^2\varkappa^2}\right]; \quad (12.4.14)$$

$$R^2\Pi^{\mathrm{b}}_{\varkappa} = \frac{(m^2-1)^2}{2\sin\alpha\cos^2\alpha}\ln\varkappa^{-1}, \quad \varkappa = \frac{s_1}{s_2}. \quad (12.4.15)$$

Formula (14) is obtained for bending (11) and (12) where in case (11) $\zeta = 1$, and in case (12) $\zeta = m^2\left(m^2 - \sin^2\alpha\right)^{-1}$. If simple support conditions apply on the wide edge ($s = s_2$), we assume in (14) that $k = 2$, and if they apply on the narrow edge we assume that $k = 1$. Formula (15) corresponds to bending (13).

We can now begin to calculate the potential energy due to the edge effect solutions by assuming that

$$R = B(s_2) = s_2\sin\alpha. \quad (12.4.16)$$

Using formulae (3.12-14) we find the expressions for $Z^{\mathrm{e}}_{1,2}$ for different supports

(they are marked in the parentheses)

$$
\begin{aligned}
Z_1^{\mathrm{e}} &= \frac{1}{\sqrt{2}}\mu\varkappa^{-1/2}w^2(\cos\alpha)^{3/2}+\mu^2Rww'\cos\alpha+ \\
&\quad +\frac{1}{\sqrt{2}}\varkappa^{1/2}\mu^3R^2w'^2(\cos\alpha)^{1/2}, \\
Z_2^{\mathrm{e}} &= \frac{1}{\sqrt{2}}\mu w^2(\cos\alpha)^{3/2}+\mu^2Rww'\cos\alpha+ \\
&\quad +\frac{1}{\sqrt{2}}\mu^3R^2w'^2(\cos\alpha)^{1/2} \quad (0111); \\
Z_1^{\mathrm{e}} &= \frac{1}{2\sqrt{2}}\varkappa^{-1/2}\mu w^2(\cos\alpha)^{3/2}, \\
Z_2^{\mathrm{e}} &= \frac{1}{2\sqrt{2}}\mu w^2(\cos\alpha)^{3/2} \quad (0110); \\
Z_1^{\mathrm{e}} &= \frac{1}{2\sqrt{2}}\varkappa^{1/2}\mu^3R^2w'^2(\cos\alpha)^{1/2}, \\
Z_2^{\mathrm{e}} &= \frac{1}{2\sqrt{2}}\mu^3R^2w'^2(\cos\alpha)^{1/2} \quad (0101,\ 0011,\ 1001,\ 0001); \\
Z_1^{\mathrm{e}} &= Z_2^{\mathrm{e}}=0 \quad (0100,\ 0010,\ 0000,\ ,\ 1000).
\end{aligned}
\tag{12.4.17}
$$

We calculate the value of Z^{e} for each edge separately and that is why the contribution of the edge effect in the neighbourhood of shell edges $s = s_1$ and $s = s_2$ are denoted respectively by Z_1^{e} and Z_2^{e}. Here w and w' are taken by formulae (11)-(13).

For the parameter λ by virtue of (4) we have

$$
\lambda(m) = \left(Z_1^{\mathrm{e}} + Z_2^{\mathrm{e}} + \mu^4R^2\Pi_{\varkappa}^{\mathrm{b}}\right)\left(\mu^2\Pi_T^{\mathrm{b}}\right)^{-1}, \tag{12.4.18}
$$

where Π_T^{b} is calculated by formula (3.15), and also $\Pi_T^{\mathrm{b}} \sim m^2$.

If $Z_1^{\mathrm{e}} = Z_2^{\mathrm{e}} = 0$ then estimates (2) are valid.

If $w = 0$ in (17) then $Z_1^{\mathrm{e}} + Z_2^{\mathrm{e}} \sim \mu^3$ and we come to estimates (9).

In cases 0111 and 0110 we have $w(s_k) \neq 0$ and by virtue (11) $w(s_k) = m^{-2}\tan\alpha$. Minimizing λ by m we come to the estimates

$$
\lambda_0 \sim \mu^{5/4}, \qquad m_0 \sim \mu^{-3/8}, \tag{12.4.19}
$$

which fall between estimates (1) and (9).

Estimates (19) are not valid for cylindrical shells since, by virtue of (2.6) for the bending of a cylindrical shell, $w(0) = 0$ for $v(0) = 0$.

Example 12.3. Consider the buckling of a cylindrical shell and two conical shells, the top edge of which are free and on the lower edge one condition from the simple support group is introduced (see Figure 12.2).

The radius R of the cylindrical shell is equal to the radius of the larger base of the conical shell. The height of all of the shells is equal to L. The shells are loaded by an external hydrostatic pressure $q^*(s)$ which linearly changes with the height. We assume that $q^*(s) = q^{**}q^0(s)$, where q^{**} is the maximum value of the pressure on the lower shell edge.

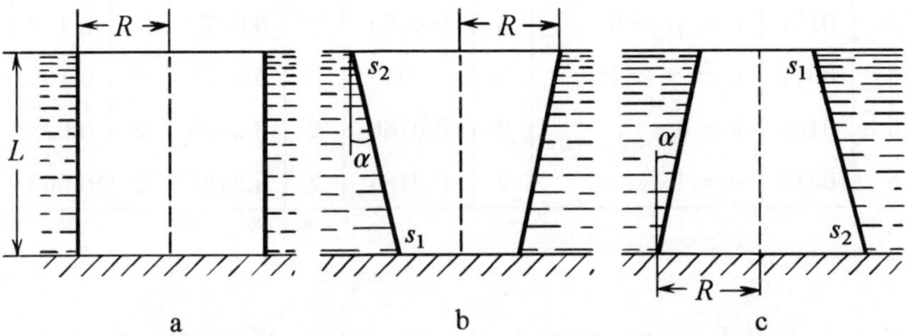

Figure 12.2: Cylindrical and conical shells in a liquid.

For the cylindrical shell, case "a"

$$q^0 = 1 - \frac{s}{L}, \quad t_1 = 0, \quad t_2 = q^0, \tag{12.4.20}$$

for the conical shell, case "b"

$$q^0 = \frac{s_2 - s}{s_2 - s_1}, \quad t_1 = -\frac{(s_2 - s)^2(s_2 + 2s)}{6\, s s_2\,(s_2 - s_1)\cos\alpha}, \quad t_2 = \frac{q^0 s}{s_2 \cos\alpha} \tag{12.4.21}$$

and in case "c"

$$q^0 = \frac{s - s_1}{s_2 - s_1}, \quad t_1 = \frac{(s - s_1)^2(2s + s_1)}{6\, s s_2\,(s_2 - s_1)\cos\alpha}, \quad t_2 = \frac{q^0 s}{s_2 \cos\alpha}. \tag{12.4.22}$$

The following relations are valid

$$\begin{aligned} q^{**} &= \frac{E\,h^2\lambda}{R^2\sqrt{12\,(1-\nu^2)}}, & R &= s_2 \sin\alpha, \\ L &= \frac{R\,(1-\varkappa)}{\tan\alpha}, & \varkappa &= \frac{s_1}{s_2}. \end{aligned} \tag{12.4.23}$$

We suppose that $R/h = 400$, $\nu = 0.3$, $\varkappa = 0.5$, $\tan\alpha = 0.5$. Then $L = R$.

Table 12.1: The critical values of λ_0

No.	Boundary Conditions		Cylinder (a)		Cone (b)		Cone (c)	
			m	λ_0	m	λ_0	m	λ_0
1	0111	$v = w = \gamma_1 = 0$	4	0.0888	3	0.0798	4	0.372
2	0110	$v = w = 0$	2	0.0186	3	0.0350	3	0.229
3	0101	$v = \gamma_1 = 0$	3	0.0650	2	0.0472	2	0.132
4	0011	$w = \gamma_1 = 0$	3	0.0650	3	0.0531	2	0.105
5	0100	$v = 0$	2	0.0186	2	0.0168	2	0.073
6	0010	$w = 0$	2	0.0186	2	0.0181	2	0.063

The critical values of λ_0 and the corresponding values of m for different boundary conditions are represented in Table 12.1.

For the cylindrical shell we take bending (2.10) which exactly satisfies supports numbers 2, 5 and 6 and the corresponding value of λ_0 may be found by (2.4). The introduction of the additional support $\gamma_1 = 0$ increases λ_0, and buckling occurs for $m \geqslant 3$ (see formulae (5) and (8)).

For the conical shell we take bending (11) or (12) depending on the boundary conditions and $s_k = s_1$ corresponds to case "b" (support at the narrow edge). We find the parameter λ_0 by means of formula (18). Here, pure bending takes place only in cases number 5 and 6.

In the other cases, the appearance of the deformation energy (Z_1^e or Z_2^e in (18)) associated with the edge effect leads to an increase in the value of λ_0.

Condition numbers 1 and 2 for the conical shell lead to estimates (19), and condition numbers 3 and 4 lead to estimates (9). In the example under consideration $\mu^{1/4} = 0.4$ and the difference between these estimates is not revealed since the other parameters of the problem are dominant.

The order of the critical load for the cylindrical shell and for the conical shell supported at the narrow edge (case "b"), are approximately the same. The conical shell supported at the wide edge is more rigid.

It can be seen from a comparison of numbers 3 and 4, and also 5 and 6 in Table 12.1 that in some cases, support $v = 0$ is more rigid, while in the other cases support $w = 0$ is more rigid.

12.5 Weakly Supported Shells of Negative Gaussian Curvature

It is a characteristic of shells of revolution with negative Gaussian curvature that if only one support at each edge is applied the shell may have bending which satisfies these supports. If this is the case the size of the shell is termed the *eigensize.* This property of shells with negative curvature has been extensively analyzed (see [4, 31, 51, 52, 56, 57, 91, 119]). Below we will consider the less often studied problem of the effect of non-tangential supports and small variations from the eigensize on the critical load (see also [50, 90, 118, 128, 129]).

Since there exist many different boundary condition variants we will consider only the case when conditions similar to the simple support group are applied at both shell edges, $s = s_1$ and $s = s_2$. For this group of supports the condition where $T_1 = 0$ (see (8.4.14)) is characteristic. As was noted above, we can represent the additional stress-strain state as a sum of three terms (3.1) and use variational formula (3.5) to determine the critical load.

Let us discuss the estimate of the value $\Pi_\varepsilon^{\mathrm{m}}$, in formula (3.5) by considering the auxiliary system of equations which is equivalent to equation (2.3)

$$\frac{dx}{ds} - m\,\frac{R_2^2}{BR_1}\,y = 0, \qquad \frac{dy}{ds} - m\,\frac{x}{BR_2} = 0. \tag{12.5.1}$$

We can represent the displacements u^{b}, v^{b} and w^{b} of the shell under bending and stress-resultants T_1, T_2 and S (in accordance with the static-geometrical analogy), which are the solutions of system (3.2), in terms of solutions x and y of system (1)

$$u^{\mathrm{b}} = \frac{B\,x}{R_2}, \quad v^{\mathrm{b}} = B\,y, \quad w^{\mathrm{b}} = B'x + m\,R_2 y; \tag{12.5.2}$$

$$\frac{T_1}{E_0 h_0} = -\frac{R_2 y}{B^2}, \quad \frac{T_2}{E_0 h_0} = \frac{R_2^2 y}{R_1 B^2}, \quad \frac{S}{E_0 h_0} = \frac{x}{B^2}. \tag{12.5.3}$$

The displacements u^{m}, v^{m} and w^{m}, which are generated by stress-resultants (3), may be found from the same formulae (2)

$$u^{\mathrm{m}} = \frac{B\,x^{\mathrm{m}}}{R_2}, \quad v^{\mathrm{m}} = B\,y^{\mathrm{m}}, \quad w^{\mathrm{m}} = B'x^{\mathrm{m}} + m\,R_2 y^{\mathrm{m}}, \tag{12.5.4}$$

where, by virtue of (3.2) x^{m} and y^{m} satisfy non-homogeneous system (1) and are equal to

$$x^{\mathrm{m}} = C_1(s)\,x_1 + C_2(s)\,x_2, \quad y^{\mathrm{m}} = C_1(s)\,y_1 + C_2(s)\,y_2,$$

$$C_1 = -\int_{s_1}^{s} \left(\frac{2\,(1+\nu)\,x x_2}{B^3} + \left(1 - \frac{2\,\nu\,R_2}{R_1} + \frac{R_2^2}{R_1^2}\right) \frac{R_2^2 y y_2}{B^3} \right) g\,ds, \tag{12.5.5}$$

$$C_2 = \int_{s_1}^{s} \left(\frac{2\,(1+\nu)\,x x_1}{B^3} + \left(1 - \frac{2\,\nu\,R_2}{R_1} + \frac{R_2^2}{R_1^2}\right) \frac{R_2^2 y y_1}{B^3} \right) g\,ds.$$

Here x and y are the solutions of system (1) determining stress-resultants (3). The fundamental set of the solutions of system (1) is formed by x_k and y_k, which satisfies (for definiteness) at $s = s_1$ the conditions

$$x_1 = y_2 = 1, \qquad x_2 = y_1 = 0. \tag{12.5.6}$$

The multiplier g is determined by (4.3.2), and for constants E, ν and h we have $g = 1$.

After integration by parts the expression

$$\Pi_\varepsilon^{\mathrm{m}} = \frac{1}{2E_0 h_0} \int_{s_1}^{s_2} (T_1\varepsilon_1 + T_2\varepsilon_2 + S\,\omega)\,B\,ds \tag{12.5.7}$$

we obtain by virtue of (3.2)

$$\Pi_\varepsilon^{\mathrm{m}} = \frac{B}{2E_0 h_0}\,(T_1 u^{\mathrm{m}} + S\,v^{\mathrm{m}})\,\big|_{s=s_2} = \frac{1}{2}\,(y^{\mathrm{m}} x - x^{\mathrm{m}} y)\,\big|_{s=s_2}, \tag{12.5.8}$$

since $u^{\mathrm{m}} = v^{\mathrm{m}} = 0$ for $s = s_1$.

Let the support $v = 0$ be introduced at $s = s_1$ and $s = s_2$ and let for some $m \geqslant 2$ be $y_1(s_2) = \delta$ and $|\delta| \ll 1$.

Then bending (2) for $x = m^{-1}x_1$, $y = m^{-1}y_1$ exactly satisfies the conditions $v^{\mathrm{b}}(s_1) = 0$ and approximately satisfies the condition $v^{\mathrm{b}}(s_2) = 0$ (the multiplier m^{-1} is introduced to get $w^{\mathrm{b}} \sim 1$).

To compensate for the error in condition $v\,(s_2) = 0$ we introduce the membrane deformation determining by (3) and (5)

$$x = D_1 x_1 + D_2 x_2, \qquad y = D_1 y_1 + D_2 y_2, \tag{12.5.9}$$

where D_1 and D_2 are unknown constants. We take $D_2 = 0$, then by virtue of (3) and (6), condition $T_1 = 0$ is fulfilled (exactly at $s = s_1$ and approximately at $s = s_2$). We can determine the constant D_1 from the condition $y^{\mathrm{b}}(s_2) + y^{\mathrm{m}}(s_2) = 0$. With an error of order δ we then obtain [128].

$$D_1 = -\frac{y_1(s_2)}{\alpha\, y_2(s_2)}, \quad \Pi_\varepsilon^{\mathrm{m}} = \frac{c}{m^2}\, y_1^2(s_2), \quad c = \frac{x_1^2(s_2)}{2\alpha}, \tag{12.5.10}$$

where $\alpha = C_2(s_2)$ for $x = x_1$ and $y = y_1$.

Let edges $s = s_1$ and $s = s_2$ be free in the tangential directions ($T_1 = S = 0$), however, support $w = 0$ is introduced, i.e. due to (2)

$$\zeta_k = y + \beta_k x = 0, \quad \beta_k = -\frac{B'}{mR_2} \quad \text{for} \quad s = s_k, \quad k = 1, 2. \tag{12.5.11}$$

We take the bending given by (2) for

$$x = m^{-1}(x_1 - \beta_1 x_2), \quad y = m^{-1}(y_1 - \beta_1 y_2), \tag{12.5.12}$$

then $w(s_1) = 0$.

Let $\zeta_2 = \delta/m$ and $|\delta| \ll 1$, i.e. the condition $w(s_2) = 0$ is fulfilled approximately. As above, to compensate for the error in satisfying this condition we introduce the membrane deformation determined by (3), (5) and (9). In order to determine constants D_1 and D_2 in (9) we have equation $w^{\mathrm{b}} + w^{\mathrm{m}} = 0$ or

$$y^{\mathrm{m}} + \beta_2 x^{\mathrm{m}} + \zeta_2 = 0 \quad \text{for} \quad s = s_2. \tag{12.5.13}$$

The second equation for D_1 and D_2 we obtain, accounts for the requirement that the energy, $\Pi_\varepsilon^{\mathrm{m}}$, be minimal. This condition has the form $D_2 + \beta_1 D_1 = 0$. Finally, we obtain

$$\begin{aligned} &\Pi_\varepsilon^{\mathrm{m}} = \frac{c_1\delta^2}{m^2}; \quad c_1 = \left[2\left(y_2 + \beta_2 x_2\right)^2\left(\alpha_{21} + 2\beta_1\alpha_{11} - \beta_1^2\alpha_{12}\right)\right]^{-1}, \\ &\delta = y_1 - \beta_1 y_2 + \beta_2\left(x_1 - \beta_1 x_2\right), \\ &\alpha_{ik} = C_i(s_2) \quad \text{for} \quad x = x_k, \quad y = y_k, \quad i, k = 1, 2, \end{aligned} \tag{12.5.14}$$

at the same time, in calculating c_1 and δ we take the functions x_k and y_2 at $s = s_2$. For $\beta_1 = \beta_2 = 0$ formula (14) turns into (10).

Consider formula (3.5)

$$\lambda(m) = \left(\Pi_\varepsilon^{\mathrm{m}} + Z_1^{\mathrm{e}} + Z_2^{\mathrm{e}} + \mu^4 R^2 \Pi_\varkappa^{\mathrm{b}}\right)\left(\mu^2 \Pi_T^{\mathrm{b}}\right)^{-1}. \tag{12.5.15}$$

For all of the boundary condition variants considered

$$\Pi_\varepsilon^{\mathrm{m}} \sim \frac{\delta^2}{m^2}, \quad \Pi_\varkappa^{\mathrm{b}} \sim m^4, \quad \Pi_T^{\mathrm{b}} \sim m^2, \tag{12.5.16}$$

however, the order $Z_i^{\rm e}$ depends on the boundary condition variant and by virtue of (3.12-14) is equal to

$$\begin{aligned}
&Z_i^{\rm e} = 0 && (0100,\,0010);\\
&Z_i^{\rm e} \sim \mu^3 m^2 && (0101,\,0011);\\
&Z_i^{\rm e} \sim \mu\, m^{-2} && (0110);\\
&Z_i^{\rm e} \sim \mu\, m^{-2} + \mu^3 m^2 && (0111), \quad i = 1,2.
\end{aligned} \tag{12.5.17}$$

We use here the estimates $w^{\rm b} \sim 1$ and $\gamma_1^{\rm b} \sim m$.

For these boundary condition variants, the orders of the terms in the expression for $\lambda(m)$ (see (15)) are presented in Table 12.2.

Table 12.2: The orders of the terms

	$\lambda\,(m)$	$\lvert\delta\rvert \lesssim \mu^2$		$\mu^2 \ll \delta \ll 1$	
		λ_0	m	λ_0	m
1	2	3	4	5	6
0100, 0010	$\dfrac{\delta^2}{\mu^2 m^4} + \mu^2 m^2$	μ^2	2	$(\mu\delta)^{\frac{2}{3}}$	$\left(\dfrac{\delta}{\mu^2}\right)^{\frac{1}{3}}$
0101, 0011	$\dfrac{\delta^2}{\mu^2 m^4} + \mu + \mu^2 m^2$	μ	2	$\mu + (\mu\delta)^{\frac{2}{3}}$	$\left(\dfrac{\delta}{\mu^2}\right)^{\frac{1}{3}}$
0110, 0111	$\dfrac{\delta^2}{\mu^2 m^4} + \dfrac{1}{\mu m^4} + \mu^2 m^2$	μ	$\dfrac{1}{\sqrt{\mu}}$	$\mu^{\frac{2}{3}}(\delta^2+\mu)^{\frac{1}{3}}$	$\left(\dfrac{\delta^2+\mu}{\mu^4}\right)^{\frac{1}{6}}$

For fixed values of m, the value of δ is a measure of the deviation of the shell size from the eigensize. It is clear that δ depends on m and that is why the minimization of $\lambda(m)$ by m may be carried out only in sorting by $m = 2, 3, \ldots$. We recall that $\lambda \sim \mu^{2/3}$ and $m \sim \mu^{-2/3}$, if for all $m \lesssim \mu^{-2/3}$ the shell sizes are not close to the eigensizes ($\delta \sim 1$).

We formally assume that δ does not depend on m, and minimize the expressions in the second column of Table 12.2 by m. For $\delta = 0$ or $|\delta| \lesssim \mu^2$ the results of the minimization are presented in columns 3 and 4 of Table 12.2, and for larger values of δ the results are given in columns 5 and 6 of that table.

It follows from Table 12.2 that if only one support, $v = 0$ or $w = 0$, is applied at each of the edges, and the shell size is close to the eigensize ($|\delta| \lesssim \mu^2$) for $m = 2$, then buckling occurs for $m = 2$, and also $\lambda \sim \mu^2$.

The introduction of the additional support $\gamma_1 = 0$ keeps the wave number ($m = 2$) under the buckling value. However, it leads to an increase in the order of the critical load ($\lambda \sim \mu$). If, simultaneously the order of the quantity δ increases ($|\delta| \gg \mu^2$) and (or) supports are introduced of type $v = w = 0$, then buckling will occur for $m > 2$.

Remark 12.2. The term $\Pi_\varepsilon^{\mathrm{m}} / \left(\mu^2 \Pi_T^{\mathrm{b}}\right)$ for $m \gg 1$ becomes the first term in (11.5.12). A value close to zero indicates that the size of the shell is close to the eigensize.

If, instead of the condition of simple support, we impose clamped conditions (see (8.4.13)), then in this case the existence of the possible tangential bending does not lead to a decrease in the order of the critical load compared to $\lambda \sim \mu^{2/3}$.

For example, if for some shell eigensizes there exists bending satisfying the supports, $u = 0$, at both edges, then the introduction (even at one edge) of an additional non-tangential support $w = 0$ immediately increases the order of the critical load from $\lambda \sim \mu^2$ to $\lambda \sim \mu^{2/3}$.

Example 12.4. Consider the buckling of a hyperboloid of revolution under a homogeneous external pressure (see Figure 12.3). Let it be formed by the revolution of hyperbola $x^2 a^{-2} - z^2 b^{-2} = 1$ around the z axis. We have [119]

$$R_1 = -a^{-4} c^2 R_2^3, \quad R_2 = a^2 \left(a^2 - \left(a^2 + c^2\right) \cos^2 \theta\right)^{-1/2}. \qquad (12.5.18)$$

After change of the independent variables due to formula

$$\eta = -\frac{c}{a^2} \int_{\pi/2}^{\theta} \frac{R_2(\theta)\, d\theta}{\sin \theta} = \frac{1}{2} \left(\arcsin \frac{1 + \tau \cos \theta}{\tau \left(1 + \cos \theta\right)} - \right.$$

$$\left. - \arcsin \frac{1 - \tau \cos \theta}{\tau \left(1 - \cos \theta\right)} \right), \quad \tau = 1 + \frac{c^2}{a^2} \qquad (12.5.19)$$

the coefficients of system of equations (1) become constant.

Let the supports

$$v\left(\theta_1\right) = v\left(\theta_2\right) = 0 \qquad (12.5.20)$$

be introduced. Then the shell eigensize (for which system (1) has a non-zero solution satisfying the conditions $y\left(\theta_1\right) = y\left(\theta_2\right) = 0$) are determined by the equation

$$\sin m \left(\eta_2 - \eta_1\right) = 0, \qquad \eta_k = \eta\left(\theta_k\right), \qquad (12.5.21)$$

from which it follows in particular that if the shell sizes are the eigensizes for some m, then they are also the eigensizes for $2m$, $3m$, ...

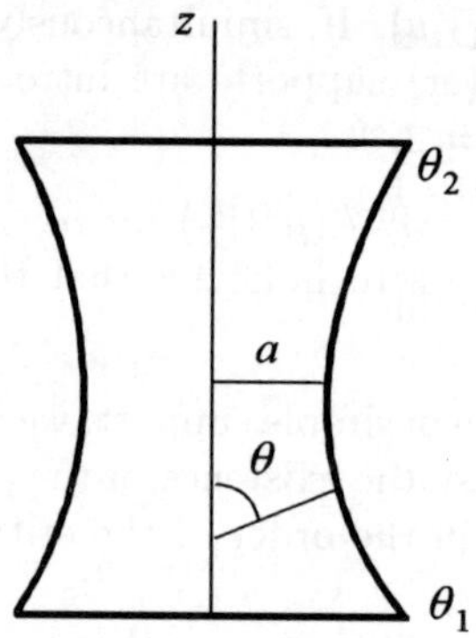

Figure 12.3: The hyperboloid of revolution.

Let us consider a numerical example [128]. As the *eigensize* of the surface we will take $R = c$. Let $\nu = 0.3$, $c/a = 2$, $\cos\theta_1 = 1/3$ and $q^0 = 1$ (in (1.11)). Then, due to (19) and (20) $\eta_1 = \pi/4$ and the shell sizes are the eigensezes for $m = 2$, if $\theta_2 = \pi - \theta_1 = \theta_2^*$, $\eta_2 = -\pi/4$.

Let only supports (20) be introduced, however, the shell differs from the eigensize ($\theta_2 = \theta_2^* + \Delta$). In Figure 12.4 the dependence of λ on Δ is shown for $m = 2$ and $m = 4$ and for $h/R = 0.003$ (solid lines) and $h/R = 0.01$ (dotted lines).

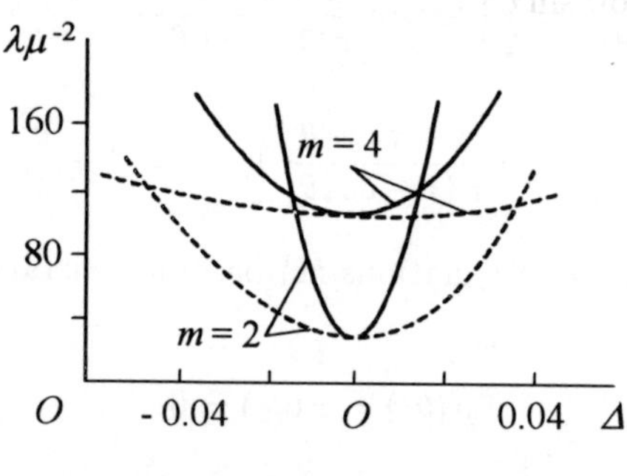

Figure 12.4: The load parameter for hyperboloid buckling.

These graphs agree with the results represented in the first line of Table 12.2. Namely, for small deviations in shell geometry from its eigensize, buckling occurs for $m = 2$, while for large deviations, buckling occurs for $m = 4$. The value of Δ which separates these cases, increases with the dimensionless thickness.

12.6 Problems and Exercises

12.1. Construct for $m = 2, 3, 4, 5$ bendings of the neutral surface of the shell of revolution that has the form of the elliptic cupola.

Hint. Construct numerically solutions of equation (2.3) regular at the apex of the cupola. Use geometric functions from (4.4.8).

12.2. Construct the solution that has a form of bending for the spherical cupola.

Hint Use relations (2.2) and (2.3) for $R_1 = R_2 = R$, $B = R\sin\theta$.

Answer

$$u = (1 - \cos\theta)\tan^{m+1}\frac{\theta}{2}, \qquad v = \sin\theta\,\tan^{m}\frac{\theta}{2}, \qquad u = (m + \cos\theta)\tan^{m}\frac{\theta}{2}.$$

12.3. Consider buckling of the weightless semi-spherical shell of the radius R and thickness h entirely immersed in a container of liquid of the density γ. The shell edge is free. The force $P = (2/3)\pi R^3\gamma$ balancing the Archimedean force is applied at the apex of the shell.

Hint. Use relation (2.4) and the bending obtained in Problem 12.2 for $m = 2$. Under calculation stresses t_1 and t_2 in (1.14) use relations (1.4.6) for $q^0 = \gamma R(1 - \sin\theta)$ and $P = (2/3)\pi R^3\gamma$. The rest loads are equal to zero.

12.4. Find the exact eigensizes of the hyperboloid of revolution of one sheet described in Problem 11.3 and Problem 11.5 for the boundary conditions $v(\theta_1) = v(\theta_2) = 0$. Compare with the results with the approximate values obtained from equation $\sin\psi_2 = 0$, where ψ_2 is given by (11.5.6).

Chapter 13

Buckling Modes Localized at an Edge

In this Chapter we will discuss the buckling modes of shells of revolution under a membrane stress state localized in the neighbourhood of the shell edge. The influence of moments on the initial stress state and pre-buckling deformations will be considered in Chapter 14. Weak edge support and variability of the determining parameters cause the appearance of these buckling modes.

Modes of this type are possible for convex shells and also for shells of zero Gaussian curvature under axial pressure. Localization of buckling modes in the neighbourhood of an edge for shells of zero Gaussian curvature under other types of loading (external pressure, torsion) and for shells of negative Gaussian curvature does not take place (see Chapters 7–12). As will be shown below, a weak edge support may significantly lower the critical load while at the same time, the variability of determining parameters changes it only slightly.

13.1 Rectangular Plates Under Compression

As an example illustrating the effect of localization of the buckling mode due to a weak edge support, consider the buckling of a rectangular plate under pressure. The buckling equation has the form [16]

$$D \Delta \Delta w + T_2 \frac{\partial^2 w}{\partial y^2} = 0, \quad D = \frac{E h^3}{12 (1 - \nu^2)}. \tag{13.1.1}$$

Edges $y = 0$ and $y = b$ (see Figure 13.1) are assumed to be simply supported

$$w = \frac{\partial^2 w}{\partial y^2} = 0 \quad \text{at} \quad y = 0, \quad y = b. \tag{13.1.2}$$

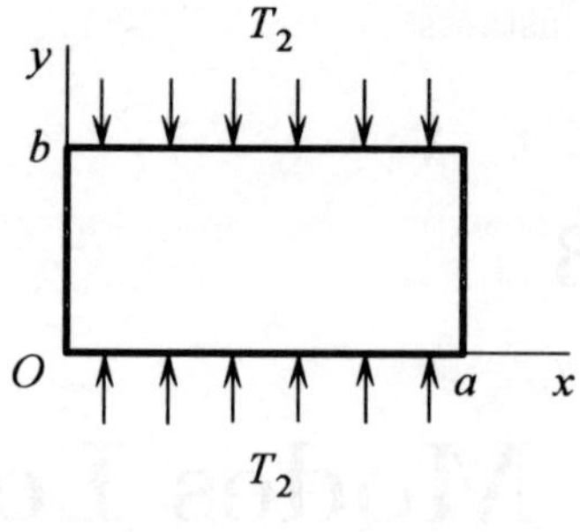

Figure 13.1: A rectangular plate under compression.

Under cylindrical bending of the plate, the deflection does not depend on x and has the form $w = w_0 \sin \dfrac{\pi y}{b}$. The critical value of the load is equal to

$$T_2 = T = \frac{\pi^2 D}{b^2}. \tag{13.1.3}$$

If edges $x = 0$, $x = a$ are also simply supported, then the deflection (for $b < a$) has the form $w = w_0 \sin \frac{\pi x}{a} \sin \frac{\pi y}{b}$, and the critical value of T_2 is equal to

$$T_2 = \lambda T, \qquad \lambda = \left(1 + \frac{b^2}{a^2}\right)^2, \tag{13.1.4}$$

where T is defined in (3). As the plate length a increases ($b/a \to 0$) $\lambda \to 1$, i.e. the critical load approaches (from above) value (3) under cylindrical bending. The same result takes place for all other type of supports of edges $x = 0$ and $x = a$ except in the case when at least one of the edges is free.

For example, let edge $x = 0$ be free, and $x = a$ be clamped, i.e.

$$\begin{aligned} &\frac{\partial^2 w}{\partial x^2} + \nu \frac{\partial^2 w}{\partial y^2} = 0, \\ &\frac{\partial^3 w}{\partial x^3} + (2 - \nu) \frac{\partial^3 w}{\partial x \, \partial y^2} = 0 \qquad \text{for} \quad x = 0, \\ &w = 0, \qquad \frac{\partial w}{\partial x} = 0 \qquad \text{for} \quad x = a. \end{aligned} \tag{13.1.5}$$

We seek the buckling mode in the form

$$w(x, y) = \left(C_1 e^{-\frac{\pi s x}{b}} + C_2 e^{-\frac{\pi r x}{b}} + C_3 e^{\frac{\pi s x}{b}} + C_4 e^{\frac{\pi r x}{b}}\right) \sin \frac{\pi y}{b}, \tag{13.1.6}$$

where C_i are arbitrary constants and

$$s = \sqrt{1 - \sqrt{\lambda}}, \qquad r = \sqrt{1 + \sqrt{\lambda}} = \sqrt{2 - s^2} \quad (\lambda < 1). \tag{13.1.7}$$

The substitution of these expressions into boundary conditions (5) leads to an equation for evaluation of λ

$$s\left(r^2 - \nu\right)^2 - r\left(\nu - s^2\right)^2 + \frac{r+s}{r-s}\Big(s\left(r^2 - \nu\right)^2 + {} + r\left(\nu - s^2\right)^2\Big)e^{-\frac{2as\pi}{b}} = 0. \tag{13.1.8}$$

We study equation (8) for $a \gg b$. That is why we neglected terms of order $\exp\{-2a/b\}$ in deriving this equation. If we express r through s by formula (7), we find that for sufficiently large a/b, equation (8) has the single root $s > 0$. Parameter λ can be expressed over this root due to (7) by the expression $\lambda = (1 - s^2)^2$. The values of s and λ for different ratios a/b for $\nu = 0.3$ are given in Table 13.1.

Table 13.1: The values of s and λ for different ratios a/b

$\pi a/b$	25	30	35	40	∞
s	0.0186	0.0324	0.0375	0.0400	0.0436
λ	0.9993	0.9979	0.9972	0.09968	0.9962

For $\nu = 0.3$, equation (8) has a real root if $\pi a/b > 22.7$. If $\pi a/b < 22.7$, the solution should be sought in a form which differs from (6). This case will not be considered here. It may be notes that the buckling of the rectangular plates under different boundary conditions and values of ratio a/b is considered, for example, in [16, 175].

The buckling of a plate with two free edges $x = 0$, $x = a$ is considered in [65] and the effect of decreasing the load compared to (3) is revealed. In spite of the difference between the boundary conditions at $x = a$ (clamped edge in our case and free edge in [65]), for $a/b = \infty$ the results ($\lambda = 0.9962$, see Table 13.1) are identical. This fact is related to the localization of the deflection near the free edge $x = 0$. As a result, the boundary conditions at $x = a$ do not effect the critical load.

Indeed, for sufficient large values of a/b we may assume that $C_3 = C_4 = 0$ in (6) and satisfy only the first two conditions (5) at $x = 0$. As a result we come to the equation

$$s\left(r^2 - \nu\right)^2 - r\left(\nu - s^2\right)^2 = 0, \tag{13.1.9}$$

which follows from (8) as $a/b \to \infty$.

It follows from Table 13.1 that the localization mentioned above takes place only for sufficiently long plates, since in this case s is small compared to unity. Indeed, from (9) we can approximate that

$$s \simeq \frac{\sqrt{2}\,\nu^2}{(2-\nu)^2}, \qquad 1-\lambda \simeq \frac{4\,\nu^4}{(2-\nu)^4}, \tag{13.1.10}$$

from which it follows that the decrease in the load is connected with the inequality $\nu \neq 0$.

Note that more significant decreasing of the critical load (from 43% to 75%) takes place for the plate, when the buckling mode is localized at the free edge $y = 0$ (see [43]).

In the work below, considering shell buckling we reveal the same localization of the deflection with a simultaneous decrease of load. However, for shells, this effect is more important than it is for plates.

13.2 Cylindrical Shells and Panels Under Axial Compression

Let us consider the influence of the boundary conditions on the buckling of a circular cylindrical shell of radius R under a homogeneous membrane stress state caused by axial pressure. As an initial system of equations for moderately long shells we take system (3.1.1)

$$\mu^2 \Delta\,\Delta\,w + 2\lambda\frac{\partial^2 w}{\partial x^2} - \frac{\partial^2 \Phi}{\partial x^2} = 0, \quad \mu^2 \Delta\,\Delta\,\Phi + \frac{\partial^2 w}{\partial x^2} = 0, \tag{13.2.1}$$

where

$$\Delta = \frac{\partial^2}{\partial x^2} + \frac{\partial^2}{\partial y^2}, \quad T_1^0 = -2\lambda\,E\,h\,\mu^2, \quad \mu^4 = \frac{h^2}{12\,(1-\nu^2)\,R^2}, \tag{13.2.2}$$

$$0 \leqslant x \leqslant l = L/R.$$

Here $x = x'/R$ is the dimensionless length of the generatrix and φ is the angle in the circumferential direction.

It was noted (see Section 3.4), that the classical critical value of the load parameter $\lambda_{\rm cl} = 1$, obtained by Lorenz and Timoshenko, is valid for many buckling modes, including axisymmetric modes. These modes satisfy the conditions of simple support at $x = 0$ and $x = l$.

For more rigid support of the shell edges, the critical load increases, but only slightly, due to the local character of the buckling pits. The critical load increases only for rather short shells (see [61, 66]) and we will not consider the short shell case here.

Let us consider cases when weaker support leads to a decrease in load. We will begin with the case of axisymmetric buckling modes of a shell with free edge $x = 0$. We seek the non-trivial solutions of the equation

$$\mu^4 \frac{\partial^4 w}{\partial x^4} + 2\lambda \mu^2 \frac{\partial^2 w}{\partial x^2} + w = 0, \tag{13.2.3}$$

satisfying at $x = 0$ the conditions

$$Q_1 - T_1^0 \gamma_1 = 0, \qquad M_1 = 0,$$

that may be written as

$$\mu^2 \frac{\partial^3 w}{\partial x^3} + 2\lambda \frac{\partial w}{\partial x} = 0, \qquad \frac{\partial^2 w}{\partial x^2} = 0. \tag{13.2.4}$$

For that we assume that the external pressure at edge $x = 0$ does not change direction under shell deformation, and that the term $-T_1^0\gamma_1$ in the first of equations (4) gives the projection of this loading on the (internal) shell normal after deformation.

We seek the solution satisfying the condition

$$w \to 0 \quad \text{for} \quad x \to \infty, \tag{13.2.5}$$

which replaces the boundary conditions at the other shell edge.

The direct verification shows that for $\lambda = 1/2$ the function

$$w(x) = e^{-\frac{2x}{\mu}} \sin\left(\frac{\sqrt{3}\,x}{2\mu} - \frac{\pi}{3}\right) \tag{13.2.6}$$

satisfies both (3) and also (4) and (5). This solution is constructed in [74]. Due to the presence of μ in the denominator, deflection (6) damps fast away from edge $x = 0$.

Under at least one of the supports $w = 0$ or for $\gamma_1 = 0$ there are no axisymmetric buckling modes localized in the neighbourhood of this edge and $\lambda = 1$.

Let us try to find the non-axisymmetric buckling modes localized in the neighbourhood of edge $x = 0$. We will consider 16 boundary condition variants in which generalized displacements (1.2.9) or corresponding to them generalized forces (1.2.10) are equal to zero except that the deflection, w, corresponds to the generalized shear stress resultant

$$Q_{1*} = Q_1 - \frac{1}{R}\frac{\partial H}{\partial \varphi} - T_1^0\gamma_1. \tag{13.2.7}$$

We seek the solution $w(x, \varphi)$ of system (1) in the form

$$w(x, \varphi) = \sum_{k=1}^{4} C_k w_k \exp\left\{\frac{i}{\mu}(p_k x + q\varphi)\right\}, \tag{13.2.8}$$

where C_k are arbitrary constants, w_k are fixed constants (below we assume that $w_k = 1$) and p_k are unknown constants satisfying the relations

$$\left(p_k^2 + q^2\right)^4 - 2\lambda p_k^2 \left(p_k^2 + q^2\right)^2 + p_k^4 = 0, \quad \Im p_k > 0. \tag{13.2.9}$$

The last of the above relations provides the damping of solution (8) away from edge $x = 0$. For $\lambda < 1$ equation (9) has four solutions, satisfying inequality (9).

For a shell which is closed in the circumferential direction $q = \mu m$. For a cylindrical panel with simply supported rectilinear edges $\varphi = 0$, $\varphi = \varphi_0$, $q = \frac{\mu m \pi}{\varphi_0}$ (m is an integer). The value of q may be found from the minimum conditions for λ.

Other unknown functions (Φ, u, v, γ_1, T_1, S, Q_{1*} and M_1) we seek in the same form as (8) with corresponding substitutions for w. We write the values Φ_k, u_k, ... in (8) (subscript k is omitted)

$$\begin{aligned}
\Phi &= \frac{p^2 w}{\left(p^2 + q^2\right)^2}, \quad u = \frac{\mu p \left(\nu p^2 - q^2\right) w}{i \left(p^2 + q^2\right)^2}, \\
v &= \frac{\mu q \left(q^2 + (2 + \nu) p^2\right) w}{i \left(p^2 + q^2\right)^2}, \\
T_1 &= -\frac{E h}{R} q^2 \Phi, \quad T_2 = -\frac{E h}{R} p^2 \Phi, \quad S = \frac{E h}{R} p q \Phi, \\
\gamma_1 &= \frac{p w}{i \mu R}, \quad \gamma_2 = \frac{q w}{i \mu R}, \quad M_1 = -E h \mu^2 \left(p^2 + \nu q^2\right) w, \\
M_2 &= -E h \mu^2 \left(q^2 + \nu p^2\right) w, \\
Q_{1*} &= \frac{E h \mu i}{R}\left(p^3 + (2 - \nu) p q^2 - 2\lambda p\right), \\
Q_{2*} &= \frac{E h \mu i}{R}\left(q^3 + (2 - \nu) p^2 q - 2\lambda q t_2\right),
\end{aligned} \tag{13.2.10}$$

which are included in the formulation of the boundary conditions.

If we substitute solutions (8) into the boundary conditions at $x = 0$ then we come to a system of four linear equations in C_k. If the determinant of the system is equal to zero then

$$\Delta(\lambda, q) = 0 \tag{13.2.11}$$

we get the equation for the evaluation of λ.

The case, when $q^2 \ll 1$ is the simplest. For the seven boundary condition variants

$$\begin{array}{lll} T_1 = S = Q_{1*} = M_1 = 0 & (0000), & \lambda = 1/2; \\ u = S = Q_{1*} = M_1 = 0 & (1000), & \lambda = 1/2; \\ T_1 = v = Q_{1*} = M_1 = 0 & (0100), & \lambda = 1/2; \\ T_1 = S = w = M_1 = 0 & (0010), & \lambda = 1/2; \\ T_1 = S = Q_{1*} = \gamma_1 = 0 & (0001), & \lambda = 1/2; \\ u = v = Q_{1*} = M_1 = 0 & (1100), & \lambda = 1/2; \\ u = S = w = M_1 = 0 & (1010), & \lambda = 1/2 \end{array} \tag{13.2.12}$$

equations (11) has a root $\lambda < 1$, moreover in all cases this root is equal to 1/2 (the designations for the boundary conditions introduced in Section 8.4 and in Table 1.1 are in the parentheses). For supports more rigid than (12), equation (11) for $q^2 \ll 1$ has no roots (except for boundary conditions (15), see below).

It follows from the above that for boundary conditions (12), a decrease in the critical load takes place. However, it remains unclear whether this decrease would be larger than a factor of two at $q \sim 1$.

We turn to the case $q \sim 1$. Let $\lambda = \cos\psi$ and $0 < \psi < \frac{\pi}{2}$. Then roots of (9) in which we are interested are the following

$$\begin{gathered} p_{1,3} = \frac{1}{2}\left(r\cos\frac{\theta}{2} \pm \cos\frac{\psi}{2}\right) + \frac{i}{2}\left(r\sin\frac{\theta}{2} \pm \sin\frac{\psi}{2}\right), \\ r = \left(\left(\cos\psi - 4q^2\right)^2 + \sin^2\psi\right)^{1/2}, \quad \tan\theta = \frac{\sin\psi}{\cos\psi - 4q^2}, \\ 0 < \theta < \pi. \end{gathered} \tag{13.2.13}$$

The roots $p_{2,4}$ differs from $p_{1,3}$ by the sign of the real part.

If $q^2 \ll 1$, then roots p_k are equal approximately to

$$p_{1,2} = \pm\cos\frac{\psi}{2} + i\sin\frac{\psi}{2}, \quad p_{3,4} = q^2 p_{1,2}. \tag{13.2.14}$$

The use of expression (14) give us double the decrease in λ for boundary conditions (12).

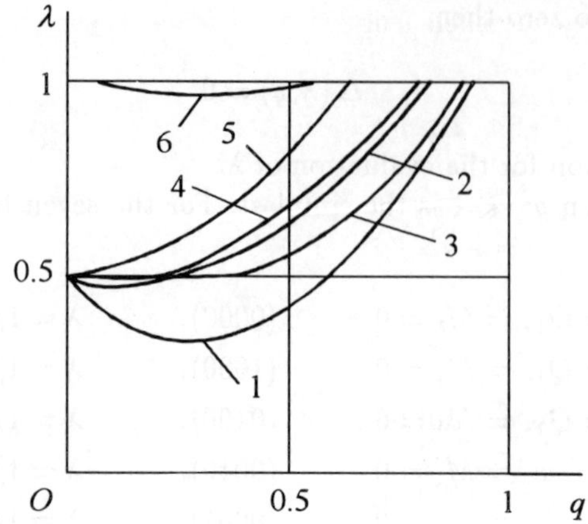

Figure 13.2: The loading parameter for a cylindrical shell with a weakly supported curvilinear edge under axial compression (1 - 0000, 2 - 0100, 3 - 1000, 4 - 1100, 5 - 0010, 1010, 0001, 6 - 0101).

The roots $\lambda = \lambda(q)$ of equation (11) for $q \neq 0$ are given in Figure 13.2. In addition we also have found a new type of boundary condition

$$T_1 = v = Q_{1*} = \gamma_1 = 0 \quad (0101), \tag{13.2.15}$$

for which a decrease in the critical load is also possible (but by not more than 3 %). It should be noted, that $\lambda \to 1$ as $q \to 0$, therefore this case does not fit (12).

Curves 1, 2 and 6 in Figure 13.2 have minimums at q=0.31, 0.20 and 0.36 respectively. Other curves have minimums at $q = 0$. The greatest decrease in the critical load takes place in the free edge case (curve 1) and is equal to 38 % of the classical value. This result was obtained in [117]. We note that curve 5 corresponds to three different boundary condition variants.

Remark 13.1. The applicability domain for the results of Section 13.2 is restricted to two circumstances.

First, the shell is assumed to be semi-infinite. That is why it is necessary to determine whether the shell length is sufficient for the necessary damping of the deflection away from edge $x = 0$. At the same time, the shell cannot be too long, since for very long shells under pressure, the beam buckling mode can occur.

Second, curves 3, 4 and 5 increase monotonically with q and for shells which are closed in the circumferential direction, the minimum is attained at $m = 2$, where m is the number of the half-waves in the circumferential direction. However, for small values of m, the accuracy of system (1) is insufficient.

Let us estimate the shell length for which the deflection decreases by N times. We have

$$l \geqslant l_0 = \mu \ln N \left(\Im p_3\right)^{-1}. \tag{13.2.16}$$

For that, due to (14) $\Im p_3 = q^2 \sin \frac{\psi}{2} \simeq 1/2\, q^2$, which corresponds to $\lambda \simeq 1/2$.

For example, for $q = 0.31$, $\nu = 0.3$ and $N = 10$ we have

$$l_0 = \frac{L_0}{R} \simeq 26 \left(\frac{h}{R}\right)^{1/2}. \tag{13.2.17}$$

If condition (16) is not fulfilled then we cannot neglect the mutual influence of shell edges $x = 0$ and $x = l$ and we must satisfy the boundary conditions at both edges. These cases will not be considered here.

Many works have been devoted to the problem of the influence of the boundary conditions and shell length on the value of the critical load for cylindrical shells under axial compression. The references may be found in [21, 22, 61, 149].

Let us consider a cylindrical panel which is simply supported at the rectilinear edges. For this panel $q = \mu\pi/\varphi_0$. If angle φ_0 is sufficiently small (for example, $\varphi_0 \leqslant \pi/6$), then we can use (after verifying condition (16)) equations (1) and the results following from these equations, which are shown in Figure 13.2.

The results for a shell which is closed in the circumferential direction qualitatively corresponding to curves 3, 4, 5 (the curve number depends on the introduced boundary condition) with minimums at $q = 0$ are shown in Figure 13.2. Quantitatively, the results in Figure 13.2 must be made more precise. However, the fact that the critical load decreases remains true.

13.3 Cylindrical Panel with a Weakly Supported Edge

Here the problem is solved under the same assumptions as in Section 13.2. It is assumed that the curvilinear edges $x = 0$ and $x = l$ are simply supported and various boundary conditions at $\varphi = 0$ are considered. Attention is concentrated on the boundary conditions for which the buckling is localized in the neighbourhood of this edge and also for which the critical load decreases.

We seek the solution of system (2.1) in a form similar to (2.8)

$$w(x, \varphi) = \sum_{k=1}^{4} C_k w_k \exp\left\{\frac{i}{\mu}\left(px + q_k y\right)\right\} \tag{13.3.1}$$

and formulae (2.10) are still valid.

The roles of parameters p and q are different than they were in Section 13.2. Here, $p = \frac{n\pi\mu}{l}$, $n = 1, 2, \ldots$, and q_k are the roots of equation (2.9), for which $\Im q_k > 0$. We then have

$$q_{1,2} = r_1 \left(\pm \cos \frac{\theta_1}{2} + i \sin \frac{\theta_1}{2} \right), \quad q_{3,4} = r_3 \left(\pm \cos \frac{\theta_3}{2} + i \sin \frac{\theta_3}{2} \right), \tag{13.3.2}$$

$$r_{1,3} = \left(p^2 \mp p^3 \cos \frac{\psi}{2} + p^4 \right)^{1/4}, \quad \tan \theta_{1,3} = \frac{\sin \dfrac{\psi}{2}}{\pm \cos \dfrac{\psi}{2} - p},$$

where the upper sign corresponds to the first subscript and $\lambda = \cos \psi$.

For $p \ll 1$ in formulae (2) we may assume approximately that

$$r_1 = r_3 = \sqrt{p}, \quad \theta_1 = \frac{\psi}{2}, \quad \theta_3 = \pi - \frac{\psi}{2}. \tag{13.3.3}$$

As in Section 13.2 we search for the values of λ by equating the fourth order determinant to zero

$$\Delta_1 (\lambda, p) = 0 \tag{13.3.4}$$

Let first $p \ll 1$. We then take approximate formulae (3) and find that the decrease in the critical load takes place for six boundary condition variants at $\varphi = 0$

$$\begin{array}{lll}
T_2 = S = Q_{2*} = M_2 = 0 & (0000), & \lambda_0 = 0.113; \\
T_2 = u = Q_{2*} = M_2 = 0 & (0100), & \lambda_0 = 0.223; \\
T_2 = S = Q_{2*} = \gamma_2 = 0 & (0001), & \lambda_0 = 0.223; \\
T_2 = u = Q_{2*} = \gamma_2 = 0 & (0101), & \lambda_0 = 0.419; \\
T_2 = S = w = M_2 = 0 & (0010), & \lambda_0 = 0.809; \\
v = S = Q_{2*} = M_2 = 0 & (1000), & \lambda_0 = 0.809.
\end{array} \tag{13.3.5}$$

Let root $\lambda (p)$ of equation (4) converge to λ_0 as $p \to 0$. Then for the other 10 boundary condition variants, equation (4) has no roots, for which $\lambda_0 < 1$.

It follows from (5), that the free edge $\varphi = 0$ generates a decrease in the critical load by a factor of nine. The least rigid are supports $u = 0$ and $\gamma_2 = 0$, and the most rigid are supports $v = 0$ and $w = 0$. If we simultaneously introduce two or more supports (except the case when $u = \gamma_2 = 0$ and one more case, which is shown below) it prevents the appearance of the buckling mode which is localized in the neighbourhood of edge $\varphi = 0$.

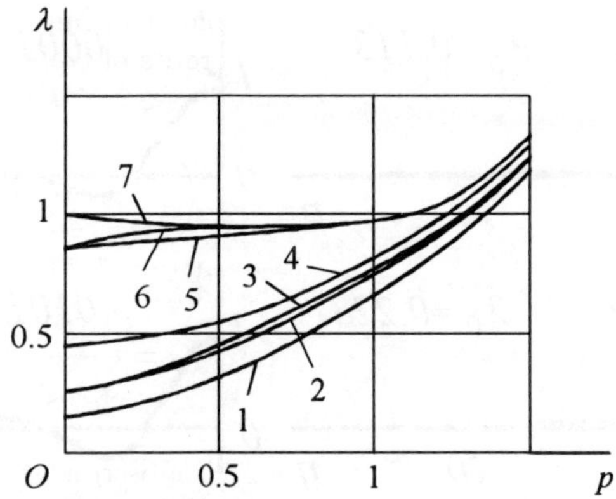

Figure 13.3: The load parameter for a cylindrical panel with a weakly supported rectilinear edge under axial compression (1 - 0000, 2 - 0001, 3 - 0100, 4 - 0101, 5 - 0010, 6 - 1000, 7 - 1100).

The numerical solution of equation (4) for $p > 0$ is shown in Figure 13.3. As in Section 13.2, we find here one more variant of the boundary condition (curve 7)

$$u = v = Q_{2*} = M_2 = 0 \quad (1100), \tag{13.3.6}$$

for which a buckling mode localized near $\varphi = 0$ is possible and $\lambda(0) = 1$. The maximum decrease in the critical load is 3.7 % and is obtained for $p \simeq 0.5$. For the other six boundary condition variants which are listed in (5), the functions $\lambda(p)$ (curves 1-6 in Figure 13.3) increase monotonically with p. Hence, the buckling mode occurs for the least possible value of p, namely for

$$p = \frac{\pi \mu}{l}. \tag{13.3.7}$$

For shells of moderate length, $p \ll 1$, and we take from the left side of the graph the value of $\lambda(p)$. The value $p \sim 1$ corresponds to a short shell.

In the case $p \ll 1$ we represent the buckling mode in the form

$$w(x, y) = Y(\eta) \sin \frac{\pi x}{l}, \qquad \eta = y\sqrt{p}.$$

For six variants of the weak support of the rectilinear edge $y = 0$ listed in (5) the functions $Y(\eta)$ are shown in Figure 13.4. One can see that for the larger values of λ_0 the deflection decrease slower away from the shell edge.

Let us consider the problem of buckling mode localization for $p \ll 1$. By

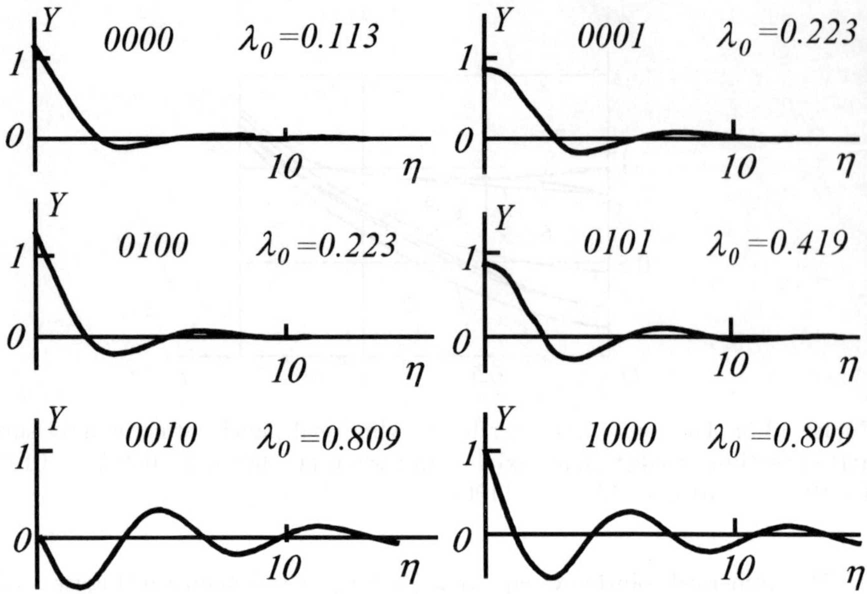

Figure 13.4: Functions $Y(\eta)$.

virtue of (2) and (3)

$$\min_k \{\Im\, q_k\} = \sqrt{p}\, \sin\frac{\psi}{4} \tag{13.3.8}$$

and the deflection decreases by N times for

$$\varphi > \varphi^0 = \frac{\mu \ln N}{\sqrt{p}\, \sin\dfrac{\psi}{4}} = 0.41\,(1-\nu^2)^{-1/8} \left(\frac{L^2 h}{R^3}\right)^{1/4} \frac{\ln N}{\sin\dfrac{\psi}{4}}, \tag{13.3.9}$$

where $\cos\psi = \lambda_0$ and the value of λ_0 is defined by (5) depending on the boundary conditions at $\varphi = 0$. The buckling modes which decrease more slowly with the growth of φ correspond to the larger values of λ_0.

Remark 13.2. It is easy to satisfy the boundary conditions at both edges $\varphi = 0$ and $\varphi = \varphi_0$ taking in solution (1) all eight roots q_k of equation (2.9). However, the results given here for a shell, which is in fact, semi-infinite in the circumferential direction, allow us to find the main part of the support influence on the critical load.

In [35] the dependence of the critical value λ on the panel width in the circumferential direction, φ_0, for various boundary conditions on the rectilinear shell edges is studied. In this work the case of the clamped curvilinear edges is also studied. It is interesting to note that in the last case the rectilinear edge is considered to be weakly supported only for four first boundary conditions (5) and the critical load is ~ 1.5 times larger than in (5).

13.4 Shallow Shell with a Weak Edge Support

Let us consider the buckling of a convex shallow shell under a membrane stress state using the same assumptions and designations as in Section 3.1. As an initial system we will take (3.1.1)

$$\mu^2 \Delta\,\Delta\, w + \lambda\,\Delta_t w - \Delta_k \Phi = 0, \quad \mu^2 \Delta\,\Delta\,\Phi + \Delta_k w = 0, \tag{13.4.1}$$

where

$$\begin{gathered}\Delta\, w = \frac{\partial^2 w}{\partial x^2} + \frac{\partial^2 w}{\partial y^2}, \quad \Delta_k w = k_2\,\frac{\partial^2 w}{\partial x^2} + k_1\,\frac{\partial^2 w}{\partial y^2}, \\ \Delta_t w = t_1\,\frac{\partial^2 w}{\partial x^2} + 2\,t_3\,\frac{\partial^2 w}{\partial x\,\partial y} + t_2\,\frac{\partial^2 w}{\partial y^2}.\end{gathered} \tag{13.4.2}$$

We will consider edge $x = 0$ and seek the boundary condition variants for which buckling modes localized in the neighbourhood of this edge are possible.

As before, we seek the solution of system (1) in the form

$$w = w_0 \exp\left\{\frac{i}{\mu}\,(p\,x + q\,y)\right\}. \tag{13.4.3}$$

In the same form we seek other unknown functions, which are included in the boundary conditions. We have

$$\begin{aligned}\Phi &= \frac{(k_2 p^2 + k_1 q^2)\,w}{(p^2+q^2)^2}, \quad T_1 = -\,\frac{Eh}{R}\,q^2\Phi, \quad S = \frac{Eh}{R}\,p\,q\,\Phi, \\ Q_{1*} &= \frac{Eh\mu\,i}{R}\Big(p^3 + (2-\nu)\,p\,q^2 - 2\lambda\,(t_1 p + t_3 q)\Big)w, \\ M_1 &= -\,Eh\mu^2\,(p^2+\nu q^2)\,w, \qquad u = \frac{\mu}{ip}\Big(k_1 w - (q^2 - \nu p^2)\,\Phi\Big), \\ v &= \frac{\mu}{iq}\Big(k_2 w - (p^2 - \nu q^2)\,\Phi\Big), \quad \gamma_1 = -\,\frac{ip}{\mu}\,w,\end{aligned} \tag{13.4.4}$$

where Q_{1*} is the generalized shear stress resultant

$$Q_{1*} = Q_1 - \frac{1}{R}\frac{\partial H}{\partial y} - T_1^0 \gamma_1 - S^0 \gamma_2 . \tag{13.4.5}$$

Condition $Q_{1*} = 0$ is a natural condition for functional (3.6.1) in the case when the deflection w is not given.

Parameter $q > 0$ in (3) is fixed, and for parameter p we get the eighth order equation

$$\left(p^2 + q^2\right)^4 - \lambda \left(t_1 p^2 + 2\, t_3 p\, q + t_2 q^2\right) \left(p^2 + q^2\right)^2 + \left(k_2 p^2 + k_1 q^2\right)^2 = 0. \tag{13.4.6}$$

We will consider the values of λ, for which equation (6) has no real roots. We seek unknown functions in the form of (2.8), where $\Im\, p_k > 0$. As in Section 13.2 we come to the equation for λ

$$\Delta_0(\lambda, q) = 0, \tag{13.4.7}$$

by equating the fourth order determinant to zero.

The roots of equation (7), depend on parameter q and also on the parameters k_i and t_i which characterize the shell curvature and membrane initial stresses, and also on the boundary condition variant introduced at edge $x = 0$.

A numerical study of the roots of equation (7) shows that for eight boundary condition variants there are domains of parameters k_i and t_i for which buckling mode localization occurs in the neighbourhood of edge $x = 0$. For that $\lambda < \lambda_0$, where λ_0 is the value of the load parameter found in Section 3.1 assuming that p and q in (3) are real and the boundary conditions are neglected.

It follows from (2) that we can choose λ or one of the values of t_i (terms of the form $\lambda\, t_i$ are unknown). If we assume, in particular, that $\lambda_0 = 1$ then the value $\lambda < 1$ is a measure of the decrease of the critical load caused by the weak support of edge $x = 0$.

The eight boundary condition variants mentioned above are the following: free edge (0000), conditions with one support (1000, 0100, 0010, 0001), and three boundary condition variants each of which contains two supports (1100, 1010, 0101).

The problem under consideration contains five parameters t_1, t_2, t_3, k_1 and k_2. We can fix one of the values of t_i by choosing λ, and we can fix one of the k_i by choosing a characteristic linear size. Three essential parameters remain as a result. A full item-by-item examination of these parameters is beyond the scope of this book, although the numerical solution of equation (7) is not difficult. Therefore, we will limit the discussion to the consideration of three load cases.

Example 13.1. Let us consider the buckling of a convex shell of revolution under axial pressure. Taking into account the expected local character of buckling, we assume that the initial stress-resultants t_1, and t_2 and curvatures k_1 and k_2 are approximately constant. We take

$$t_1 = k_2 = 1, \qquad t_3 = 0, \qquad t_2 = -k_1, \tag{13.4.8}$$

where $k_1 \geqslant 0$ is the only parameter of the problem.

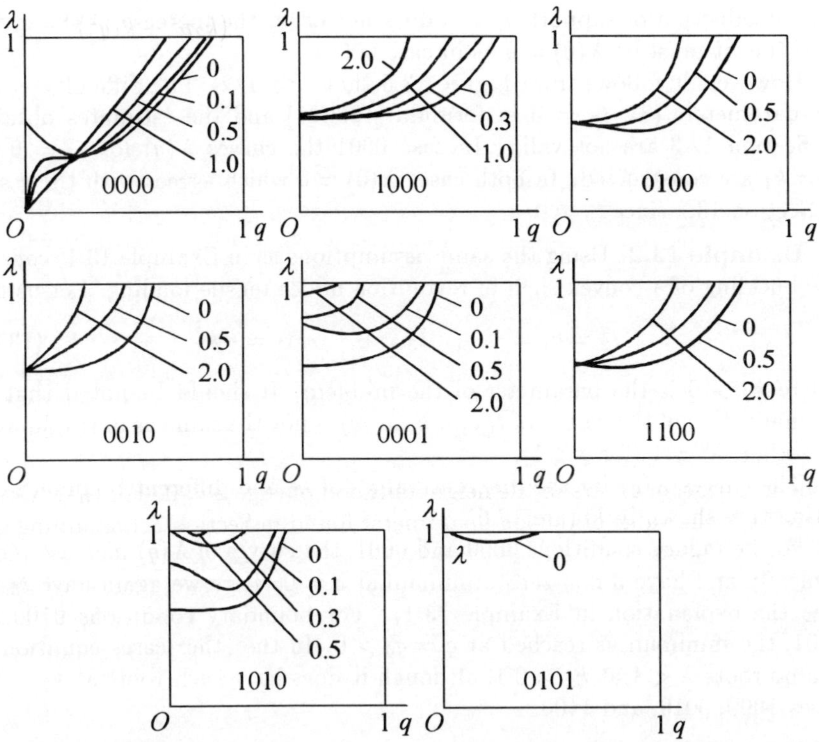

Figure 13.5: The load parameter for a convex shell of revolution under axial compression (the values of k_1 are given in the Figure).

We can see in Figure 13.5 a number of curves of $\lambda\,(q)$ which depend on the values of k_1 and the boundary conditions (shown in the Figure).

For the free edge (0000) and $k_1 > 0$, all of the curves have a minimum $\lambda = 0$ at $q = 0$, which tells us about the decrease in the critical load (see Section 12.3, where we constructed pseudo-bending for a convex shell. The existence of pseudo-bending leads to a decrease in the critical load).

For boundary conditions 1000, 0100, and 1100 (as in Section 13.2) the curves $\lambda(q)$ have a minimum $\lambda = 1/2$ at $q = 0$.

In the case of boundary conditions 0001 and 1010, the curves of $\lambda(q)$ may have a minimum at $q = q_0 > 0$. In these cases equation (7) has a root $\lambda < 1$ only for $k_1 < k_1^0$ (see Figure 13.5).

For condition 0101 equation (7) has a root at $k_1 = 0$, which is close to unity (see curve 6 in Figure 13.2). For $k_1 \geqslant 0.1$ equation (7) has no roots such that $\lambda \leqslant 1$.

Let us return to the discussion of case 0001. As was shown in Section 12.3, the introduction of support $\gamma_1 = 0$ does not effect the increase of the critical load (i.e. it must be $\lambda(0) = 0$ as in case 0000).

However, it follows from Figure 13.5 that $\lambda \geqslant 1/2$. The difficulty is that for parameters (8), $t_0 = 0$ in formula (12.3.21) and the estimates obtained in Section 12.3 are not valid. In case 0001 the curves $\lambda(q)$ for $t_2 = 0$ and $t_2 = k_1$ are constructed. In both cases $\lambda(0) = 0$ which agrees with the results of Section 13.2 since $t_0 \neq 0$.

Example 13.2. Using the same assumptions as in Example 13.1, consider the buckling of a convex shell of revolution under tensile loading. Let us take

$$t_2 = k_1 = 1, \qquad t_3 = 0, \qquad t_1 = -k_2, \tag{13.4.9}$$

where $k_2 > 0$ is the parameter of the problem. It should be noted that the normalization of the values of t_i and k_i in (9) is not the same as in (8), however as earlier, $\lambda^0 = 1$ (for $q < 1$).

The curves $\lambda(q)$ for the different values of k_2 and different boundary conditions are shown in Figure 13.6.

For boundary conditions 0000 and 0001, the curves of $\lambda(q)$ increase monotonically and have a non-zero minimum at $q = 0$. Here we again have $t_0 = 0$ (see the explanation in Example 13.1). For boundary conditions 0100 and 0101, the minimum is reached at $q = q_0 > 0$. In the other cases equation (7) has no roots $\lambda < 1$ at $k_2 \geqslant 0.1$, although it does have such roots at $k_2 = 0$ in cases 1000, 0010, and 1100.

Example 13.3. Let us consider the buckling of a convex shell of revolution under torsion. We take

$$t_1 = t_2 = 0, \qquad t_3 = \sqrt{k_1}, \qquad k_2 = 1, \tag{13.4.10}$$

where $k_1 > 0$ is the parameter of the problem. Again, $\lambda^0 = 1$, i.e. $\lambda < 1$ is a measure of the decrease in the critical load connected with the weak support of the edge.

The decrease of the critical load occurs only for two boundary condition variants (0000 and 0001) (see Figure 13.7). The minimum is reached at $q = 0$ in both cases.

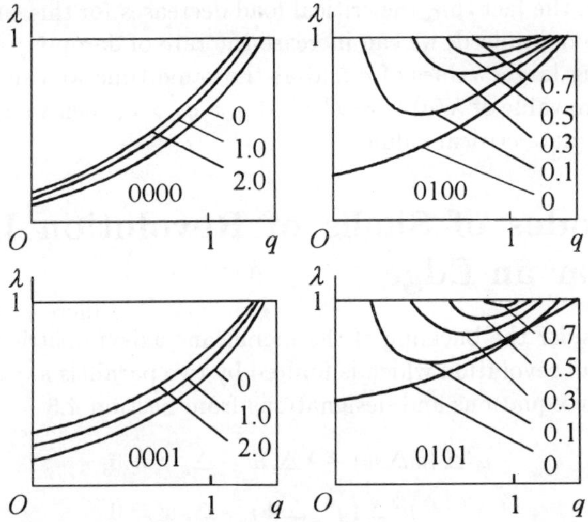

Figure 13.6: The loading parameter for a convex shell of revolution under axial tension (the values of k_2 are given in the Figure).

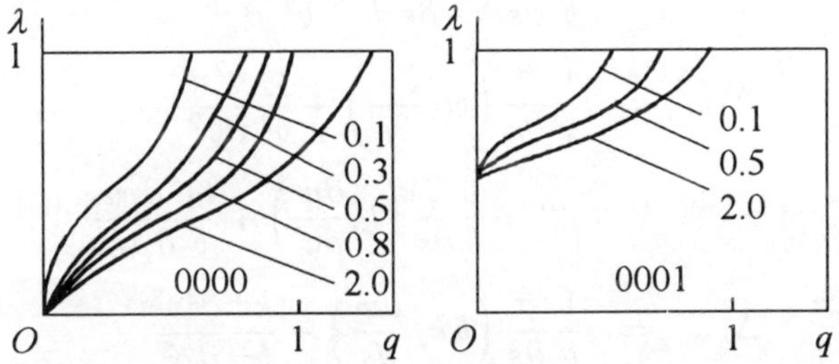

Figure 13.7: The load parameter for a convex shell of revolution under torsion (the values of k_1 are given in the Figure).

Let us make a remark similar to Remark 13.1 in connection with the results of this Section.

Remark 13.3. The minimum value of $\lambda(q)$ is reached at $q = 0$ in most of the cases considered. This means that we should take the minimum possible value of q to evaluate the critical load. However, equation (6) has the root $p = iq\sqrt{\frac{k_1}{k_2}}$ for small values of q and this root determines the slow damping of the buckling mode, which may lead to necessity to consider the boundary conditions at the other edge $x = x_2$.

Therefore it is not a reliable practice to take the value $\lambda(0)$ as the critical load. However, the fact that the critical load decreases for this support at edge $x = 0$ remains and indeed, we can increase the rate of damping of the buckling mode by taking larger values of q and at the same time we can get $\lambda(q) < 1$. If the minimum value of $\lambda(q)$ is reached at $q = q_0 > 0$, then we can take $\lambda(q_0)$ as the approximate critical value.

13.5 Modes of Shells of Revolution Localized near an Edge

Let us consider the buckling of the membrane axisymmetric stress state of a convex shell of revolution which is limited by two parallels $s = s_1$ and $s = s_2$. We will use the equations and designations from Section 4.3

$$\begin{aligned} \mu^2\Delta\,(d\,\Delta\,w) + \lambda\,\Delta_t w - \Delta_k\Phi = 0, \\ \mu^2\Delta\left(g^{-1}\Delta\,\Phi\right) + \Delta_k w = 0, \end{aligned} \tag{13.5.1}$$

where

$$\begin{aligned} \Delta w &= \frac{1}{b}\frac{\partial}{\partial s}\left(b\,\frac{\partial w}{\partial s}\right) + \frac{1}{b^2}\frac{\partial^2 w}{\partial \varphi^2}, \\ \Delta_t w &= \frac{1}{b}\frac{\partial}{\partial s}\left(bt_1\frac{\partial w}{\partial s}\right) + \frac{t_2}{b^2}\frac{\partial^2 w}{\partial \varphi^2} + \\ &\quad + \frac{1}{b}\frac{\partial}{\partial s}\left(t_3\frac{\partial w}{\partial \varphi}\right) + \frac{t_3}{b}\frac{\partial^2 w}{\partial \varphi\,\partial s}, \\ \Delta_k w &= \frac{1}{b}\frac{\partial}{\partial s}\left(b\,k_2\,\frac{\partial w}{\partial s}\right) + \frac{k_1}{b^2}\frac{\partial^2 w}{\partial \varphi^2}. \end{aligned} \tag{13.5.2}$$

As in Section 4.3, we assume that the determining functions b, d, g, k_i and t_i do not depend on φ, but may depend on s.

In Chapter 4 we developed the buckling modes for which the weakest parallel is situated completely inside the shell. Here we will consider cases in which the buckling mode is localized in the neighbourhood of edge $s = s_1$. This localization may occur due to the weak support of an edge or if the weakest parallel coincides with the edge.

Let us show first that the results of Section 13.4 are applicable to a discussion of the weak support of an edge. Indeed, the solution of equation (1) we seek in the form of a formal series in μ

$$w(s,\varphi) = \sum_{k=1}^{4} C_k w_k^* \exp\left(\frac{i}{\mu}\int_{s_1}^{s} p_k ds + i\,m\,\varphi\right), \tag{13.5.3}$$

where

$$w_k^* = \sum_{n=0}^{\infty} \mu^n w_k^{(n)}(s), \quad \Im p_k(s_1) > 0, \tag{13.5.4}$$

and m is the number of waves in the circumferential direction.

Other unknowns we seek in the form (3), and for which have the lowest order in μ terms in (4) we have the same relationship (4.4). For that case $p_k = p_k(s)$ are the roots of the equation

$$d\left(p^2 + \frac{\rho^2}{b^2}\right)^4 - \lambda\left(t_1 p^2 + 2\,t_3 p\,\frac{\rho}{b} + t_2\,\frac{\rho^2}{b^2}\right)\left(p^2 + \frac{\rho^2}{b^2}\right)^2 + \\ + g\left(k_2 p^2 + k_1\,\frac{\rho^2}{b^2}\right)^2 = 0, \qquad \rho = \mu\, m. \tag{13.5.5}$$

If we normalize the parameters E and h in such a way that $d = g = 1$ at $s = s_1$ and assume that $q = \rho\big(b(s_1)\big)^{-1}$, then equation (5) coincides with equation (4.6). As before, if we substitute solution (3) into the boundary condition at $s = s_1$, then the fourth order determinant is equal to zero

$$\Delta(\lambda, q, \mu) = \Delta_0(\lambda, q) + O(\mu) = 0, \tag{13.5.6}$$

The elements of this determinant differ by values order of μ from those of determinant (4.7). If we neglect this difference we get the same results as in Section 13.4, namely that the same boundary conditions allow the desired buckling modes. Concerning the critical value of λ, we have

$$\lambda = \lambda_0 + \mu\,\lambda_1(\mu), \tag{13.5.7}$$

where λ_0 is the root of equation (4.7), and the correction term $\mu\,\lambda_1$ depends on the variability of the determining functions and may have any sign.

Now let the edge support at $s = s_1$ not be weak, i.e. the boundary conditions at $s = s_1$ are such that for $0 \leqslant q < \infty$ equation (6) has no roots λ for $\lambda < \lambda_0$, where $\lambda_0 = \min\limits_{p,q} f$ at $s = s_1$ (see below). Let the weakest parallel be when $s = s_1$. This will be discussed below (see also Section 4.3).

If we resolve equation (5) in λ, we get

$$\lambda = f(p, \rho, s). \tag{13.5.8}$$

We seek the smallest value of this function $\lambda_0 > 0$ in the domain $-\infty < p < \infty$, $0 \leqslant \rho < \infty$, $s_1 \leqslant s \leqslant s_2$. Let the value of $\lambda = \lambda_0$ be reached for real $p = p_0$, $\rho = \rho_0$ and $s = s_0$. The weakest parallel is s_1 if $s_0 = s_1$. We additionally assume that the parallel s_0 is unique and

$$f_s^0 = \frac{\partial f}{\partial s} > 0 \quad \text{for} \quad p = p_0,\ \rho = \rho_0,\ s = s_1. \tag{13.5.9}$$

For $\lambda = \lambda_0$, $\rho = \rho_0$ equation (5) has real root $p = p_0$ at $s = s_1$ and has no real roots for $s \neq s_1$. Then we take

$$\lambda = \lambda_0 + \Delta\,\lambda, \qquad \Delta\,\lambda > 0. \tag{13.5.10}$$

and equation (5) has real roots for $s_1 \leqslant s \leqslant s_*$ and has no real roots for $s > s_*$. Indeed, if we fix $\rho = \rho_0$ and expand (8) into a Taylor series we obtain approximately

$$\Delta\,\lambda = f_s^0\,(s - s_1) + \frac{1}{2}\,f_{pp}^0\,(p - p_0)^2\,, \quad f_{pp}^0 = \frac{\partial^2 f}{\partial p^2}, \tag{13.5.11}$$

where f_{pp}^0 is calculated for the same variables as in (9). From (10) and (11) we get

$$s_* = s_1 + \frac{\Delta\,\lambda}{f_s^0}. \tag{13.5.12}$$

Point $s = s_*$ is called a turning point. For $s_1 \leqslant s < s_*$ by virtue of (3) and (11), system (1) has oscillating integrals, and for $s > s_*$ all of its solutions increase or decrease exponentially. The existence of the oscillations in direction s means that buckling is possible.

Selecting $\Delta\lambda$ in (10) one must select the phase for the oscillating solution such that together with the other solutions of system (1) for which $\Im\, p_k > 0$ satisfies the boundary conditions introduced at $s = s_1$. For that we take the oscillating solution which transforms to a damped solution for $s > s_*$ and as a result we get the buckling mode localized in the neighbourhood of edge $s = s_1$. The critical value of λ does not depend on the boundary conditions at $s = s_2$, if this edge is not weakly supported.

The technical realization of the outlined plan proves to be rather complex and is discussed in Section 13.6. Here we only note that the critical value of λ can be expanded in powers of $\mu^{1/3}$

$$\lambda = \lambda_0 + \Delta\,\lambda, \quad \Delta\,\lambda = \mu^{2/3}\,\lambda_2 + \mu\,\lambda_3 + \cdots\,, \tag{13.5.13}$$

where λ_0 is defined by formulae (4.3.7) and (4.3.9), and λ_2, λ_3 are obtained in Section 13.6.

In direction s the size of the buckling domain is of order $\mu^{2/3}$. The last circumstance may be some justification for not taking into account pre-buckling displacements and moment stresses, since the boundary effect solutions penetrate only to a depth of order μ (smaller than $\mu^{2/3}$). Their influence on the critical load is discussed in Section 14.4.

13.6 Buckling Modes with Turning Points

The difficulty of developing analytical solutions of system (1) with a turning point, s_*, is caused by the fact that in the neighbourhood of $s = s_*$ exponential expression (5.3) becomes invalid. This is related to the fact that at $s = s_*$ equation (5.5) has multiple roots and the coefficients $w_k^{(n)}$ in (5.4) approach infinity at $s = s_*$.

We note that using the orthogonal sweep method of numerical integration of the system of equations obtained from (1) after separation of the angle coordinate φ, the presence of the turning point does not bring additional difficulties. Therefore, the aim of the derivations given below is to clarify qualitatively the phenomenon and to obtain formula (5.13).

Let us limit the discussion to two cases, when $t_3 \neq 0$ at $s = s_1$ if at least one of the inequalities $t_1 > 0$, $t_2 > 0$, $t_3^2 > t_1 t_2$ (case 1) is valid or $t_3 = 0$ and simultaneously $t_2 > 0$, $t_2 k_2 > t_1 k_1$ (case 2).

In case 1 the pits are inclined at an acute angle to the edge (see (3.1.12)), and in case 2 the pits are orthogonal to the edge. The case defined by formula (4.3.10) ($t_3 = 0$, $t_1 > 0$, $t_1 k_1 > t_2 k_2$), in which the pits are stretched along the edge is not considered here.

Under the above assumptions, equation (5.5) has two roots which coincide at $s = s_*$: p_1 and p_2. By virtue of (5.11) we have approximately

$$p_{1,2} = p_0 \pm \left(\frac{2 f_s^0 (s_* - s)}{f_{pp}^0} \right)^{1/2} . \tag{13.6.1}$$

The solution of system (5.1) is damped for $s > s_*$ and has the form similar to (11.7.2) (see also [157])

$$w(s, \varphi, \mu) = \left[w^{(1)} \mathrm{Ai}(\eta) + w^{(2)} \mathrm{Ai}'(\eta) \right] e^{\dfrac{i}{\mu} \displaystyle\int_{s_*}^{s} \frac{p_1 + p_2}{2} ds + im\varphi} , \tag{13.6.2}$$

where $\mathrm{Ai}(\eta)$ is the Airy function, $\mathrm{Ai}(\eta) = U(-\eta)$, $U(\eta)$ is the solution of equation (11.7.3)

$$\eta = \left(\frac{3}{2\mu} \int_{s_*}^{s} \frac{p_2 - p_1}{2} ds \right)^{2/3} , \tag{13.6.3}$$

$w^{(1)}$, $w^{(2)}$ are the formal power series in μ,

$$w^{(k)} = \mu^{\gamma_k(w)} \sum_{n=0}^{\infty} \mu^n w^{(k,n)}(s), \qquad k = 1, 2. \tag{13.6.4}$$

The other variables included in the boundary conditions have the same form as (2).

The substitution of the linear combination of solution (2) and the other three solutions of type (5.3) into the boundary conditions leads to the approximate equation for the determination of λ

$$\mathrm{Ai}\,(\eta_1) = \mu^{1/3} F\,(\lambda)\,\mathrm{Ai}'(\eta_1), \quad \eta_1(\lambda) = \eta\,(s_1) < 0, \tag{13.6.5}$$

where the rather awkward function $F\,(\lambda)$ depends on the boundary conditions at $s = s_1$. For a clamped edge, the function F is given in [157].

The solution of equation (5) leads to formula (5.13) where

$$\Delta\,\lambda = \left(\frac{\mu^2\,(f_s^0)^2\,f_{pp}^0}{2}\right)^{1/3}\left[\zeta_0 - \mu^{1/3} F\,(\lambda_0)\right], \quad \zeta_0 = 2.338. \tag{13.6.6}$$

Here ζ_0 is the first root of the equation $\mathrm{Ai}\,(-\zeta_0) = 0$. It should be noted that only the second term in the brackets of (6) depends on the specific form of the boundary conditions.

Example 13.4. Let us consider the torsion of a convex shell of revolution by moments M applied to shell edges $s = s_1$ and $s = s_2$. The parameters E, ν and h are assumed to be constant.

Then, due to (4.3.3) and Section 4.5

$$\begin{aligned} &T_1^0 = T_2^0 = 0, \quad S^0 = \frac{M}{2\pi\,R^2 b^2} = -\,\lambda\,T\,t_3, \quad T = E\,h\,\mu^2, \\ &t_3 = -\frac{1}{b^2}, \qquad \lambda_0 = \min_s \{\gamma\,(s)\}\,, \qquad \gamma\,(s) = 2\,b^2\sqrt{k_1 k_2}. \end{aligned} \tag{13.6.7}$$

Let us assume that the function $\gamma(s)$ attains its minimum value at $s = s_1$. Then, by virtue of (3.1.14) and (4.5.5)

$$p_0 = \frac{k_1\sqrt{2\,k_2}}{k_1 + k_2}, \quad m_0 = -\frac{b\,k_2\sqrt{2\,k_1}}{\mu\,(k_1 + k_2)}. \tag{13.6.8}$$

The derivatives in (6) are equal to

$$f_s^0 = \frac{\partial\gamma}{\partial s_1}, \quad f_{pp}^0 = 2\,b^2 k_1^{-3/2} k_2^{-1/2}\left(4\,k_1^2 + (k_1 - k_2)^2\right). \tag{13.6.9}$$

The critical value for the torsional moment

$$M = \frac{2\pi\,E\,h^2}{\sqrt{3\,(1-\nu^2)}}\,\frac{B^2\,(s_1)}{\sqrt{R_1(s_1)\,R_2(s_1)}}\left(1 + \frac{\Delta\,\lambda}{\lambda_0}\right). \tag{13.6.10}$$

Let us consider, for example, the torsion of an oblate ellipsoid of revolution with semi-axes $a = R$ and $b = R/\sqrt{2}$ (see Figure 4.4).

Let $R/h = 500$, $\nu = 0.3$, $\theta_1 \leqslant \theta \leqslant \theta_2$, $\theta_1 = \pi/4$ and $\theta_2 = \pi/2$, where θ is an angle between the shell normal and the axis of symmetry. Taking into account formulae (4.4.8) $\gamma(s) = 2\sqrt{2}\sin^2\theta$, hence the weakest parallel is $\theta = \theta_1$ and condition (5.9) is fulfilled. The critical value for the torsional moment is equal to

$$M = \frac{2\pi E h^2 R}{\sqrt{6(1-\nu^2)}}\left[1 + 6.1\mu^{2/3} - 1.2\mu\right] = 1.48\,\frac{2\pi E h^2 R}{\sqrt{6(1-\nu^2)}}. \qquad (13.6.11)$$

Here the second term in the brackets does not depend on the boundary conditions at $s = s_1$, and the third term corresponds to the case of a clamped edge.

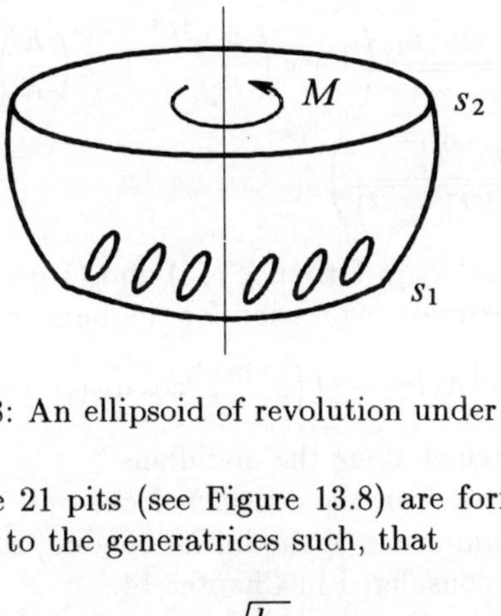

Figure 13.8: An ellipsoid of revolution under torsion.

On the shell surface 21 pits (see Figure 13.8) are formed. These pits are inclined at an angle ψ_0 to the generatrices such, that

$$\tan\psi_0 = \sqrt{\frac{k_1}{k_2}} = 1.2. \qquad (13.6.12)$$

The shape of the pits is approximately defined by the expression

$$w(s,\varphi) = \left[\mathrm{Ai}(\eta) + O\left(\mu^{1/3}\right)\right]\cos m_0\left(\sqrt{\frac{k_1}{k_2}}\,\frac{s}{b} - \varphi + \varphi_0\right), \qquad (13.6.13)$$

which is obtained after simplifying the real part of function (2). The values k_1, k_2 and b are calculated at $\theta = \theta_1$.

Example 13.5. Let us consider the buckling of a convex shell of revolution under tensile loading by an axial force P. The parameters E, ν and h are assumed to be constant. As in 4.4 we have

$$P = 2\pi R E h \mu^2 \lambda, \qquad t_1 = -\frac{1}{k_2 b^2}, \quad t_2 = \frac{k_1}{k_2^2 b^2},$$
$$\lambda_0 = \min_s \{\gamma(s)\}, \quad \gamma(s) = 2 b^2 k_2^2 = 2\sin^2\theta, \tag{13.6.14}$$
$$p_0 = 0, \quad q_0 = \sqrt{k_1}, \quad m_0 = \frac{b\, q_0}{\mu}.$$

The minimum $\gamma(s)$ for a convex shell is reached at one of the edge parallels. Let it be parallel $s = s_1$. We then obtain

$$f_s^0 = \gamma_s' = 4 k_1 \sin\theta \cos\theta = 4 k_1 k_2 b\, b', \quad f_{pp}^0 = \frac{8 b^2 k_2^3}{k_1^2}. \tag{13.6.15}$$

Now we obtain the critical value by formulae (5.13) and (6)

$$P = \frac{2\pi E h^2 \sin^2\theta_1}{\sqrt{3(1-\nu^2)}} \left(1 + c\left(\frac{h}{R}\right)^{1/3} + O\left(\frac{h}{R}\right)^{1/2}\right),$$
$$c = \left(\frac{8 k_2 (b')^2}{b^2\sqrt{12(1-\nu^2)}}\right)^{1/3} \zeta_0. \tag{13.6.16}$$

The values of b, k_1 and k_2 in formulae (15) and (16) are calculated at $s = s_1$.

We get the approximate expression for the buckling mode

$$w(s,\varphi) = \left[\mathrm{Ai}(\eta) + O\left(\mu^{1/3}\right)\right] \cos m_0(\varphi - \varphi_0), \tag{13.6.17}$$

and the pits are stretched along the meridians.

The initial stress state in the considered example unlike Example 13.4, is not membrane. The influence of the initial bending stresses and pre-buckling deformations will be considered in Chapter 14.

Here, as in (13), the O-term includes the sum of the edge effect integrals damping away from edge $s = s_1$ faster than the first term in the brackets of (17).

13.7 Modes Localized near the Weakest Point on an Edge

As in Section 6.1, we will consider the buckling of a convex shell under a membrane stress state. Let us fix the coefficients of the system of equations (6.1.1) and seek a solution in the form of (6.1.4)

$$w = w_0 \exp\left[i\mu^{-1}(p_1\alpha_1 + p_2\alpha_2)\right]. \tag{13.7.1}$$

Then, the parameters p_1, p_2 and λ are related by expression (6.1.5)

$$\lambda = f(p_i, \alpha_i) = \frac{d\left(q_1^2+q_2^2\right)^4 + g\left(k_2 q_1^2 + k_1 q_2^2\right)^2}{\left(q_1^2+q_2^2\right)^2\left(t_1 q_1^2 + 2\,t_3 q_1 q_2 + t_2 q_2^2\right)}, \tag{13.7.2}$$

where $q_i = p_i/A_i$. The approximate value of the load parameter, λ_0, the wave numbers p_1^0, p_2^0 and the weakest point α_1^0, α_2^0 are evaluated by minimizing (6.1.6)

$$\lambda_0 = \min_{p_i,\alpha_i}\{f(p_i, \alpha_i)\} = f\left(p_i^0, \alpha_i^0\right). \tag{13.7.3}$$

In Sections 6.1-6.3 it was assumed that the weakest point is placed sufficiently far from the shell edge. Here we consider the case when this point lies at one of the shell edges.

Let us first make some simplifying assumptions. Let the shell edge in which we are interested coincide with the line of curvature. Instead of α_1, α_2 we introduce a system of the curvilinear coordinates s, φ coinciding with the lines of curvature such that line $s = 0$ coincides with the shell edge, and the weakest point is $s = \varphi = 0$. We denote $p_1 = p$, $p_2 = q$ and introduce the matrix

$$A = \begin{pmatrix} f_{pp} & f_{pq} & f_{p\varphi} \\ f_{qp} & f_{qq} & f_{q\varphi} \\ f_{\varphi p} & f_{\varphi q} & f_{\varphi\varphi} \end{pmatrix}, \tag{13.7.4}$$

where the partial derivatives calculated at $p = p^0$, $q = q^0$ and $s = \varphi = 0$ are denoted by $p_{pp}, \ldots$. Let us assume that $f_s > 0$ for the same values of the variables, and that matrix A is positive definite. Under these assumptions functions f have a strict minimum at the considered point.

Then we assume that edge $s = 0$, and also the other shell edges are not weakly supported (see Section 13.4), i.e. that the edge itself can not cause buckling.

Let us consider the problem with the initial stresses of one of the types which were defined at the beginning of Section 13.6.

Under these assumptions the buckling mode and the load parameter are found in [96] and for the buckling mode the following formula is obtained

$$\begin{aligned} w(s,\varphi) = \exp\left[\frac{i}{\mu}\left(p^0 s + q^0\varphi\right) - \frac{a\varphi^2}{2\mu}\right]\Big(\mathrm{Ai}(\eta)w_0 + \mu^{1/3}\mathrm{Ai}'(\eta)w_1\Big) + \\ + \mu^{1/3}\sum_{j=2}^{4} w_j \exp\left[\frac{i}{\mu}\left(p^{(j)}s + q^0\varphi\right)\right], \end{aligned} \tag{13.7.5}$$

where

$$w_j = \sum_{l=0}^{\infty} \mu^{l/6} w_j^{(l)}, \quad j = 0, 1, \ldots, 4; \quad \eta = \zeta_0 + ks\mu^{-2/3}. \tag{13.7.6}$$

Here $w_0^{(l)}$, $w_1^{(l)}$ are polynomials in the powers of $s\mu^{-2/3}$, $\varphi\mu^{-1/2}$, and $w_2^{(l)}$, $w_3^{(l)}$, $w_4^{(l)}$ are polynomials in powers $s\mu^{-1}$, $\varphi\mu^{-1/2}$; Ai (η) is the Airy function, damping as $\eta \to \infty$ (see (6.2), (11.7.7)), $\zeta_0 = 2.338$ is the root of the equation Ai $(-\zeta_0) = 0$.

The second term in (5) is a sum of the edge effect solutions adjusted with the first term at $s = 0$.

The approximate representation for the buckling mode we get by taking

$$w(s, \varphi) = \exp\left[\frac{i}{\mu}(p^0 s + q^0 \varphi) - \frac{a\,\varphi^2}{2\,\mu}\right] \mathrm{Ai}\,(\eta). \tag{13.7.7}$$

Function (7) describes the system of elongated pits with width of order μ, inclined to the edge by angle δ where ($\tan\delta = -q^0/p^0$).

The domain occupied by the pits is localized in the neighbourhood of the weakest point $s = \varphi = 0$ and has size of order $\mu^{2/3} \times \mu^{1/2}$ (see Figure 13.9 below, where $\delta = \pi/2$). In direction s, function (7) is similar to functions (6.2) and (6.13) describing the localization of the deflected shape in the neighbourhood of the weakest line coinciding with the shell edge. In direction φ function (7) coincides with function (4.2.6) describing the localization near the line lying completely inside the shell.

We can give the expression for k and a in (6) and (7)

$$k = \left(\frac{2\,f_3}{f_{pp}}\right)^{1/3}, \quad a = \frac{i\,(f_{pq} f_{p\varphi} - f_{pp} f_{q\varphi}) - \sqrt{f_{pp}\ \det A}}{f_{pp} f_{qq} - f_{pq}^2}. \tag{13.7.8}$$

The parameter λ can be expanded in powers of $\mu^{1/6}$

$$\lambda = \lambda_0 + \mu^{2/3}\lambda_1 + \mu\,\lambda_2 + \mu^{7/6}\lambda_3 + \ldots, \tag{13.7.9}$$

where λ_0 is defined in (3),

$$\lambda_1 = \zeta_0 \left(\frac{f_s^2\, f_{pp}}{2}\right)^{1/3}, \tag{13.7.10}$$

and $\mu^{2/3}\lambda_1$ coincides with the first term in $\Delta\lambda$ (see (6.6)) and does not depend on the boundary conditions at $s = 0$. The next terms $\mu\lambda_2$, $\mu^{7/6}\lambda_3$ and ... in (9) do however depend on the boundary conditions at $s = 0$.

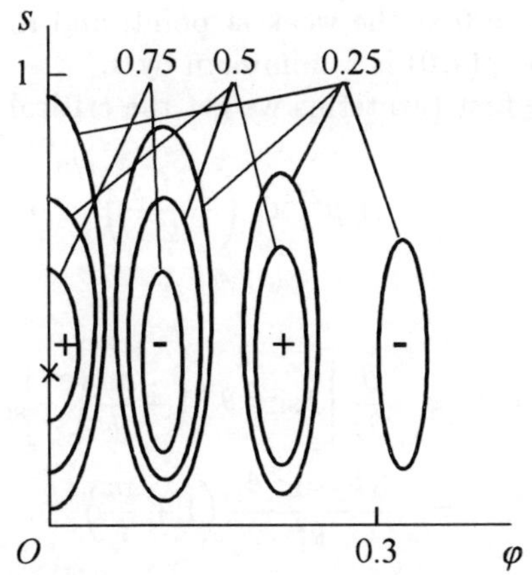

Figure 13.9: The buckling mode of a convex shell in the case when the weakest point lies on the edge of the shell.

As earlier (see Section 4.2, the case B; Section 6.1), the eigenvalues are asymptotically double.

Example 13.6. Let us consider the buckling of a convex shell of revolution with constant parameters E, ν and h being stretched by an axial force, P, and a bending moment, M_1 (see Figure 1.2).

For only one force P, one of the edge parallels is the weakest parallel (see Example 13.5). If we add the moment M_1 then we have the situation when the points of this parallel are loaded non-homogeneously and buckling occurs in the neighbourhood of the weakest point. The initial stress-resultants T_i^0, S^0 are determined by formulae (1.4.6). We then assume

$$t_1 = -\frac{1}{b^2 k_2} - \frac{m\cos\varphi}{b^3 k_2}, \quad t_2 = \frac{k_1}{b^2 k_2^2} + \frac{m\,k_1\cos\varphi}{b^3 k_2^2}, \quad t_3 = 0,$$
$$M_1 = -\frac{PRm}{2}, \qquad P = 2\pi\,R\,E\,h\,\mu^2\lambda. \tag{13.7.11}$$

If we calculate the minimum of (3) in p and q, we find that the weakest point is the point where the function (see (4.4.2))

$$\gamma\,(s,\varphi) = \frac{2\,k_1}{t_2} = 2\sin^2\theta\left(1 + \frac{m\cos\varphi}{B}\right)^{-1} \tag{13.7.12}$$

attains a minimum value. For that $p^0 = 0$ and $q^0 = b\sqrt{k_1}$.

Let $m > 0$, then $\varphi = 0$ at the weakest point, and this point is situated at an edge parallel where $\gamma(s, 0)$ is a minimum by s.

Keeping in (9) the first two terms we get the critical value

$$\lambda = 2\sin^2\theta_1 \left(1 + \frac{m}{b}\right)^{-1} + \mu^{2/3}\zeta_0 \left(\frac{f_s^2 f_{pp}}{2}\right)^{1/3} + O(\mu), \tag{13.7.13}$$

where

$$\begin{aligned} f_s &= \frac{\partial}{\partial s}\left[2\sin^2\theta \left(1 + \frac{m}{b}\right)^{-1}\right], \\ f_{pp} &= \frac{8\,k_2 \sin^2\theta_1}{k_1^2}\left(1 + \frac{m}{b}\right)^{-1}. \end{aligned} \tag{13.7.14}$$

For $m = 0$ formula (13) transforms to formula (6.16).

Parameter a determines the rate of decrease of the depth of the pits in the circumferential direction and is equal to

$$a = \left(\frac{f_{\varphi\varphi}}{f_{qq}}\right)^{1/2} = \left(\frac{m\,k_1}{4\,(m + b)}\right)^{1/2}. \tag{13.7.15}$$

In [96] for an ellipsoid of revolution the term $\mu\lambda_2$ in (9) taking into account the edge support at $s = 0$ is obtained. For a shell with parameters $b_0/a_0 = \sqrt{5}$, $\pi/3 \leqslant \theta \leqslant \pi/2$, $R/h = 500$, $\nu = 0.3$ for different m, it occurs that this effect does not exceed 1 %.

The contour lines for the buckling deflection, $w = \pm 0.75$, ± 0.50 and ± 0.25 for $m = 4$ are shown in Figure 13.9. The point with the largest deflection, $w = 1$, is marked by a small cross. The deflection damps very fast away from generatrix $\varphi = 0$ and therefore there is no contour line for the third pit $w = 0.75$, and, likewise, there is no contour line for the fourth pit $w = -0.75$ and $w = -0.50$.

13.8 Problems and Exercises

13.1. Consider a long tube with square cross-sectional area, each four faces of which are rectangular plates of width a and length b ($a \ll b$). Both plate edges $y = 0$ and $y = b$ are subjected to the uniformly distributed load $h\sigma$, where h is the plate thickness, and σ is a pressure. Using asymptotic method find critical axial pressure.

Answer

$$\sigma_{\rm cr} = \frac{\pi^2 E h^2}{12 a^2}\frac{3 + \nu}{1 + \nu}. \tag{13.8.1}$$

13.2. Consider buckling of cylindrical panel with weakly supported rectilinear edge $y = 0$ and clamped edge $y = y_0$ under axial compression. The curvilinear edges are simply supported. For different boundary conditions on the edge $x = 0$ analyze the possibility of existence of the localized buckling modes near the edge $x = 0$ and find the critical buckling resultants T_1^0. Compare results for $y_0 = R$ with given in formulae (3.5) for the semi-infinite in circular direction panel.

13.3. Plot graphs $\lambda(y_0)$ for 6 variants of the weak boundary conditions (3.5) found in problem 13.2.

13.4. Find the critical load for the paraboloid of revolution under external normal pressure. The initial axisymmetric stress-strain state is determined by the nondimentional stresses

$$t_1 = \frac{1}{2k_2} \qquad t_2 = \frac{1 - t_1 k_1}{k_2}$$

where $k_2 = \frac{1}{\sqrt{1+b^2}}$ and $k_1 = k_2^3$.

13.5. Determine the cases of the weak support of the edge $x = 0$ of the convex shell of revolution under combined axial compression and torsion for the particular case, when $T_1^0 = S^0$. The results represent is the form similar to Figure 13.4.

Hint. Under solution of (4.6) and (4.7) assume $t_1 = t_3 = k_2 = 1$, $t_2 = -k_1$.

Chapter 14

Shells of Revolution under General Stress State

In contrast to the previous Chapters, here we will assume that the initial shell stress-strain state is a sum of a membrane state and an edge effect. We will refer to this as a general stress-strain state.

We will also assume that the shell is sufficiently long, i.e. that we may neglect the mutual influence of the edge effects. In cases when the influence of the initial edge effect and pre-buckling deformations is small, the order of this influence on the critical load will be found. If the influence of these factors is significant then for the critical load parameter in the zeroth approximation the standard boundary value problem which does not contain the relative shell thickness is developed.

14.1 The Basic Equations and Edge Effect Solutions

Assume that the initial general shell stress-strain state is a sum of a membrane state w^0, T_i^0 and S^0 and an edge effect w^e, T_i^e and S^e, then we can write system (2.4.2) in the form

$$\begin{aligned} \mu^2\Delta\,(d\,\Delta\, w) + \lambda\,(\Delta_t w + \delta\Delta_t^e w) - \Delta_k\Phi - \lambda\,\delta\,\Delta_k^e\Phi &= 0, \\ \mu^2\Delta\,(g^{-1}\Delta\Phi) + \Delta_k w + \lambda\,\delta\,\Delta_k^e w &= 0, \end{aligned} \tag{14.1.1}$$

where the multiplier $\delta = 1$ is introduced for the convenience of discussion.

Below, we will study only the case of an axisymmetrically loaded shell of revolution. Then, for the operators Δ, Δ_t and Δ_k we may use expressions

(4.3.2), (4.3.3)

$$\begin{aligned} \Delta w &= \frac{1}{b}\frac{\partial}{\partial s}\left(b\frac{\partial w}{\partial s}\right)+\frac{1}{b^2}\frac{\partial^2 w}{\partial \varphi^2}, \\ \Delta_k w &= \frac{1}{b}\frac{\partial}{\partial s}\left(b\,k_2\frac{\partial w}{\partial s}\right)+\frac{k_1}{b^2}\frac{\partial^2 w}{\partial \varphi^2}, \\ \Delta_t w &= \frac{1}{b}\frac{\partial}{\partial s}\left(b\,t_1\frac{\partial w}{\partial s}\right)+\frac{1}{b}\frac{\partial}{\partial s}\left(t_3\frac{\partial w}{\partial \varphi}\right)+ \\ &\quad +\frac{t_3}{b}\frac{\partial^2 w}{\partial s\,\partial \varphi}+\frac{t_2}{b^2}\frac{\partial^2 w}{\partial \varphi^2}, \end{aligned} \tag{14.1.2}$$

$$(T_1^0, T_2^0, S) = -\lambda\, E_0 h_0 \mu^2\,(t_1, t_2, t_3), \quad \mu^4 = \frac{h^2}{12\,(1-\nu^2)\,R^2}. \tag{14.1.3}$$

Here the generatrix length s and the deflection w are referred to R.

The terms with the multiplier δ take into account the influence of the initial edge effect stress-resultants and the corresponding pre-buckling deformations

$$\begin{aligned} \Delta_t^e w &= \frac{1}{b}\frac{\partial}{\partial s}\left(b\,t_1^e\frac{\partial w}{\partial s}\right)+\frac{t_2^e}{b^2}\frac{\partial^2 w}{\partial \varphi^2}, \quad T_i^e = -\lambda\, E_0 h_0 \mu^2 t_i^e, \\ \Delta_k^e w &= \frac{1}{b}\frac{\partial}{\partial s}\left(b\,(\varkappa_2^0+\varkappa_2^e)\frac{\partial w}{\partial s}\right)+\frac{\varkappa_1^0+\varkappa_1^e}{b^2}\frac{\partial^2 w}{\partial \varphi^2}, \end{aligned} \tag{14.1.4}$$

where the t_i^e are the dimensionless initial edge effect stress-resultants.

As before, the loading is assumed to have only one parameter, $\lambda > 0$. We represent the pre-buckling deflection W^0 as a sum $W^0 = \lambda w^0 + \lambda w^e$ where λw^0 and λw^e are the membrane and the edge effect deflection components. The pre-buckling change of curvatures are also in the form of sums, $\lambda\,(\varkappa_i^0 + \varkappa_i^e)$, $i = 1, 2$. For high intensity (large) loads ($\lambda \sim 1$), the functions w^e, $\varkappa_i^e$, and t_i^e each depend non-linearly on λ (see equation (6) below).

We find the membrane initial stress-resultants T_i^0, and S^0 by formulae (4.3.8). Let $\max\{|t_i|\} \sim 1$, and the intensity of the loading be defined by the parameter λ. Taking into account that for the membrane state, unknown functions do not increase in differentiation we obtain estimates

$$w^0,\, \varkappa_i^0 \sim \mu^2. \tag{14.1.5}$$

For the edge effect solutions we use equation (1.5.8) which we write in the form

$$\mu^4 d\,\frac{d^4 w^e}{ds^4} + \lambda\,\mu^2 t_1\,\frac{d^2 w^e}{ds^2} + g\,k_2^2 w^e = 0, \tag{14.1.6}$$

where d, g, t_1, and k_2 are functions in s. The applicability domain of equation (6) is discussed below.

We seek solutions w^e in the neighbourhood of edge $s = s_0$ ($s_1 \leqslant s \leqslant s_2 = s_0$) in the form

$$w^e(s,\mu) = \sum_{j=1}^{2} C_j \sum_{n=0}^{\infty} \mu^n w_{jn}^e(s) \exp\left\{ \frac{1}{\mu} \int_{s_0}^{s} q_j(s)\, ds \right\}, \tag{14.1.7}$$

where C_1 and C_2 are arbitrary constants which are defined below from the boundary conditions. The functions $q_1(s)$ and $q_2(s)$ satisfy equations

$$d\, q^4 + \lambda t_1 q^2 + g\, k_2^2 = 0, \tag{14.1.8}$$

and

$$\Re\, q_j > 0, \qquad j = 1, 2. \tag{14.1.9}$$

Condition (9) which guarantees the damping of the solutions (6) away from the edge is fulfilled if for $t_1 > 0$ and

$$\lambda < \min_s \{\gamma(s)\} = \lambda^0, \quad \gamma(s) = \frac{2\, k_2}{t_1}\, (g\, d)^{1/2}. \tag{14.1.10}$$

For $t_1 \leqslant 0$ condition (9) is always fulfilled.

With an error of order μ solutions (7) may be replaced by

$$w^e(s,\mu) = \sum_{j=1}^{2} C_j \exp\left\{ \mu^{-1} q_j(s_0)(s - s_0) \right\}. \tag{14.1.11}$$

The tangential boundary conditions are fulfilled when we develop the membrane initial stress state. To find the constants C_1 and C_2 in (11) we use non-tangential boundary conditions

$$\begin{aligned} w^e &= w_* \quad \text{or} \quad Q_1^e = Q_{10}, \\ \gamma_1^e &= \gamma_{1*} \quad \text{or} \quad M_1^e = M_{10} \quad \text{at} \quad s = s_0, \end{aligned} \tag{14.1.12}$$

in which λQ_{10} and λM_{10} are the shear stress resultant and the bending stress-couple introduced at edge $s = s_0$ and λw_* and $\lambda \gamma_{1*}$ are the differences between the given and the membrane deflection and the angle of rotation.

We can give the explicit expressions for w^e for various boundary conditions (12)

$$w^e = \alpha\, w_* (C - S) - \gamma\, S \tag{14.1.13}$$

for $w^e = w_*$, $\gamma_1^e = \gamma_{1*} = \mu^{-1}\gamma$;

$$w^e = w_* \left(\alpha\, C - \frac{\alpha^2 - \beta^2}{2\alpha}\, S\right) + \frac{m}{2\alpha}\, S \tag{14.1.14}$$

for $w^e = w_*$, $M_1^e = M_{10} = E_0 h_0 R\, d\, \mu^2 m$;

$$w^e = \frac{1}{2k_2}\left(\frac{d}{g}\right)^{1/2}\left[\gamma\Big(\left(\beta^2 - 3\alpha^2\right) C + \left(\alpha^2 - 3\beta^2\right) S\Big) + q\,(C - S)\right] \tag{14.1.15}$$

for $\gamma_1^e = \gamma_{1*}$, $Q_1^e = Q_{10} = E_0 h_0 \mu\, d\, q$;

$$w^e = \frac{d}{gk_2^2}\left[m\Big(\left(3\alpha^3 - \alpha\beta^2\right) C - \left(\alpha^3 - 3\alpha\beta^2\right) S\Big) + \right.$$
$$\left. + q\Big(2\alpha^2 C - \left(\alpha^2 - \beta^2\right) S\Big)\right] \tag{14.1.16}$$

for $Q_1^e = Q_{10}$, $M_1^e = M_{10}$;

$$w^e = \frac{d}{gk_2^2}\left(1 - \frac{\lambda t_1}{k_2\sqrt{gd}}\right)^{-1}\left[q\Big(2\alpha^2 C - \left(\alpha^2 - \beta^2\right) S\Big) + \right.$$

$$\left. + m\left(\left(3\alpha^3 - \alpha\beta^2 + \alpha\frac{\lambda t_1}{d}\right) C - \left(\alpha^3 - 3\alpha\beta^2 + \alpha\frac{\lambda t_1}{d}\right) S\right)\right] \tag{14.1.17}$$

for $Q_1^e - T_1^0\gamma_1^e = Q_{10}$, $M_1^e = M_{10}$,
where $q\,(s_0) = \alpha + i\beta$, $\alpha > 0$,

$$C = \alpha^{-1} e^{\alpha_1} \cos\beta_1, \qquad S = \beta^{-1} e^{\alpha_1} \sin\beta_1,$$
$$\alpha_1 = \frac{\alpha\,(s - s_0)}{\mu}, \qquad \beta_1 = \frac{\beta\,(s - s_0)}{\mu}. \tag{14.1.18}$$

For large tensile stress-resultants ($t_1 < 0$) the value β is imaginary but the formulae given above are still valid in this case and give real values of $w^e(s)$ (see Example 14.5).

In formulae (13)-(18) the values w_*, γ, q, and m have the dimension of a length while the rest of the quantities are dimensionless.

By choosing the values R, E_0, h_0, and λ it is possible (without loss of generality at $t_1 > 0$) to get

$$d = g = k_2 = 1, \quad t_1 = 2 \quad \text{at} \quad s = s_0, \tag{14.1.19}$$

and as a result the formulae given above are simplified. In particular, inequality (10) takes on the form $\lambda < 1$.

As $\lambda \to \lambda^0$ (see (10)) we have $\alpha \to 0$ and the edge effect loses its local character. The critical load (3.1.16) which is found above without taking into account the boundary conditions and the variability of the determining parameters corresponds to the value $\lambda = \lambda^0$.

Boundary conditions (17) correspond to edge $s = s_0$, at which the deflection, w, and the angle of rotation, γ_1, are not introduced and the edge loading maintains its direction under deformation. In Section 13.4 (see Example 13.1) it was shown that for such boundary conditions the critical load decreases by a factor of two compared to the value $\lambda = \lambda^0$. It could be seen in formula (17) according to which $w^e \to \infty$ as $\lambda \to \lambda^0/2$. For boundary conditions (17) we must modify restriction (10) by $\lambda < \lambda^0/2$.

In [61] the expression for w^e is given for the case when a reinforcing elastic symmetric rib is fixed at shell edge $s = s_0$.

The functions t_i^e and $\varkappa_i^e$ in (1) and (3) one can find by using approximate formulae (1.5.3)

$$
\begin{aligned}
t_2^e &= \frac{g k_2}{\mu^2}\, w^e, \qquad & t_1^e &= \frac{b'}{b}\int_{-\infty}^{s} t_2^e ds \sim \mu\, b' t_2^e, \\
\varkappa_1^e &= \frac{d^2 w^e}{ds^2}, \qquad & \varkappa_2^e &= \frac{b'}{b}\frac{dw^e}{ds}.
\end{aligned}
\tag{14.1.20}
$$

To estimate the applicability domain for equation (6) and subsequent formulae (7) and (11) we write equation (1.5.8) in the form

$$
\begin{aligned}
&\mu^4 d\,\frac{d^4 W^0}{ds^4} + \lambda\mu^2\left(t_1 + t_1^e\right)\frac{d^2 W^0}{ds^2} + g\, k_2^2 W^0 + \\
&\qquad + \lambda\mu^2\left(t_1 + t_1^e\right)\left(k_1 + \nu\, k_2\right) = \lambda\, p, \quad p = \frac{q^* R}{E_0 h_0 \lambda}.
\end{aligned}
\tag{14.1.21}
$$

Here the initial axial stress-resultant, T_1, is represented as a sum of the membrane term and the edge effect, $T_1 = T_1^0 + T_1^e$, (see formulae (2) and (3)). It also is possible to represent the solution, W^0, of this equation in form $W^0 = \lambda w^0 + \lambda w^e$. The values t_1, w^0, and p in (21) are slowly varying functions and near the edge they may be considered as constants. Then from (21) we get

$$
w^0 = \left(p - \mu^2 t_1\left(k_1 + \nu k_2\right)\right)\left(g k_2^2\right)^{-1}. \tag{14.1.22}
$$

Equation (21) was, however, obtained under the assumption that the unknown solution is a fast varying function (with an index of variation equal to

1/2) and it is not acceptable for the development of the membrane part of the deflection. In particular, for $b' \neq 0$, formula (22) gives the wrong value of deflection w^0.

From the system of equations

$$\begin{aligned} \frac{du}{ds} - k_1 w &= \varepsilon_1^0 = \nu t_2 - t_1, \\ \frac{b'}{b} u - k_2 w &= \varepsilon_2^0 = \nu t_1 - t_2 \end{aligned} \qquad (14.1.23)$$

we find the expressions for u^0 and w^0

$$\begin{aligned} u^0 &= b k_2 \left(\int\limits_{s_0}^{s} \frac{k_2 \varepsilon_1^0 - k_1 \varepsilon_2^0}{b k_2^2} \, ds + C \right), \\ w^0 &= -\frac{\varepsilon_2^0}{k_2} + b' \left(\int\limits_{s_0}^{s} \frac{k_2 \varepsilon_1^0 - k_1 \varepsilon_2^0}{b k_2^2} \, ds + C \right), \end{aligned} \qquad (14.1.24)$$

where C is constant. For $b' = 0$, according to the third equilibrium equation $T_1^0 k_1 + T_2^0 k_2 + Rq^* = 0$ expression (24) for w^0 coincides with (22) and for $b' \neq 0$, we ought to use only formulae (24).

By excluding w^0 we find from (21) the equation for w^e

$$\mu^4 d \frac{d^4 w^e}{ds^4} + \lambda \mu^2 (t_1 + t_1^e) \frac{d^2 w^e}{ds^2} + g\, k_2^2 w^e + \mu^2 t_1^e (k_1 + \nu\, k_2) = 0, \qquad (14.1.25)$$

in which the first approximation as $\mu \to 0$, all values except w^e and t_1^e are considered to be given and constant. We seek the solution w^e that is decreasing for $s < s_0$. The value t_1^e is related to w^e by equation (20).

We will now try to determine the conditions at which equation (25) transforms into linear equation (6). Let the values of d, g, t_1, and k_2 be of order 1 and $|k_1| \lesssim 1$ in (25).

The last term in (25) may be neglected with an error of order μ for $b' \neq 0$ and with an error of order μ^2 for $b' = 0$.

The order of the non-linear term in (25)

$$z = \lambda\, \mu^2 t_1^e \frac{d^2 w^e}{ds^2} \qquad (14.1.26)$$

coincides for $b' \neq 0$ with the order of the value of $\lambda b' \mu^{-1} (w^e)^2$. That is why term (26) is of the same order as the rest of the terms in equation (25), if $w^e \sim \mu$ for $b' \neq 0$ and if $w^e \sim 1$ for $b' = 0$. According to formulae (13)-(17)

the order of w^e depends on the boundary conditions and on the values of w_*, γ, m, and q and it is equal to

$$w^e \sim \max\{|w_*|, |\gamma|, |m|, |q|\}, \tag{14.1.27}$$

where the maximum is calculated by two of the four values listed above, which appear in boundary conditions at $s = s_0$.

Using equation (6) and formulae (13)-(17) incurs an error of order μ if the order of the value of μz does not exceed the orders of the remaining terms in equation (25) (namely if $w^e \lesssim \mu^2$ for $b' \neq 0$ and if $w^e \lesssim \mu$ for $b' = 0$). We will assume later that these conditions are fulfilled.

But now let us consider an example in which these conditions are violated. We will study the buckling of a truncated circular conical shell with angle 2α at the vertex under axial compression in the case when its edge is free in the radial direction. In this case

$$b' = \sin\alpha, \quad Q_{10} = P\tan\alpha, \quad q \sim \mu, \quad w^e \sim \mu \tag{14.1.28}$$

and term (26) has the same order as the remaining terms in (25) and so it may not be neglected. Here, we expect buckling of the limiting point type (see Section 2.1).

At present we will not consider such cases since $\gamma_1^e \sim 1$ for $w^e \sim \mu$ and equation (6) and equations (25) and (21) and some of the relations of Chapter 1 are not applicable since in their derivation we assumed that $\gamma_i^2 \ll 1$. The problems in which geometric non-linearity is modelled exactly are studied for example in [95, 170].

14.2 Buckling with Pseudo-bending Modes

In this Section we will consider problems in which the influence of the edge effect initial stress resultants and pre-buckling deformations on the critical load is small and may be found (or estimated) by the perturbation method. These are buckling problems of a shell of revolution with negative Gaussian curvature ($\lambda \sim \mu^{2/3}$) and for shells with zero Gaussian curvature under an external normal pressure ($\lambda \sim \mu$) and under torsion ($\lambda \sim \mu^{1/2}$). For all of these problems a relatively low level of the membrane stress-resultants ($\lambda \ll 1$) is typical.

Let us temporarily assume that the parameter δ is small and we consider the self-adjoint boundary value problem L_δ consisting of two equations (1.1) and of four boundary conditions at each of the shell edges $s = s_1$ and $s = s_2$. We will refer to the problem L_δ as the general problem.

For $\delta = 0$ we get the problem L_0 (discussed above, see Chapters 7–11) on the membrane stress state buckling without taking into account any pre-buckling deformations or edge effect stress-resultants (shortly to be a membrane problem). We seek a solution of the problem L_δ in the form of a formal series

$$\begin{aligned} w_\delta &= w_0 + \delta\, w_1 + \ldots\,, \qquad \Phi_\delta = \Phi_0 + \delta\, \Phi_1 + \ldots\,, \\ \lambda_\delta &= \lambda_0 + \delta\, \lambda_1 + \ldots\,, \end{aligned} \tag{14.2.1}$$

where λ_0, w_0, and Φ_0 are respectively the eigenvalue and the eigenfunction of problem L_0.

To find the first approximation for w_1 and Φ_1 we get the non-homogeneous problem "on spectrum". From its compatibility condition we find the value of λ_1 which is the main part of the correction to λ and we represent it in the form

$$\lambda_1 = \eta\, \lambda_0, \qquad \eta = (I_2 - I_1)\, I_0^{-1}, \tag{14.2.2}$$

where

$$\begin{aligned} I_0 &= \iint \left[t_1 \left(\frac{\partial w_0}{\partial s} \right)^2 + 2\, \frac{t_3}{b}\, \frac{\partial w_0}{\partial s}\, \frac{\partial w_0}{\partial \varphi} + \frac{t_2}{b^2}\, \frac{\partial^2 w_0}{\partial \varphi^2} \right] b\, ds\, d\varphi, \\ I_1 &= \iint \left[t_1^e \left(\frac{\partial w_0}{\partial s} \right)^2 + \frac{t_2^e}{b^2} \left(\frac{\partial w_0}{\partial \varphi} \right)^2 \right] b\, ds\, d\varphi, \\ I_2 &= 2 \iint \left[(\varkappa_2^0 + \varkappa_2^e)\, \frac{\partial w_0}{\partial s}\, \frac{\partial \Phi_0}{\partial s} + \frac{\varkappa_1^0 + \varkappa_1^e}{b^2}\, \frac{\partial w_0}{\partial \varphi}\, \frac{\partial \Phi_0}{\partial \varphi} \right] b\, ds\, d\varphi. \end{aligned} \tag{14.2.3}$$

Here the domain of integration is the entire shell neutral surface. In evaluating I_1 and I_2 we take into account that the pre-buckling deformation is axisymmetric.

The eigenfunction for problem L_0 we represent in the form of a sum

$$w_0 = w_0^0 + w_0^e, \qquad \Phi_0 = \Phi_0^0 + \Phi_0^e, \tag{14.2.4}$$

where w_0^0 and Φ_0^0 describe the main stress state and w_0^e and Φ_0^e correspond to the edge effect. Let us have the following estimates of the orders of the terms in (4) and of their indices of variation

$$\begin{aligned} & w_0^0,\, \Phi_0^0 \sim 1; \quad w_0^e,\, \Phi_0^e \sim \mu^{\alpha_e}; \quad w^0 \sim \mu^{\alpha_{00}}; \quad w^e \sim \mu^{\alpha_{0e}}; \\ & \frac{\partial w_0^0}{\partial s} \sim \mu^{-\alpha_s} w_0^0; \quad \frac{\partial w_0^0}{\partial \varphi} \sim \mu^{-\alpha_\varphi} w_0^0. \end{aligned} \tag{14.2.5}$$

For shells with zero Gaussian curvature under external pressure or torsion we obtain $\alpha_s = 0$ and $\alpha_\varphi = 1/2$ and for shells with negative Gaussian curvature we have $\alpha_s = \alpha_\varphi = 2/3$. For the edge effect solutions, w_0^e and w^e, we get $\partial w_0^e/\partial s \sim \mu^{-1} w_0^e$, $\partial w^e/\partial s \sim \mu^{-1} w^e$.

The variables in (3) due to (1.20) are of order

$$\begin{gathered} t_2^e \sim \mu^{\alpha_{0e}-2}, \qquad t_1^e \sim \mu\, t_2^e \quad (b' \neq 0), \\ \varkappa_1^0,\, \varkappa_2^0 \sim \mu^{\alpha_{00}}, \quad \varkappa_1^e \sim \mu^{\alpha_{0e}-2}, \quad \varkappa_2^e \sim \mu^{\alpha_{0e}-1} \quad (b' \neq 0). \end{gathered} \tag{14.2.6}$$

If $b'(s_0) = 0$ then $t_1^e \sim \mu^2 t_2^e$ and $\varkappa_2^e \sim \mu^{\alpha_{0e}}$.

We represent the integral I_j in the form

$$I_j = I_j^0 + I_j^e, \qquad j = 0, 1, 2, \tag{14.2.7}$$

where I_j^0 does not contain the edge effect solutions and I_j^e contains the edge effect solutions describing the initial stress state and/or eigenfunction (4) of problem L_0. In particular, I_0^0 is calculated by formula (3) for $w_0 = w_0^0$ and I_2^0 is calculated for $w_0 = w_0^0$, $\Phi_0 = \Phi_0^0$, and $\varkappa_i^e = 0$, and $I_1^0 = 0$.

Now we will consider buckling specifically. First, we will study the buckling of a cylindrical or a conical shell of revolution of moderate length under external normal pressure. We assume that the shell is "well" supported and that the load parameter λ_0 satisfies the estimate

$$\lambda_0 \sim \mu, \tag{14.2.8}$$

and the possible variants of "well" supported edges are given in the first six lines of Table 8.4.

Under these assumptions the following estimates

$$\alpha_{00} = 2, \quad I_0^0 \sim \mu^{-1}, \quad I_2^0 = O(\mu), \quad \alpha_e \geqslant 1. \tag{14.2.9}$$

are valid.

The intensity, α_e, of the edge effect solutions, w_0^e, coincides with the error order that occurs in the transition from the full boundary value problem, L_0, to the simplified problem (see Chapter 8). In particular, $\alpha_e = 1$ if $w = 0$ is one of the boundary conditions.

The order of I_2^0 depends on whether or not the equality $\varkappa_1^0 = 0$ is valid. If $\varkappa_1^0 = 0$ then $I_2^0 \sim \mu^2$, otherwise $I_2^0 \sim \mu$.

The initial edge effect intensity, α_{0e}, depends on the type of edge $s = s_0$ support. For example, by introducing a pre-stressed support or with edge stress-resultants and/or stress-couples we may obtain arbitrary values of the right sides of boundary conditions in (1.13)-(1.17). Below we assume that

$$\alpha_{0e} = 2, \tag{14.2.10}$$

i.e. the deflection corresponding to the edge effect is of the same order as the membrane initial deflection. In particular, estimate (10) is valid if the initial edge effect is only caused by the fact that the initial membrane deflection did not satisfy the condition $w = 0$ (and may be the condition $\gamma_1 = 0$).

We will neglect the mutual influence of the edge effects, that is why the integrals, I_j^e, are the sums of terms corresponding to shell edges $s = s_1$ and $s = s_2$. We will study one of these edges $s = s_2 = s_0$.

The integrals I_j^e are sums of terms of the form

$$F(\mu) = \int_{s_1}^{s_2} f(s,\mu)\, g\left(\frac{s - s_0}{\mu}\right) ds, \qquad f = \mu^\alpha \Big(f_0(s) + \mu f_1(s) + \cdots\Big), \tag{14.2.11}$$

where the function $f(s,\mu)$ may be represented as the asymptotic series (11) and $g(\zeta) \to 0$ as $\zeta \to -\infty$. By expanding functions f_i in a Taylor series at $s = s_0$ we get

$$F(\mu) = \mu^{\alpha+1}\Big[f_0(s_0)\, c_0 + \mu\Big(f_1(s_0)\, c_0 + f_0'(s_0)\, c_1\Big) + \\ + \mu^2\Big(f_2(s_0)\, c_0 + f_1'(s_0)\, c_1 + \frac{1}{2} f_0''(s_0)\, c_2\Big) + \cdots\Big], \tag{14.2.12}$$

where

$$c_k = \int_{-\infty}^{0} \zeta^k g(\zeta)\, d\zeta. \tag{14.2.13}$$

Studying the boundary condition variants from the clamped and simple support groups (see Section 8.4), we find that at $s = s_0$ the following estimates are valid

$$w_0^0(s_0) \sim \mu, \qquad w_0^e = O(\mu), \\ \frac{\partial w_0^0}{\partial s_0} \sim \mu \quad \text{(clamped)}, \qquad \frac{\partial w_0^0}{\partial s_0} \sim 1 \quad \text{(simple support)}. \tag{14.2.14}$$

It is important to recall that the $\sim$ symbol indicates the exact order of the variable and the symbol O gives the upper order estimate. In particular, for some boundary condition variants we get $w_0^e \sim \mu^2$ or $w_0^e \sim \mu^{5/2}$ (see Section 8.4).

In estimating the integrals I_j^e we assume that $t_2 \sim 1$; t_1, t_3, and $b' = O(1)$. Then according to (12), (6), (10) and (14) we find

$$I_0^e = O(\mu), \quad I_1^e \sim \mu^2, \quad I_2^e = O(\mu), \tag{14.2.15}$$

and then due to (2) we get

$$\eta = O\left(\mu^2\right), \tag{14.2.16}$$

i.e. the relative order of the initial moment stress-resultants and the influence of pre-buckling deformations on the critical load under the above assumptions does not exceed μ^2. In addition, the term

$$z = z^0 + z^e = 2 \iint \left(\varkappa_1^0 + \varkappa_1^e\right) \frac{\partial w_0^0}{\partial \varphi} \frac{\partial \Phi_0^0}{\partial \varphi} \frac{1}{b}\, ds\, d\varphi, \tag{14.2.17}$$

in I_2^0 is the most significant. For boundary conditions of the clamped support group we have $w_0^0\left(s_0\right) \sim \mu$, $\Phi_0^0\left(s_0\right) \sim 1$ and

$$z^e \sim \mu, \qquad \eta = \left(z + O\left(\mu^2\right)\right)\left(I_0^0\right)^{-1} \sim \mu^2. \tag{14.2.18}$$

For simple support group boundary conditions we obtain $w_0^0\left(s_0\right)$, $\Phi_0^0\left(s_0\right) \sim \mu$ and for $\varkappa_1^0 = 0$ we have

$$z \sim \mu^2, \qquad \eta \sim \mu^3. \tag{14.2.19}$$

For moderately long shells the calculation of η by formulae (2) is scarcely practical since the correction is very small. However, for short shells the correction, η, becomes sufficient. In [22, 61, 149] one may find a survey of the numerical results for a cylindrical shell with various boundary conditions which agree with the estimates obtained here.

Let us now study the buckling of an axisymmetrically loaded shell of revolution with negative Gaussian curvature. Let there be no torsion ($t_3 = 0$) and let the boundary conditions be of the clamped or simple support group. The case when both edges are simply supported is not considered. Then the estimates

$$\lambda_0 \sim \mu^{2/3}, \quad \alpha_s = \alpha_\varphi = \frac{2}{3}, \quad \alpha_{00} = \alpha_{0e} = 2, \quad \alpha_e = \frac{2}{3}\alpha, \tag{14.2.20}$$

are valid (see (5)). Here the value α depends on the boundary conditions at $s = s_0$ and it is given in Table 11.2.

We can find the value of η by formula (2). Keeping the main terms (with respect to μ) we get

$$\eta = I_2'\left(I_0^0\right)^{-1}, \quad I_2' = 2 \iint \varkappa_1^e \frac{\partial w_0^0}{\partial \varphi} \frac{\partial \Phi_0^0}{\partial \varphi} \frac{1}{b}\, ds\, d\varphi. \tag{14.2.21}$$

For all boundary condition variants considered we have $I_0^0 \sim \mu^{-4/3}$. As before, we calculate the value of I_2' separately for each edge, $s = s_1$ and $s = s_2$, and denote the edge under consideration by $s = s_0$. By taking into account buckling modes (11.4.2) and (11.4.6) and by using an expansion of type (12) we obtain the estimates

$$\begin{aligned} \eta &= O\left(\mu^{5/3} + \mu^{1+\frac{2\alpha}{3}}\right) & &\text{(clamped)}, \\ \eta &= O\left(\mu^{2} + \mu^{\frac{4+2\alpha}{3}}\right) & &\text{(simple support)}. \end{aligned} \tag{14.2.22}$$

where the values of α are given in Table 11.2.

A comparison with estimates (18) and (19) reveals that for shells of negative Gaussian curvature, the use of a membrane formulation is accompanied by a greater error than for cylindrical and conical shells under external pressure.

14.3 The Cases of Significant Effect of Pre-buckling strains

The buckling of convex shells of revolution under arbitrary axisymmetric loading and the buckling of shells of revolution of zero Gaussian curvature under axial compression is studied in this Section for the case where $\lambda \sim 1$. The corresponding membrane problems have already been discussed in Chapters 3 and 4 and it was shown that in the zeroth approximation, the critical value of λ is a function in $\gamma(s)$ (see formulae (4.3.7), (4.3.9), and (4.3.10)). We will discuss several related cases here.

Case 1. Let $\gamma(s) = \lambda^0 = \text{const}$.

Then, in the membrane statement, all points on the neutral surface are equally predisposed to buckling and buckling pits cover the entire surface. This case is important particularly for the buckling of cylindrical and conical shells under axial pressure and for the buckling of a spherical shell subject to a homogeneous external pressure. The initial moment stress-resultants and the pre-buckling deformations infringe on this homogeneity near the edges and buckling may occur for $\lambda < \lambda^0$ and the buckling mode is localized near one of the shell edges.

Case 2. Let there be no membrane stress ($\gamma \equiv 0$) or let it not contain compressive membrane stress-resultants in any direction (see (3.1.11)).

Then, only the general statement of the buckling problem is valid. The initial edge effect stress-resultants and the pre-buckling deformations which are generated by the local loads at $s = s_0$, are the only cause for buckling and the buckling mode is localized near the line $s = s_0$.

Case 3. Let the function $\gamma(s)$ be variable and attain its minimum value at $s = s_0 = s_2$.

The edge parallel $s = s_0$ is the weakest line and the buckling mode is again localized in its neighbourhood. In the membrane statement the solution of this problem for $\gamma'(s_0) \neq 0$ is given in Section 13.6 and in the general statement it is given in Section 14.4.

We assume that the characteristic values of the functions describing the initial state satisfy the following estimates (see (1.20))

$$w^0,\, w^e \sim \mu^2; \quad \varkappa_1^e,\, t_2^e \sim 1; \quad \varkappa_2^e,\, t_1^e = O(\mu). \tag{14.3.1}$$

Now we will give the solution algorithm for this problem in Cases 1 and 2.
The critical value λ we seek from the bifurcation condition into the adjacent non-symmetric mode with m waves in a circumferential direction in the form

$$w(s,\varphi) = \sum_{j=0}^{\infty} \mu^j w_j(\zeta)\, e^{im\varphi},$$
$$\Phi(s,\varphi) = \sum_{j=0}^{\infty} \mu^j \Phi_j(\zeta)\, e^{im\varphi}, \tag{14.3.2}$$
$$\zeta = \mu^{-1}(s - s_0), \qquad \lambda = \lambda_0 + \mu\lambda_1 + \cdots,$$

where in the neighbourhood of the parallel $s = s_0$ the scaled variable ζ is introduced.

To find λ_0 we come to the system

$$d_0 \left(\frac{d^2}{d\zeta^2} - \rho_1^2\right)^2 w_0 + \lambda_0 \left(t_{10}\frac{d^2 w_0}{d\zeta^2} + 2i\rho_1 t_{30}\frac{dw_0}{d\zeta} - \rho_1^2\left(t_{20} + t_2^e(\zeta)\right) w_0\right) -$$
$$- k_{20}\frac{d^2\Phi_0}{d\zeta^2} + \rho_1^2\left(k_{10} + \lambda_0 \varkappa_1^e(\zeta)\right)\Phi_0 = 0, \qquad \rho_1 = \frac{\mu m}{b_0},$$
$$g_0^{-1}\left(\frac{d^2}{d\zeta^2} - \rho_1^2\right)^2 \Phi_0 + k_{20}\frac{d^2 w_0}{d\zeta^2} - \left(k_{10} + \lambda_0 \varkappa_1^e(\zeta)\right)\rho_1^2 w_0 = 0, \tag{14.3.3}$$

where by d_0, g_0, b_0, k_{i0}, and t_{i0} we denote the values of the corresponding functions at $s = s_0$.

Functions w_0 and Φ_0 satisfy the four homogeneous boundary conditions at $s = s_0$ (see below) and also the damping conditions

$$w_0 \longrightarrow 0, \quad \Phi_0 \longrightarrow 0 \quad \text{at} \quad \zeta \longrightarrow -\infty. \tag{14.3.4}$$

We can express by w and Φ the variables which are contained in the boundary conditions

$$
\begin{aligned}
u &= \frac{b\omega}{im} + \frac{b^2}{m^2}\left(\varepsilon_2' - \frac{b'}{b}\varepsilon_1 + bk_2\left(\frac{w}{b}\right)'\right),\\
v &= \frac{b}{im}\left(\varepsilon_2 + k_2 w - \frac{b'}{b}u\right), \qquad M_1 = \mu^4 d\,(\varkappa_1 + \nu\varkappa_2),\\
T_1 &= -\frac{\mu^2 m^2}{b^2}\Phi + \frac{\mu^2 b'}{b}\Phi', \qquad S = -\frac{i\mu^2 m}{b}\left(\Phi' - \frac{b'}{b}\Phi\right),\\
Q_{1*} &= -\frac{\mu^4}{b}\Big(db\,(\varkappa_1 - \nu\varkappa_2)\Big)' - \frac{2\,(1-\nu)\,\mu^4 imd}{b}\tau - \\
&\quad -\frac{\mu^4 db'}{b}(\varkappa_2 + \nu\varkappa_1) - \mu^2\,(t_1 + t_1^e)\,\lambda w' + \lambda\,(w^e)'\,T_1 - \mu^2\lambda t_3 imw,
\end{aligned}
\tag{14.3.5}
$$

where

$$
\begin{aligned}
\varepsilon_1 &= \frac{1}{g}(T_1 - \nu T_2) - \lambda\,(w^e)'\,w', \qquad \varepsilon_2 = \frac{1}{g}(T_2 - \nu T_1),\\
\omega &= \frac{2\,(1+\nu)}{g}S - \lambda\,(w^e)'\,\frac{im}{b}w, \qquad T_2 = \mu^2\Phi'',\\
\varkappa_1 &= w'', \qquad \varkappa_2 = -\frac{m^2}{b^2}w + \frac{b'}{b}w', \qquad \tau = \frac{im}{b}\left(w' - \frac{b'}{b}w\right).
\end{aligned}
\tag{14.3.6}
$$

Here with the $(')$ symbol we denote the derivative with respect to s. The error in formulae (5) is of order μ^2. In the right sides of (5) in the formulation of the boundary conditions for the system we pass to the variable ζ and keep the main terms with respect to μ.

Then we get

$$
\begin{aligned}
u &= \mu\left[\frac{1}{\rho_1^2 g_0}\frac{d^3\Phi_0}{d\zeta^3} - \frac{2+\nu}{g_0}\frac{d\Phi_0}{d\zeta} + \frac{k_{20}}{\rho_1^2}\frac{dw_0}{d\zeta} - \lambda_0\left(\frac{dw^e}{d\zeta}\right)_0 w_0\right],\\
v &= \mu\left[\frac{1}{i\rho_1 g_0}\left(\frac{d^2\Phi_0}{d\zeta^2} + \nu\rho_1^2\Phi_0\right) + \frac{k_{20}}{i\rho_1}w_0\right],\\
T_1 &= -\rho_1^2\Phi_0, \quad S = -i\rho_1\frac{d\Phi_0}{d\zeta}, \quad M_1 = \mu^2 d_0\left(\frac{d^2 w_0}{d\zeta^2} - \nu\rho_1^2 w_0\right),\\
Q_{1*} &= -\mu\left[d_0\left(\frac{d^3 w_0}{d\zeta^3} - (2-\nu)\,\rho_1^2\frac{dw_0}{d\zeta}\right) + \right.\\
&\quad \left. + \lambda_0\left((t_{10} + t_1^e)\frac{dw_0}{d\zeta} + \rho_1^2\left(\frac{dw^e}{d\zeta}\right)_0\Phi_0 + i\rho_1 t_{30} w_0\right)\right],
\end{aligned}
\tag{14.3.7}
$$

where we take $\left(\frac{dw^e}{d\zeta}\right)_0$ for $\zeta = 0$.

The value of λ_0 may be found after minimization by the number of waves, m, in the circumferential direction ($\rho_1 = \mu\, m\, b_0^{-1}$).

The next approximation w_1, Φ_1 satisfies the non-homogeneous system of equations with the same left sides as in system (3) and the non-homogeneous boundary conditions with the left sides of form (7). The decreasing conditions (4) are still valid. From the existence condition for the solution w_1, Φ_1 we can find λ_1.

The construction method for λ_1 proposed here is interesting only to validate the expansion (2) for the load parameter $\lambda = \lambda(\mu)$. If we want to find the numerical value of $\lambda(\mu)$ then it is more convenient to solve system (1.1) numerically with boundary conditions of type (5).

From system (3) it follows that the initial stress-resultant, t_2^e, and the pre-buckling change of curvature, $\varkappa_1^e$, affect equally λ_0 (at least from the asymptotic point of view) and the functions, t_2^e and $\varkappa_1^e$, ought to be found from the equation of the non-linear edge effect (1.6).

If we neglect one of the values, t_2^e or $\varkappa_1^e$, or evaluate them from the simple edge effect equation (equation (1.6) for $\lambda = 0$) we incur an error which does not go to zero with the shell thickness. In [114] for the buckling problem of a hemisphere under external pressure (see Example 14.3) with various boundary conditions, the numerical results for three statements of the problem are given:

(i) for $\varkappa_1^e$ and t_2^e which are found from the non-linear edge effect equation,

(ii) for $\varkappa_1^e = 0$,

(iii) for $\varkappa_1^e$ and t_2^e which are found from the simple edge effect equation.

For all boundary conditions considered, the minimum value of λ is obtained under the exact formulation of problem (i). In the approximate formulations, (ii) and (iii), the values of λ are greater by 10 % and 20 % respectively.

Remark 14.1. The term $\mu\lambda_1$ in formula (2) takes into account the effect of some factors which are neglected in the zeroth approximation (3) (namely the variability of coefficients b, d, g, k_i, and t_i in system (1.1) and the effect of relatively small values of t_1^e and $\varkappa_2^e$).

If, at $s = s_0$ the derivatives of the listed coefficients are equal to zero then $\lambda_1 = 0$ in formula (2) and the error in system (3) and of the boundary conditions which are introduced by formulae (7) is of order μ^2.

In order to solve system (1.1) and approximate system (3), which have variable coefficients, various numerical methods (such as numerical integration, finite element methods or variational methods) may be used. We note also the method of invariant immersion [179] which is a type of numerical integration technique in which the decreasing conditions (4) may be satisfied very conveniently.

Examples 14.1–14.3 considered below illustrate Case 1 and Example 14.4

illustrates Case 2. In these Examples $t_3 = 0$ and E, ν, and h are constant and therefore $d = g = 1$.

Example 14.1. Consider the buckling of a cylindrical shell under axial compression. In this case $k_1 = t_2 = 0$ and we assume that $b = k_2 = 1$ and $t_1 = 2$, then due to (1.2)

$$T_1^0 = -2\,E\,h\,\mu^2\lambda, \qquad P = 2\pi\,R\,T_1^0. \tag{14.3.8}$$

The value $\lambda = 1$ gives the classical Lorenz–Timoshenko critical load (see (3.4.3)). The value of $\lambda < 1$ is a measure of the decrease of the critical load. The edge effect stresses and deformations near shell edge $s = s_0$ depend on the type of edge support. Let

$$u = v = w = M_1 = 0 \quad \text{(that is 1110)} \quad \text{at} \quad s = s_0. \tag{14.3.9}$$

Then by formulae (1.14), (1.20) we find

$$\begin{aligned} &w^e = 2\,\nu\,\mu^2\left(\alpha\,C - \frac{\alpha^2-\beta^2}{2\,\alpha}\,S\right), \\ &\alpha = \left(\frac{(1-\lambda_0)\,k_2}{2}\right)^{1/2}, \qquad \beta = \left(\frac{(1+\lambda_0)\,k_2}{2}\right)^{1/2}, \\ &t_2^e = \mu^{-2}k_2 w^e, \qquad \varkappa_1^e = -\frac{\nu\left(\alpha^2+\beta^2\right)^2}{\alpha}\,S, \end{aligned} \tag{14.3.10}$$

where we use the designations from (1.18).

Now, if we let $\nu = 0.3$, we obtain $\lambda_0 = 0.867$ (see [2]).

The problem under consideration has been extensively studied [2, 21, 22, 44, 61, 112, 149] for the various relationships among the parameters and for various boundary conditions.

In all cases except 0010 and 1010 the buckling mode occurs with a large number of waves, m, in the circumferential direction ($m \sim \mu^{-1}$). Since λ_0 is close to 1, the parameter α, which causes the decrease of functions t_2^e, $\varkappa_1^e$ is small (see formulae (10) and (1.18)) and for relatively short shells the mutual influence of the edge effects at $s = s_1$ and at $s = s_2$ may be important. This question has been studied in [2, 22, 44, 61]. In cases 0010 and 1010 where the support is weak, buckling occurs for $m = 2$. The same value of $\lambda = 0.5$ is also found without taking into account initial moment stresses (see (13.2.12)).

The decrease in the critical value of λ_0 noted is related to the fact that Poisson's ratio $\nu \neq 0$ (see (10)). In Table 14.1 (the data for which has been obtained in [71]) for six boundary condition variants, the relationships of $\lambda_0(\nu)$ and $\rho(\nu)$ are given. The last of these, due to the expression $\rho = \mu m$, allows us to estimate the expected number of waves, m, under buckling.

Table 14.1: $\lambda_0(\nu)$ and $\rho(\nu)$

	ν	0.1	0.2	0.3	0.4	0.5
1111	λ_0	0.968	0.946	0.927	0.911	0.897
	ρ	0.48	0.48	0.47	0.47	0.47
1011	λ_0	0.968	0.946	0.927	0.911	0.897
	ρ	0.48	0.48	0.48	0.47	0.47
0111	λ_0	0.957	0.930	0.910	0.893	0.879
	ρ	0.45	0.45	0.44	0.44	0.44
0011	λ_0	0.956	0.929	0.909	0.892	0.877
	ρ	0.45	0.45	0.44	0.44	0.44
1110	λ_0	0.928	0.894	0.868	0.848	0.831
	ρ	0.48	0.48	0.47	0.47	0.47
0110	λ_0	0.907	0.870	0.844	0.824	0.808
	ρ	0.45	0.44	0.44	0.44	0.44

Example 14.2. Let us now study the buckling of a conical shell under axial compression. For a conical shell we have

$$k_1 = 0, \qquad k_2 = b^{-1}\cos\alpha_0, \qquad b = s\sin\alpha_0,$$
$$P = 2\pi R\, b\cos\alpha_0\, T_1^0, \qquad T_2^0 = S^0 = 0, \tag{14.3.11}$$

where $2\alpha_0$ is the cone vertex angle and Rs is the distance between a point on the shell and the vertex. In relation (1.2) $T_1^0 = -\lambda\, E\, h\, \mu^2 t_1$ we take $t_1 = 2b^{-1}\cos\alpha_0$.

Then

$$\gamma(s) = \frac{2\,k_2}{t_1} \equiv 1. \tag{14.3.12}$$

The minimization of function $\gamma(s)$ gives the weakest parallel (see formula (4.3.10)). It follows from (12) that in the membrane statement this Example (as in Example 14.1) corresponds to Case 1 and the buckling mode occupies the whole shell surface. The critical value of the force is equal to

$$P = -\frac{2\pi\, h^2\cos^2\alpha_0}{\sqrt{3\,(1-\nu^2)}}\,\lambda. \tag{14.3.13}$$

For $\lambda = 1$, formula (13) is similar to the Lorenz–Timoshenko formula (3.4.3) and has been obtained in [140].

Now we will examine the buckling mode which is localized near shell edge $s = s_0$. Let this edge be simply supported by a rigid body which may progressively move in the axial direction. The boundary conditions under buckling we take in form (9). The initial stress-strain state is determined by relations

$$u^0 = -2\,\mu^2 \cot \alpha_0 \,(\ln s + c), \quad w^0 = -2\,\mu^2\,(\nu + \ln s + c), \tag{14.3.14}$$

and w^e, t_2^e, and $\varkappa_1^e$ may be calculated by the same formulae (10) as for a cylindrical shell.

System of equations (3) for a conical shell differs from system (3) for a cylindrical shell by only two parameters: b_0 and k_{20}. We can therefore make the substitutions

$$\tilde{\zeta} = \zeta\, k_{20}^{1/2}, \qquad \tilde{\rho}_1 = \rho_1\, k_{20}^{-1/2} = \mu\, m\, b_0^{-1} k_{20}^{-1/2}. \tag{14.3.15}$$

Then, both system (3), and boundary conditions (9), which are written using formulae (7), coincide for cylindrical and conical shells. Hence, for $\nu = 0.3$, $\lambda_0 = 0.874$. Here, in contrast to a cylindrical shell, $\lambda_1 \neq 0$ i.e. the error of the zeroth approximation is greater.

If edges $s = s_1$ and $s = s_2$ are supported identically then (in the zeroth approximation) buckling in the neighbourhood of each of them corresponds to equal values of P. But the number of waves, m, in the circumferential direction are different. By virtue of (15) we can write

$$m = \tilde{\rho}_1 \Big(12\,(1-\nu^2)\Big)^{1/4} \left(\frac{B \cos \alpha}{h}\right)^{1/2}, \tag{14.3.16}$$

where the B is the radius of circle $s = s_0$. For larger radius the number m is larger.

Taking into account the noted agreement of equations (3), the values of λ_0 obtained for a cylindrical shell with various boundary conditions may be used in formula (13) also for a conical shell. Boundary condition variants 0010 and 0110 for which $\lambda = 0.5$ and the buckling mode is axisymmetric [44] are exceptions.

Example 14.3. Consider the buckling of a hemisphere under a homogeneous external pressure. For a spherical shell

$$\begin{aligned} &k_1 = k_2 = 1, \qquad T_1^0 = T_2^0 = -\frac{q\,R}{2} = -\,2\,\lambda\,E\,h\,\mu^2, \\ &t_1 = t_2 = 2, \qquad q = \frac{2\,E\,h^2 \lambda}{\sqrt{3\,(1-\nu^2)}\,R^2}, \end{aligned} \tag{14.3.17}$$

where q is the external pressure and R is the sphere radius. For $\lambda = 1$ we get formula (3.3.11).

Let the boundary conditions $T_1 = v = w = M_1 = 0$ apply at edge $s = s_0$. The initial stress state is then described by

$$w^0 = 2\mu^2(1-\nu), \qquad w^e = -2(1-\nu)\,\mu^2\left(\alpha\, C - \frac{\alpha^2-\beta^2}{2\alpha}\, S\right), \qquad t_2^e = \mu^{-2} w^e, \qquad \varkappa_1^e = \frac{(1-\nu)\left(\alpha^2+\beta^2\right)^2}{\alpha}\, S, \tag{14.3.18}$$

where we use same designations as in formulae (1.18), and (10).

System (3) gives $\lambda_0 = 0.701$ and $\rho_1 = 0.79$. The same value of λ has been obtained in [114], though in that paper other boundary condition variants were also considered. The agreement of the values of λ is noteworthy despite the fact that system (3) is approximate, i.e. function $b\,(s) = \cos s$ is replaced by the value $b_0 = 1$.

Example 14.4. We will now examine the buckling of a cylindrical shell under concentrated radial forces. This problem corresponds to Case 2 above. The initial membrane stresses are absent ($t_i = 0$) and the initial general stress state is symmetric with respect to $s = 0$ and for $s < 0$ it is described by formulae (see (1.15))

$$w^e = \frac{\mu^2\lambda}{2}(C-S), \qquad \alpha = \beta = 1/2, \qquad Q = 2\,\mu^3 E\, h\, \lambda, \qquad t_2^e = \frac{\lambda}{2}(C-S), \qquad \varkappa_1^e = -\frac{\lambda}{4}(C+S), \tag{14.3.19}$$

where functions $C\,(s)$, and $S\,(s)$ are introduced in (1.18) and Q is the intensity of the compressive radial load at circle $s = 0$. We find the load parameter λ from system (3) with boundary conditions (4) and by noting the symmetry of the problem

$$\frac{dw}{d\zeta} = \frac{d^3 w}{d\zeta^3} = \frac{d\Phi}{d\zeta} = \frac{d^3\Phi}{d\zeta^3} = 0 \quad \text{at} \quad \zeta = 0. \tag{14.3.20}$$

The solution of this problem has been obtained in [34, 80, 93, 116].

$$Q = 0.344\,\frac{E\, h}{(1-\nu^2)^{3/4}}\left(\frac{h}{R}\right)^{3/2}, \qquad m = 0.357\,\mu^{-1}. \tag{14.3.21}$$

A survey of the work on the buckling of cylindrical shells under edge couples and radial forces may be found in [21, 61, 149].

14.4 The Weakest Parallel Coinciding With an Edge

The weakest parallel, $s = s_0$, is determined from the minimum condition for the function $\gamma(s)$ which was introduced in Section 4.3. Let parallel s_0 coincide with the edge of the shell ($s_0 = s_2$). Here we will study problems with the same initial membrane stress-resultants as in Section 13.6 but in addition, the moment stress-resultants and any pre-buckling deformations will also be taken into account. We will examine, in particular, buckling problems for convex shells under torsion and under tensile load by an axial force.

Depending on the loading and type of edge support, one of two possibilities for buckling may occur: $\lambda < \lambda^0 = \gamma(s_0)$ or $\lambda > \lambda^0$. For the case when $\lambda < \lambda^0$ we may use expansion (3.2) to find λ. In this expansion λ_0 is evaluated using the same system (3.3).

Now let $\lambda > \lambda^0$ and examine the characteristic equation (13.5.5) of system (1.1)

$$F(p, s, \rho, \lambda) = 0 \tag{14.4.1}$$

assuming that $t_i^e = \varkappa_i^e = 0$.

The roots of equation (1) for which

$$\Im\, p < 0. \tag{14.4.2}$$

correspond to a solution of system (1.1) which exponentially decreases with s.

As was shown in Section 13.6, system (1.1) for $t_i^e = \varkappa_i^e = 0$ has a turning point, $s = s_*$, such that for $s < s_*$ equation (1) has four roots satisfying condition (2).

Therefore system (1.1) for t_i^e, $\varkappa_i^e \neq 0$ has four linearly independent solutions which exponentially decrease with s for $s < s_*$ (since the functions t_i^e, and $\varkappa_i^e$ decrease away from edge $s = s_2$). Four such solutions allow satisfaction of four boundary conditions introduced at $s = s_2$.

For $t_2^e = \varkappa_1^e = 0$ the characteristic equation of system (3.3) is $F(p, s_2, \rho, \lambda) = 0$. This equation for $\lambda > \lambda^0$ has only three roots satisfying condition (2) and therefore system (3.3) generally has no solutions satisfying the four conditions for $\zeta = 0$ and decreasing conditions (4) and so it cannot be used as the zeroth approximation.

In Sections 13.5 and 13.6 it was shown that in the membrane formulation for $\gamma'(s_0) < 0$, parameter λ has the expansion

$$\lambda = \lambda^0 + \mu^{2/3}\lambda_1 + O(\mu), \tag{14.4.3}$$

where

$$\lambda^0 = \gamma(s_0), \quad \lambda_1 = \left(\frac{1}{2}\left(f_s^0\right)^2 f_{pp}^0\right)^{1/3} \zeta_0, \quad \zeta_0 = 2.338, \tag{14.4.4}$$

and f_s^0 and f_{pp}^0 are described in Section 13.5. In formula (3) only terms of order μ, $(O(\mu))$, depend on the boundary conditions.

Let us solve the general formulation of the problem. As in Section 14.2 it can be shown that $\eta = O(\mu)$, where η is a measure of the relative influence of the initial moment stress-resultants and the pre-buckling deformations on the critical load (see (2.2)). In other words, in formula (3) the first two terms remain the same in the general formulation.

To prove this fact we return to formulae (2.2) and (2.3) taking in them the deflection w_0 in the form of (13.6.13)

$$w_0 = \left(Ai\left(\alpha\mu^{-2/3}(s - s_*)\right) + \right.$$
$$\left. + \mu^{1/3}\sum_{k=1}^{3} C_k e^{\frac{ip_k(s-s_2)}{\mu}}\right)\cos\left(m(\beta s - \varphi + \varphi_0)\right), \tag{14.4.5}$$

where α, C_k, and β do not depend on μ and s, for that $\beta = 0$ if there is no torsion. Formula (5) is approximate but its accuracy is sufficient to obtain the following estimates. By substituting (5) into (2.3) we find

$$I_0 \sim \mu^{-4/3}, \qquad I_1, I_2 \sim \mu^{-1/3}, \tag{14.4.6}$$

and due to (2.2) it follows that $\eta \sim \mu$.

Example 14.5. Let us consider the buckling of a spherical belt, (a portion of a hemisphere with edges parallel but not on the equator), $s_1 \leqslant s \leqslant s_2$ with constant values of E, ν, and h under an axial tensile force P. By using formulae (13.6.14) we obtain

$$\begin{aligned} P &= 2\pi R E h \mu^2\lambda, \quad t_2 = -t_1 = b^{-2}, \quad b = \cos s, \\ \gamma(s) &= 2\sin^2 s, \qquad m = b_0\mu^{-1}, \qquad k_1 = k_2 = 1, \end{aligned} \tag{14.4.7}$$

where s is the angle between the point and the equator.

Let $s_2 > |s_1|$, then the parallel $s = s_2$ is the weakest.

Let $s_2 = \pi/4$. Then, by formulae (3) and (13.6.16) we find

$$P = \frac{\pi E h^2 \lambda}{\sqrt{3(1-\nu^2)}}, \qquad \lambda = 1 + 2\zeta_0\mu^{2/3} + O(\mu). \tag{14.4.8}$$

Table 14.2: Values of $\lambda(m)$ obtained by numerical integration

m	membrane		general		formula (8)
	simple support	clamped	simple support	clamped	
	1	2	3	4	
28	1.476	1.500	1.549	1.590	
29	1.464	1.486	1.539	1.577	
30	1.458	1.479	1.535	1.571	1.426
31	1.460	1.479	1.538	1.573	
32	1.467	1.483	1.549	1.581	
33	1.482	1.495	1.566	1.595	

In order to estimate the influence of the boundary conditions and initial moment stress state, we take $R/h = 400$ and $\nu = 0.3$. Then $\mu = 0.0275$ and the first two terms in formula (8) give $\lambda = 1.426$.

In Table 14.2 we have presented the values of $\lambda(m)$ obtained by numerical integration (with orthogonalization) of system (1.1). Columns 1 and 3 correspond to a simply supported edge $s = s_2$ ($T_1 = v = w = M_1 = 0$), and columns 2 and 4 correspond to a clamped edge ($u = v = w = \gamma_1 = 0$).

In columns 1 and 2 we place the results, obtained for the membrane formulation ($t_i^e = \varkappa_i^e = 0$), and evaluating Columns 3 and 4 we take into account the stress-resultants t_2^e and the change of the curvature changes $\varkappa_1^e$. For the simply supported edge we can write

$$t_2^e = -2(1+\nu)\left(C - \frac{\alpha^2+\beta^2}{2\alpha^2} S\right), \quad \varkappa_1^e = \frac{1+\nu}{\alpha^2} S \tag{14.4.9}$$

and for the clamped edge

$$t_2^e = -2(1+\nu)(C-S), \quad \varkappa_1^e = 2(1+\nu)(C+S), \tag{14.4.10}$$

where

$$C = \alpha^{-1} e^{\alpha_1} \operatorname{ch} \beta_1, \qquad S = \beta^{-1} e^{\alpha_1} \operatorname{sh} \beta_1,$$
$$\alpha_1 = \frac{\alpha (s - s_2)}{\mu}, \qquad \beta_1 = \frac{\beta (s - s_2)}{\mu}, \tag{14.4.11}$$
$$\alpha = \left(\frac{1+\lambda}{2}\right)^{1/2}, \qquad \beta = \left(\frac{\lambda - 1}{2}\right)^{1/2}.$$

It follows from the Table that buckling occurs for $m = 30$ which is slightly larger than the value $m = 26$ predicted by formula (7). This is because the value of b in the buckling pit centre is larger than b_0 at the shell edge.

From the results presented in the Table it follows that the comparatively low critical load depends on both the boundary conditions and whether or not we take into account the initial moment stress state. It agrees with formula (8). If we include the moment stress-resultants in the calculation then we obtain a greater value of λ since near the edge $t_2^e < 0$ the effect of the compressive membrane stress-resultants $t_2 > 0$ decreases.

14.5 Problems and Exercises

14.1 Consider an axisymmetrically loaded shell of revolution with negative Gaussian curvature. Find relative the order of the first term in expression for the critical value (14.2.2) for different boundary conditions of the free support and clamped group.

Answer

$$\eta = O\left(\mu^{4/3}\right) \qquad (1011,\ 1010),$$
$$\eta = O\left(\mu^{5/3}\right) \qquad (1111,\ 1110,\ 1101,\ 1100,\ 0111,\ 0011,\ 0010),$$
$$\eta = O\left(\mu^{2}\right) \qquad (0110,\ 0101,\ 0100),$$

where, in parenthesis, we list the boundary condition variants for which these estimates are valid.

14.2. Consider the buckling of a cylindrical shell with the Poisson ratio $\nu = 0.3$ under axial compression for the following boundary conditions at the edge $s = s_0$ ([2]):

a) 1111, 1011

b) 0111

c) 0011

d) 0110
e) 0010, 1010

Answer

Boundary conditions	λ_0
1111, 1011	0.926
0111	0.910
0011	0.908
0110	0.844
0010, 1010	0.5

14.3 Consider buckling of the semi-infinite circular cylindrical shell at the free edge of which the following load is applied
(i) radial compressive load,
(ii) radial tensile load.

14.4 Consider buckling of the infinite circular cylindrical shell under radial load distributed homogeneously along the parallel.

Bibliography

[1] N.A. Alfutov. *Background of Elastic System Stability Design.* Mashinostroenie, Moscow, 1978. 312p. (Russian).

[2] B.O. Almorth. Influence of the edge condition on stability of axially compressed cylindrical shells. *AIAA Journal*, 4(1):134–140, 1966.

[3] N.A. Alumae. Critical load for a long cylindrical shell under torsion. *Prikl. Mat. Mekh.*, 18(1):27–34, 1954 (Russian).

[4] N.A. Alumae. The determination of the critical load for shells limited by a hyperboloid of one sheet. *Prikl. Mat. Mekh.*, 20(2):397–408, 1956 (Russian).

[5] N.A. Alumae. On the basic system of solutions of equations of small stable axisymmetric vibrations of an elastic conic shell of revolution. *Izv. AN Est. SSR, Ser. tekh. & phys.-math. sci.*, 1960 (Russian).

[6] Ju.A. Amen-zade, R.M. Bergman, and T.T. Nasirov. Stability of a non-circular cylindrical shell under external pressure. *Dokl. AN AzSSR*, 34(4):38–42, 1978 (Russian).

[7] S.I. Amstislavskaya. Investigation of the stress state and stability of elliptic cisterns of mixing cameras. *Khimich. Neft. Prom*, (8):8–9, 1976 (Russian).

[8] L.V. Andreev and N.I. Obodan. Stability of cylindrical shells with variable thickness under arbitrary external loading. *Prikl. Mekh. (Kiev)*, 10(1):41–47, 1974 (Russian).

[9] J. Arbocz. *Comparison of Level-1 and Level-2 buckling and post-buckling solutions.* Rep.-LR-700. TU Delft, 1992.

[10] J. Arbocz. *The effect of initial imperfections on shell stability.* Rep.-LR-695. TU Delft, 1992.

[11] J. Arbocz and C.D. Babcock. The effect of general imperfections on the buckling of cylindrical shells. *Journal of Applied Mechanics*, 36:28–38, 1969.

[12] E.L. Axelrad. *Flexible Shells*. Nauka, Moscow, 1976. 376p.

[13] E.L. Axelrad. Shells theory and its specialized brabches. *Solids and Structures*, (37):1425–1451, 2000.

[14] S.M. Bauer. Stability of cylindrical shells with imperfections. In *Dynamics and Stability of Mechanical Systems*, volume 6, pages 172–175. Leningr. Univ., 1984. (Russian).

[15] S.M. Bauer, S.B. Filippov, A.L. Smirnov, and P.E. Tovstik. Asymptotic methods in mechanics with applications to thin shells and plates. In *Asymptotic Methods in Mechanics. CRM Proc. and Lecture Notes*, volume 3, pages 3–141. AMS, 1993.

[16] I.A. Birger and Ya.G. Panovko. *Strength. Stability. Vibrations: Directory in 3 vol.* Mashinostroenie, Moscow, 1968 (Russian).

[17] V.V. Bolotin. *Dynamic Stability of Elastic Systems.* Gostechizdat, Moscow, 1956. 600p. (Russian).

[18] G.H. Brayan. Application of an energy test to the collapse of a thin long pipe under external pressure. *Proc. Cambridge Philos. Soc*, 6:287–292, 1988.

[19] B. Budiansky. Theory of buckling and post-buckling behavior of elastic structures. *Adv. in Appl. Mech.*, (14), 1974.

[20] V.A. Bulygin. On a class of shells of gaussian curvature which changes sign. *Izv. Akad. Nauk SSSR Mekh. Tverd. Tela*, (5):97–105, 1977 (Russian).

[21] D. Bushnell. Buckling of shells — pitfalls for designers. *AIAA Journal*, 19(9):1183–1226, 1981.

[22] D. Bushnell. Computerized buckling analysis of shells. In *Mechanics of Elastic Stability*. Martinus Nijhoff Publ., Dordrecht, 1985.

[23] A.N. Cheremis. On the stability of an ellipsoidal shell under external pressure. *Proml. Prochnosti*, (7):78–81, 1975 (Russian).

[24] K.F. Chernykh. *Linear Theory of Shells*. v.1, 1962; v.2, 1964. Leningrad Univ. Press (Russian).

[25] K.F. Chernykh. Simple boundary effect and splitting of the boundary conditions in the linear theory of thin shells. *Izv. Akad. Nauk SSSR Mekhanika*, (1):89–98, 1965 (Russian).

[26] K.F. Chernykh. *Nonlinear theory of elasticity in mashine design.* Mashinostroenie, Leningrad, 1986. 336p. (Russian).

[27] W.Z. Chien. The intrinsic theory of thin shells and plates. *Quart. Appl. Math*, 1(4):297–327; 2(1) 43–59; 2(2), 120–135, 1943.

[28] P.G. Ciarlet. *Introduction to Linear Shell Theory.* Dauthier-Villars, Paris, 1998. 184p. (French).

[29] P.G. Ciarlet and P. Rabier. *Les equations de von Kármán.* Springer-Verlag, Berlin, Heidelberg, New York, 1980. (French).

[30] D.A. Danielson. Buckling and initial post-buckling behaviour of spheroidal shells under pressure. *AIAA Journal*, (5):936–944, 1969.

[31] A.I. Danilov and G.N. Chernyshev. Stiffness and flexure of shells of negative curvature used in wave transmissions. *English transl., Mech. Solids*, 13(5), 1978.

[32] L.H. Donnell. A discussion of thin shell theory. In *Proc. 5-th Intern. Congr. Appl. Mech.*, pages 66–70, 1938.

[33] L.H. Donnell. *Beams, Plates and Shells.* McGraw-Hill, Inc., 1976.

[34] V.V. Ershov and V.A. Ryabtsev. Stability of cylindrical shells with a variable generatrix thickness under radial pressure. *Prikl. Mekh. (Kiev)*, 10(4):38–46, 1974 (Russian).

[35] Z.G. Ershova. Stability of cylindrical panels with weakly restrained rectilinear edges. *English transl., Len. Univ. Mech. Bul*, (3):59–63, 1993.

[36] L. Euler. *Methodus niveniendineas curvas... Additamentum 1: De curvis elasticis.* Lausanne et Geneve, 1744.

[37] M.A. Evgrafov. *Asymptotic Estimates and Entire Functions.* Nauka, Moscow, 1979. 320p. (Russian).

[38] V. I. Feodos'ev. On the stability of a spherical shell under the action of uniform external pressure. *English transl., J. Appl. Maths Mechs*, 18(1), 1954.

[39] S.B. Filippov. Stability of cylindrical shells joined by a mitred joint under uniform external pressure. *Prikl. Mat. Me h.*, 59(1):696–702, 1995.

[40] S.B. Filippov. Stability of cylindrical shells joined by a mitred joint under uniform external pressure. *English transl., J. Appl. Maths Mechs*, 59(1), 1995.

[41] S.B. Filippov. Application of asymptotic methods for the evaluation of the optimal parameters for the ring-stiffened cylindrical shells. In *Integral Methods in Science and Engineering, vol. 2*, pages 79–83. UK, 1997.

[42] S.B. Filippov. *Theory of joined and stiffened shells.* St.Petersburg Univ. Press, St.Petersburg, 1999. 196p, (Russian).

[43] S.B. Filippov, E. Haseganu and A.L. Smirnov. Buckling analysis of axially-compressed square elastic tubes with weakly supported edges. *Technishe Mechanik.* 2(1):13–20), 2000.

[44] G. Fischer. Influence of boundary conditions on the stability of thin-walled cylindrical shells under axial load and initial pressure. *AIAA Journal*, 3(4):736–737, 1965.

[45] W. Flügge. *Statics and Dynamics of Shells.* Strojizdat, 1961. 306p. (German).

[46] A. Föppl. Vorlesungen über technische mechanik. volume 5, pages 132–144. Oldenburg Verlag, München, 1907 (German).

[47] Y.C. Fung and E.E. Sechler. Instability of thin elastic shells. In *Proc. First Symp. on Naval structural Mechanics*, Oxford, London, New York, Paris, 1960. Pergammon Press.

[48] Y.C. Fung and E.E. Sechler (eds.). *Thin-Shell Structures: Theory, Experinent, and Design.* Prentics-Hall, New Jersey, 1974. 376p.

[49] A.L. Gol'denveiser. Membrane theory of shells defined by second order surfaces. *Prikl. Mat. Me h.*, 11(2):285–290, 1947 (Russian).

[50] A.L. Gol'denveiser. *Effect of a Thin Shell Edge Support on its Stress State.* Number 669 in: Trudy ministerstva aviatsionnoi promyshlennosti SSSR. 1948. 23p. (Russian).

[51] A.L. Gol'denveiser. *Theory of Thin Elastic Shells.* Nauka, Moscow, 1953. 544p. (Russian).

[52] A.L. Gol'denveiser. *Theory of Thin Elastic Shells.* Nauka, Moscow, 1976. 512p. (Russian).

[53] A.L. Gol'denveiser. The bending of surfaces and the lowest frequencies of thin shell vibrations. *Izv. Acad. Nauk SSSR Mekh. Tverd. Tela*, (5):106–117, 1977 (Russian).

[54] A.L. Gol'denveiser. Mathematical rigidity of surfaces and physical rigidity of shells. *English transl., Mech. Solids*, 14(6):53–63, 1979 (Russian).

[55] A.L. Gol'denveiser. Geometrical theory of stability of shells. *English transl., Mech. Solids*, 18(1):135–145, 1983.

[56] A.L. Gol'denveiser and Ju.D. Kaplunov. Dynamical boundary layer in vibration problems of shells. *English transl., Mech. Solids*, 23(4):146–155, 1988.

[57] A.L. Gol'denveiser, V.B. Lidsky, and P.E. Tovstik. *Free Vibrations of Thin Elastic Shells.* Nauka, Moscow, 1979. 384p. (Russian).

[58] D.P. Goloskokov and P.A. Zhilin. Stability of cylindrical shells under torsion. In *Vodnye puti i portovye soorugeniya*, pages 24–29. Trudy Leningr. Inst. vodnogo transporta, 1983. (Russian).

[59] E.I. Grigoluk and V.V. Kabanov. *Stability of Circular Cylindrical Shells. Mechanics of Deformable Silids.* VINITI SSSR, Moscow, 1969. 348p. (Russian).

[60] E.I. Grigoluk and V.I. Mamai. *Nonlinear deformation of thin-walled constructions.* Nauka, Moscow, 1997. 272p. (Russian).

[61] E.I. Grigolyuk and V.V. Kabanov. *Stability of Shells.* Nauka, Moscow, 1978. 360p. (Russian).

[62] E.I. Grigolyuk and V.I. Shalashilin. *Problems of Non-Linear Deformation. Method of Proceeding by Parameter in Non-Linear Problems of mechanics of Deformable Solids.* Nauka, Moscow, 1988. 232p. (Russian).

[63] J. Heading. *Introduction to the Phase-Integral Method.* Methuen, Wiley, London, New York, 1962. 260p.

[64] J.W. Hutchinson and W.T. Koiter. Post-buckling theory. *Appl. Mech. Rev.*, 1970.

[65] A.Ju. Ishlinsky. On one limiting transition in the stability theory of elastic rectangular plates. *English transl., Soviet Phys. Dokl.*, 95(3), 1954.

[66] N. Jamaki and S. Kodama. Buckling of circular cylindrical shells under compression. *Sci. Repts Res. Inst. Tohocu. Univ. B: High Speed Mech.*, v.23, 99-123; v.24, 111-142, 1971.

[67] Janke, Emde, and Losch. *Tafeln hoherer funktionen.* Teuburger VG, Stuttgart, 1960 (German).

[68] V.V. Kabanov. Stability of circular cylindrical shells under nonhomogeneous compression. *Izv. Acad. Nauk SSSR. Mekhanika i Mashinistroenie*, (1):181–183, 1963 (Russian).

[69] E. Kamke. *Differentialgleichungen, 5th ed.* v.1, 1964; v.2, 1965. Academische Verlagsgesellschaft, Leipzig (German).

[70] J.D. Kaplunov, L.Yu. Kossovich, and E.V. Nolde. *Dynamics of thin walled elastic bodies.* Academic press, London, 1998. 226p.

[71] D.S. Karaulov. Critical load as a function of poisson's ratio in axial compressing of a cylindrical shell. *English transl., Len. Univ. Mech. Bul*, (4):10–13, 1991.

[72] T. Karman. Festigkeitsprobleme im maschinenban. In *Enzyklopedie der mathematischen Wissenschaften*, volume 5, n. 31 of *Bd. 4. Mechanik*, pages 695–770. B. G. Tenner, Leipzig, 1914 (German).

[73] A.V. Karmishin, V.A. Lyjaskovets, V.I. Myachenkov, and A. N. Frolov. *Statics and Dinamics of Thin–Walled Shell Structures.* Mashinostroenie, Moscow, 1975. 376p. (Russian).

[74] N.A. Kil'chevsky. On the axisymmetric deformations and elastic stability of a circular tube under the action of longitudinal compressive forces. *Prikl. Mat. Mech.*, 6(6):497–508, 1942.

[75] G. Kirchhoff. *Vorlesungen über matematische Physik. Mechanik.* Leipzig, 1877. 466p. (German).

[76] W.T. Koiter. *On the stability of the elastic equilibrium.* Thesis, Delft, Amsterdam, 1945.

[77] W.T. Koiter. On the non-linear theory of thin elastic shells. *Proc. Koninkl. Nederl. Acad. Westenschap*, 69:1–54, 1966.

[78] V.M. Kornev. On the solution of shell stability problems taking into account the density of the eigenvalues. In *Theory of Shells and Plates*, pages 132–134. Sudostroenie, 1975 (Russian).

[79] V.I. Korolev. *Elastic-Plastic Deformation of Shells.* Mashinostroenie, Moscow, 1971. 304p. (Russian).

[80] E.M. Koroleva. Stability of cylindrical shells under a bending stress state. *English transl., J. Appl. Maths*, 32(4), 1968.

[81] A.V. Krotov and P.E. Tovstik. Asymptotically double natural frequencies of thin elliptical cylindrical shell vibrations. In: *Day on Difftaction'97*, St.Petersburg, 1997.

[82] V.I. Kruglyakova. On the design of thin-walled tubes with curvilinear axes. *English transl., Mech. Solids*, 7(5), 1972.

[83] S.N. Kukudganov. Stability of cylindrical shells of variable thickness under external pressure. *Trudy Tbilisy Math. Inst*, 52(6):81–92, 1976 (Russian).

[84] R.E. Langer. The asymptotic solutions of ordinary linear differential equations of second order with special reference to a turning point. *Trans. Amer. Math. Soc.*, (67):461–490, 1949.

[85] M.A. Lavrent'ev and A.Ju. Ishlinsky. Dinamik modes of stability loss of elastic systems. *English transl., Soviet Phys. Dokl*, 6, 1949.

[86] A.I. Liapunov. *General Problems of Motion Stability*, volume 2. Akad. Nauk SSSR, Moscow, 1956. 475p. (Russian).

[87] A. Libai and J.G. Simmonds. *The Nonlinear Theory of Elastic Shells.* Cambridge Univ. Press, Cambridge, New York, Melbourne, 1998. 542p.

[88] V.B. Lidsky and P.E. Tovstik. Spectra in the theory of shells. *Adv. in Mech*, 7(2):25–54, 1984.

[89] T.V. Liiva. On free non-axisymmetric vibrations of shells on revolution of negative gaussian curvature. In *Trans. Tallinn Politech. Inst. Ser. A.*, volume 5, pages 47–60. 1970 (Russian).

[90] T.V. Liiva. On the free non-symmetric vibrations of shells of revolution with negative gaussian curvature for various boundary conditions. *Trans. Tallinn Polytech. Inst*, (345):41–52, 1973 (Russian).

[91] T.V. Liiva and P.E. Tovstik. On the linear approximation stability of shells of revolution of negative gaussian curvature. In *Problems of Mechanics of Deformable Solids*, pages 231–238. Sudostroenie, Leningrad, 1970 (Russian).

[92] T.V. Liiva and P.E. Tovstik. On the buckling of shells of revolution of negative gaussian curvature under torsion. *English transl., Mech. Solids*, 6(6):79–85, 1973.

[93] Ju.V. Lipovtsev. On the stability of elastic and viscoelastic shells in the presence of local stresses. *Ing. J. Mekh. Tverd. Tela*, (5):174–180, 1968.

[94] R. Lorenz. Die nicht achsensymmetrische knickung dunnwandiger hohlzylinder. *Physical Zeitschrift*, 12(7):241–260, 1911.

[95] N.I. Obodan L.V. Andreev and A.G. Lebedev. *Stability of Shells at the Non-Symmetric Deformation.* Nauka, Moscow, 1988 (Russian).

[96] A.L. Maiboroda. Construction of the main solution of the buckling problem for convex shells near the edge. *English transl., Len. Univ. Mech. Bull*, (1), 1986.

[97] A.L. Maiboroda. The effect of internal pressure on the stability of a circular cylindrical shell under bending. *English transl., Len. Univ. Mech. Bull*, (3), 1986.

[98] A.L. Maiboroda. Buckling of a circular conical shell under forced bending. In *Prikl. Mekh.*, volume 8, pages 209–214. Lening. Univ. Press, Leningrad, 1990 (Russian).

[99] V.P. Maslov. *Complex WKB Methods in Non-Linear Equations.* Nauka, Moscow, 1977. 384p. (Russian).

[100] G.I. Mikhasev. Local loss of stability of a shell of zero curvature with an variable thickness and elastic module. *English transl., Len. Univ. Mech. Bul.*, (2), 1984.

[101] G.I. Mikhasev. Local loss of stability of thin truncated ellipsoid acted upon by a combined load. *English transl., Len. Univ. Mech. Bul.*, (4):63–71, 1984.

[102] G.I. Mikhasev. On propagation of bending waves in a non-circular cylindrical shell. *Izv. Ross. Akad. Nauk. Mekh. Tverd. Tela*, (3):99–103, 1994.

[103] G.I. Mikhasev and P.E. Tovstik. Stability of conical shells under external pressure. *English transl., Mech. Solids*, 25(4), 1990.

[104] R. Mises. Der kritische aussendruck zylindrischer rohre. *VDI Zeitschrift*, 58(19):750–755, 1914.

[105] A.I. Molchanov. Free oscillations of noncircular shells close to shells of zero gaussian curvature. *Vestnik Leningrad. Univ. Mat. Mekh. Astronom*, (4):43–45, 1986.

[106] A.I. Molchanov. Asymptotic integration of a system of equations the free vibrations of noncircular shells close to shells of zero gaussian curvature. *Vestnik Leningrad. Univ. Mat. Mekh. Astronom*, (2):106–107, 1987.

[107] N.F. Morozov. To nonlinear theory of thin plates. *English transl., Soviet Phys. Dokl*, 114(5), 1957.

[108] N.F. Morozov. On the existence of nonsymmetric solutions in the problem of large deflections of circular plate under symmetrical load. *Izv. vuzov. Matematica*, 21(2):126–129, 1961 (Russian).

[109] H.M. Mushtari. On the elastic equilibrium of a thin shell with initial imperfections of a middle surface form. *Prikl. Mat. Mekh.*, 15(6):743–750, 1951 (Russian).

[110] H.M. Mushtari. On the elastic equilibrium of a thin shell with initial imperfections of a middle surface form. *English transl., J. Appl. Maths Mech*, 15(6), 1951.

[111] H.M. Mushtari and H.Z. Galimov. *Nonlinear theory of elastic shells.* Tatknigoizdat, Kazan', 1957. 431p. (Russian).

[112] H.M. Mushtari and S.V. Prochorov. On the stability of a cylindrical shell under nonhomogeneous compression. *Trudy Kazan. Chim. Technol. Inst*, (22):10–23, 1957 (Russian).

[113] V.I. Myachenkov. Stability of cylindrical shells under axisymmertical lateral pressure. *Prikl. Mekh. (Kiev)*, 6(1):27–33, 1970 (Russian).

[114] V.I. Myachenkov. Stability of spherical shells under simultaneous action of external pressure and local axisymmetric loads. *English transl., Mech. Solids*, (6), 1970.

[115] V.I. Myachenkov and I.V. Grigor'ev. *Design of consistent shell constructions by computer. Directory.* Mashinostroenie, Moscow, 1981. 216p. (Russian).

[116] V.I. Myachenkov and L.V. Pakhomova. Stability of cylindrical shells under concentrated circular loads. *Ing. J. Mekh. Tverd. Tela*, (6):143–147, 1968 (Russian).

[117] W. Nachbar and N.J. Hoff. The buckling of free edge of a axially compressed circular cylindrical shell. *Quart. Appl. Math*, 20(3):160–172, 1962.

[118] Ju.B. Nedeshev and G.N. Chernyshev. Transformation of rigidity of shells of reversed sign of curvature at scale changing. *English transl., Mech. Solids*, 21(5), 1986.

[119] V.V. Novozhilov. *Theory of thin shells.* Volters–Noordhoff, Groningen, 1970. 422p.

[120] V.V. Novozhilov. *Mathematical models and exactness of engineering designs.* Inst. Problem Mekh. Acad. Nauk SSSR, Moscow, 1983. 215p. (Russian).

[121] V.V. Novozhilov and R.M. Finkel'shtein. On errors of kirchhof's hypotheses in shell theory. *Prikl. Mat. Mekh.*, 7(5):331–340, 1943.

[122] Ju.I. Kaplan N.S. Kan and A.Z. Shternis. Stability of orthotropic cylindrical shells with variable thickness. *Prikl. Mekh. (Kiev)*, 10(5):129–132, 1974.

[123] I.F. Obrazstov, B.V. Nerubailo, and I.V. Andrianov. *Asymptotic methods in civil engineering of thin-wall structures.* Mashinostroenie, Moscow, 1991. 416p. (Russian).

[124] P.M. Ogibalov. *Problems of dynamics and stability of shells.* Moscow Univ. Press, Moscow, 1963. 418p. (Russian).

[125] F. Olver. *Asymptotics and special functions.* Academic Press, 1974. 528p.

[126] Ya.G. Panovko and I.I. Gubanova. *Stability and vibrations of elastic systems. Modern conceptions, antinomies and mistakes.* Nauka, Moscow, 1979. 352p. (Russian).

[127] P.F. Papkovich. Design formulas for a stability test of a cylindrical shell of a submarine strength body. *Bul. nauchno-tekh. kom. UMVS RKKA*, 6(2):113–123, 1929 (Russian).

[128] M.B. Petrov. Study of loss of stability of shells of revolution of negative gaussian curvature by pure bending modes. In *Prikl. Mekh.*, volume 4, pages 211–224. Lening. Univ. Press, Leningrad, 1979 (Russian).

[129] M.B. Petrov. On eigen and critical sizes of shells of revolution of negative gaussian curvature. In *Vibrations and stability of mechanical systems*, volume 5 of *Prikl. Mekh.*, pages 228–238. Lening. Univ. Press, Leningrad, 1981 (Russian).

[130] M.B. Petrov. Integral equations of superlow-frequency vibrations of shells of revolution of curvature which changes sign. In *Dynamics and Stability of Mechanical Systems*, volume 6 of *Prikl. Mekh.*, pages 161–168. Lening. Univ. Press, Leningrad, 1984 (Russian).

[131] W. Pietraszkiewicz. Geometrically nonlinear theories of thin elastic shells. *Advances in Mechanics*, (12), 1988.

[132] A.V. Pogorelov. *Geometric Theory of shell Stability.* Nauka, Moscow, 1966. 296p. (Russian).

[133] A.V. Pogorelov. *Bending of surfaces and stability of shells.* Nauka, Moscow, 1986. 96p. (Russian).

[134] Ju.N. Rabotnov. Local stability of shells. *English transl. Soviet. Phys. Dokl*, 52(2), 1946.

[135] E. Reissner. On some problems in shell theory. In *Proc. First Symp. on Naval structural Mechanics*, pages 71–114. Pergammon Press, Oxford, London, New York, Paris, 1960.

[136] A.V. Sachenkov. On the stability of a cylindrical shell with arbitrary boundary conditions under lateral pressure. *Izv. Kazan. fil. Akad. Nauk SSSR*, (12):127–132, 1958 (Russian).

[137] E. Schwerin. Die torsionsstabilitat des dunnwandigen rohres. *Z. Angew. Math. and Mech*, 5(3):235–243, 1925.

[138] D.I. Shil'krut. *Problems of qualitative theory of nonlinear shells.* Shtinitsa, Kishinev, 1974. 144p. (Russian).

[139] V.P. Shirshov. Local stability of shells. *Proc. 2-nd All-Union Conf. on the Theory of Shells and Plates*, pages 314–317, 1962 (Russian).

[140] A.Ya. Shtaerman. Stability of shells. *Trudy Kiev. Avia. Inst*, (1), 1936 (Russian).

[141] J.G. Simmonds. A set of simple, accurate equations for circular cylindrical elastic shells. *Int. J. Solids Struct.*, (2):525–541, 1966.

[142] V.I. Smirnov. *High Mathematics*, volume 3(2). Gostekhizdat, 1953. 676p. (Russian).

[143] H. Sobell. Effect of boundary conditions on the stability of cylinders subject to lateral and axial pressure. *AIAA Journal*, 2(8):1437–1440, 1961.

[144] R. Southwell. On the collapse of tubes by external pressure. parts 1, 2, 3. *Philos. Mag., Ser. 6*, 25(149):687–697, 1913.

[145] L.S. Srubschik. *Buckling and postbuckling behavior of shells.* Rostov. Univ. Press, 1981. 96p. (Russian).

[146] L.S. Srubschik. Energetic test of dynamical buckling of spherical shells. *English transl., Soviet Phys. Dokl*, 280(1), 1985.

[147] L.S. Srubschik. Dynamical buckling of elastic shell under action of pulse loading. *English transl., J. Appl. Maths Mechs*, 51(1), 1988.

[148] M. Stein. Recent advances in the investigation of shell buckling. *AIAA J.*, (6):2339–2346, 1968.

[149] J.G. Teng. Buckling of thin shells: Recent advances and trends. *Appl. Mech. Rev.*, 1996.

[150] I.Ju. Teterin. On stability of imperfection cylindrical and spherical shells. In *Plikl. mekh. Dynamics and Stability of Mechanical Systems*, volume 6, pages 176–184. Leningrad Univ. Press, 1984 (Russian).

[151] G.V. Timofeeva. Local buckling of cylindrical shells under nonuniform axial pressure. In *Dynamics and Stability of Mechanical Systems*, volume 6 of *Prikl. Mekh.*, pages 184–188. Leningrad Univ. Press, 1984.

[152] G.V. Timofeeva. Loss of stability of cylindrical shells under bending by force and moments. *English transl., Len. Univ. Mech. Bul.*, (4):72–76, 1984.

[153] G.V. Timofeeva and P.E. Tovstik. On a representation of solutions of problems of shell dynamics and buckling. *English transl., Len. Univ. Mech. Bul*, (1):41–50, 1985.

[154] S.P. Timoshenko. On the problem of deformation and stability of cylindrical shells. *Vestnik obshchestva tekhnologii*, 21:785–792, 1914 (Russian).

[155] S.P. Timoshenko. *Stability of elastic systems.* Gostekhizdat, Moscow Leningrad, 1946. 531p. (Russian).

[156] P. Tong and T.H. Pian. Investigation of post-buckling behavior of revolution shells by finite-elements method. In *Thin-shell structures. Theory, Experiment and Design (Y. C. Fung and E. E. Sechler, eds.)*, pages 438–453. Prentice-Hall, Englewood, New Jersey, 1980.

[157] P.E. Tovstik. Buckling of shells of revolution in the linear approximation. *Raschety prostranstvennykh konstruktsii*, (13):118–138, 1970 (Russian).

[158] P.E. Tovstik. On the finding of the minimal frequency of free vibrations of a thin shell. *Asimptot. Metody v Teorii Sistem*, 8:5–22, 1975 (Russian).

[159] P.E. Tovstik. On the problem of cylindrical shell stability under torsion. *Prikl. Mekh. (Kiev)*, 16(9):132–134, 1980 (Russian).

[160] P.E. Tovstik. On the problem of local shells buckling. *English transl., Len. Univ. Mech. Bul*, (3), 1982.

[161] P.E. Tovstik. Some problems in cylindrical and conical shell buckling. *English transl., J, Appl. Math. Mech*, 47(5):657–663, 1983.

[162] P.E. Tovstik. Two-dimensional problems of buckling and vibrations of the shells of zero gaussian curvature. *English transl., Soviet Phys. Dokl*, 28(7):593–594, 1983.

[163] P.E. Tovstik. The wkb method in two-dimensional problems of buckling and vibrations of thin shells. *Proc. of 13 All-Union Conf. on the Theory of Shells and Plates*, pages 194–199, 1983 (Russian).

[164] P.E. Tovstik. Local loss of stability by cylindrical shells under axial compression. *English transl., Len. Univ. Mech. Bul*, (1):46–54, 1984.

[165] P.E. Tovstik. On the vibrations and stability of shells of revolution with intervals of positive and negative gaussian curvature. In *Dynamics and Stability of Mechanical Systems*, volume 6 of *Prikl. Mech.*, pages 161–168. Leningrad Univ. Press, 1984 (Russian).

[166] P.E. Tovstik. On forms of local buckling of thin elastic shells. *Trans. CSME*, 15(3):199–211, 1991.

[167] P.E. Tovstik. Axisymmetric deformation of shells of revolution made of the non-linearly elastik material. *J. Appl. Math. Mech.*, (4), 1997.

[168] P.E. Tovstik. Derivation of two-dimensional equations of shells of revolution from three-dimensional equations of nonlinear theory of elasticity. *Strength problems of deformed bodies, St.Petersburg*, (1):189–201, 1997.

[169] R. Valid. *The Nonlinear Theory of Shells through Variational Principles. From elementary Algebra to Differential Geometry.* J. Wiley & Sons, New York, 1995. 477p.

[170] N.V. Valishvili. *Design of shells of revolution by computer.* Mashinostroenie, Moscow, 1976. 287p. (Russian).

[171] V.Z. Vlasov. The principal differential equations of the general shells theory. *Prikl. Mat. Mech.*, 8(2):109–140, 1944 (Russian).

[172] V.Z. Vlasov. The momentless theory of thin shells outlined by surfaces of revolution. *Prikl. Mat. Mech.*, 11(4):397–408, 1947 (Russian).

[173] V.Z. Vlasov. *General theory of shells and its applications in engineering.* Gostechizdat, Moscow, Leningrad, 1949. 784p. (Russian).

[174] V.S. Buldyrev V.M. Babich. *Asymptotic methods in short waves diffraction problems.* Nauka, Moscow, 1972. 456p. (Russian).

[175] A.S. Vol'mir. *Stability of deformable systems.* Nauka, Moscow, 1967. 984p. (Russian).

[176] A.S. Vol'mir. *Nonlinear dynamics of plates and shells.* Nauka, Moscow, 1972. 432p. (Russian).

[177] A.S. Vol'mir. *Shells in a stream of liquid and gas. Problems of aeroelasticy.* Nauka, Moscow, 1976. 416p. (Russian).

[178] I.I. Vorovich. *Mathematical problems of the non-linear theory of shallow shells.* Nauka, Moscow, 1989. 376p. (Russian).

[179] A.N. Vozianov. Stability of orthotropic shell of revolution in zone of a simple edge effect. *Izv. Vuzov. Mashinostroenie*, (4):25–28, 1978 (Russian).

[180] W. Wasow. *Asymptotic expansions for ordinary differential equations.* J. Wiley & Sons, Inc., New York, London, Sidney, 1965. 464p.

[181] J.C. Yao and W.C. Jenkins. Buckling of elliptical cylinders under normal pressure. *AIAA Journal*, 8(1):22–27, 1970.

[182] H. Ziegler. *Principles of structural stability.* Blaisdell Publ. Comp, Waltham, Massachusetts, Toronto, London, 1968.

[183] R. Zoelly. *Uber ein Knickungsproblem an der Kugelshale.* Diss., Zurich, 1915 (German).

List of Figures

List of Tables

Index

About the Authors and Editors

Petr E. Tovstik was born in Leningrad, Russia, in 1935. He presently is professor and Head of the Department of Theoretical and Applied Mechanics of the Faculty of Mathematics and Mechanics at St. Petersburg State University. He graduated in Applied Mathematics from Leningrad State University (now St. Petersburg State University - St. PSU) in 1958, obtained his Ph.D. in Mechanics of Solids from the same university in 1963 and the degree of Doctor of Science in 1968 for the thesis: *Free Vibrations and the Stability of Thin Elastic Shells*.

Prof. Petr Tovstik is a leading Russian specialist in the application of asymptotic methods in thin shell and plate theory. He continues the tradition of Russian (Soviet) study in mechanics of thin shell structures which is associated with the names Goldenveizer, Mushtari, Novozhilov, Vlasov, and Vorovich. He is the author of a monograph entitled *Stability of Thin Shells*, the co-author, with A.L. Goldenveiser and V.B. Lidsky, of a monograph entitled *Free Vibrations of Thin Elastic Shells* and with his pupils of books *Asymptotic Methods in Mechanics, Asymptotic Methods in Problems and Examples, Asymptotic Methods in Thin Shell Structures* and also numerous papers on asymptotic methods in mechanics of thin structures published in *Doklady of the Academy of Science of the USSR*, *Mechanics of Solids*, *Transactions of the Canadian Society of Mechanical Engineers*, *Differential Equations*, *Soviet Applied Mathematics*, *Technische Mechanik* and many other journals. He received First Prize for Research at Leningrad State University in 1970. In 1999 he got the highest scientific award in Russia, the State Prize. He is member of the National Committee on Theoretical and Applied Mechanics and the editor of the journal *Vestnik Sankt-Peterburgskogo Universiteta*, serie *Mathematics, Mechanics, Astronomy.* He is a member of AMS.

Andrei L. Smirnov was born in Leningrad, USSR, in 1956. He is Associate Professor at the Department of Theoretical and Applied Mechanics of the Faculty of Mathematics and Mechanics at St. Petersburg State University (formerly Leningrad State University). He graduated in Applied Mathematics from Leningrad State University in 1978 and obtained his Ph.D. in Mechanics of Solids from the same university in 1981 for the thesis: *Vibrations of the Rotating Shells of Revolution.* He is the editor and/or the author of the books *Asymptotic Methods in Mechanics, Asymptotic Methods in Problems and Examples, Asymptotic Methods in Thin Shell Structures, Problems in Stability Theory, Introduction to the Theory of Stability* and papers on asymptotic and numerical methods in mechanics of thin structures published in *Transactions of the ASME, Transactions of the CSME, Technische Mechanik, Mechanics Research Communications, Vestnik Leningradskogo Universiteta* and other journals. He is a member of AMS.

Peter R. Frise was born in Peterborough, Canada in 1958. Since 1997 he has been a Professor of Mechanical and Automotive Engineering and has held the NSERC/DaimlerChrysler Canada Industrial Research Chair in Mechanical Design Chair at the Faculty of Engineering of the University of Windsor, Canada. He graduated in Mechanical Engineering from Queen's University in Kingston in 1981 and, after working in the oil industry in West Africa, obtained his M.Sc. also in Mechanical Engineering from the same university in 1984. He spent a year and a half at Husky Injection Molding Systems Ltd. in Bolton, Canada working as an R&D engineer and engineering group leader and then left Husky to attend Carleton University where he obtained his Ph.D. in 1991 for the thesis: *Fatigue Crack Growth and Coalescence in a Notch Region.* He was presented with the Ralph R. Teetor Engineering Education Award by the Society of Automotive Engineers (SAE) in 1993. Dr. Frise is the author of papers on modelling fatigue crack growth and non-destructive testing of structural steels and zirconium-niobium alloys in *The Journal of Pressure Vessels and Piping, The International Journal of Fatigue* and *The British Journal of Non-Destructive Testing* and has spoken at a number of international conferences on Design Engineering including ICED and the ASME-DETC conferences in 1999. In 2000 he was elected a Fellow of the Canadian Academy of Engineering.

Ardeshir Guran was born in Tehran, Iran, in 1957. He is the director of the Institute of Structronics in Canada. After enjoying his childhood in the Savadkooh Mountains and the Caspian seashores, he studied Medicine at Tehran University and Structural Engineering at Aryamehr University(Tehran). In 1979 he emigrated to Canada, received his M.Eng.(in Applied Mechanics) from McGill University(Montreal) in 1982. Both his M.Sc.(in Mathematics) and PhD (for the thesis *Contributions to Study of Instabilities in a Class of Conservative Systems*) were obtained from the University of Toronto. Professor Guran served as the associate editor of the *International Journal of Modeling and Simulations* during 1993–95. His books on *Theory of Elasticity for Scientists and Engineers* (2000) and *Adaptive Control of Nonlinear Systems* (2001) have been published by Birkhäuser.

Dr. Guran chaired several international scientific conferences including: The First International Symposium on Impact and Friction of Solids, Structures and Intelligent Machines(Ottawa Congress Center, 1998), the First International Congress on Dynamics and Control of Systems (Chateau Laurier, 1999) and the First International Conference on Acoustics, Noise and Vibrations (McGill University, 2000). He is the editor-in-chief of series on Stability, Vibration and Control of Systems.

